The
Structure and
Evolution of Stars

The Structure and Evolution of Stars

J J Eldridge
University of Auckland, New Zealand

Christopher A Tout
University of Cambridge, UK

We World Scientific

NEW JERSEY · LONDON · SINGAPORE · BEIJING · SHANGHAI · HONG KONG · TAIPEI · CHENNAI · TOKYO

Published by

World Scientific Publishing Europe Ltd.

57 Shelton Street, Covent Garden, London WC2H 9HE

Head office: 5 Toh Tuck Link, Singapore 596224

USA office: 27 Warren Street, Suite 401-402, Hackensack, NJ 07601

Library of Congress Cataloging-in-Publication Data

Names: Eldridge, J. J., author. | Tout, Christopher A., author.
Title: The structure and evolution of stars / by J.J. Eldridge (University of Auckland, New Zealand)
   and Christopher A. Tout (University of Cambridge, United Kingdom).
Description: New Jersey : World Scientific, 2018. | Includes bibliographical references and index.
Identifiers: LCCN 2018033728| ISBN 9781783265794 (hc : alk. paper) |
   ISBN 9781783265800 (pbk : alk. paper)
Subjects: LCSH: Stars--Formation. | Stars--Evolution.
Classification: LCC QB806 .E53 2018 | DDC 523.8/8--dc23
LC record available at https://lccn.loc.gov/2018033728

British Library Cataloguing-in-Publication Data
A catalogue record for this book is available from the British Library.

For any available supplementary material, please visit
https://www.worldscientific.com/worldscibooks/10.1142/P974#t=suppl

Desk Editors: V. Vishnu Mohan/Jennifer Brough/Koe Shi Ying

Typeset by Stallion Press
Email: enquiries@stallionpress.com

To those who have taught us and those whom we have taught

Foreword

A star is more than a ball of gas. Stars are perhaps more accurately described as self-gravitating nuclear reactors. But stars are truly the most fundamental of all astronomical objects. Almost everything we know about the Universe comes from the analysis of starlight. The history of the Galaxy, or the Universe itself, can be traced through the history of light from stars. An alternative, and indeed complementary, approach is to trace the history of the Universe through its composition, a composition which changes in time (and space) due to the lives of stars.

The discipline of astrophysics concerns time and space, on both the largest scales and the smallest scales. There is no length-scale larger than that of the structure of the Universe, composed as it is of hundreds of billions of individual galaxies. Yet the fundamental properties of the various astronomical bodies are often driven by the behaviour of sub-atomic physics. Again, stars stand at the cross-road of these two extremes. The life cycles of stars result in them stirring the gas within their host galaxy. This has a fundamental impact on the large-scale structure and appearance of galaxies. But stars also eject gas and dust into the galaxy, thus contributing importantly to the smallest components of the galaxy. Stars shine as a result of almost instantaneous nuclear processes, acting over millions and even billions of years. Again we see stars covering the full range of scales in both space and time.

It is no wonder that a quantitative understanding of stars is an essential ingredient of most modern astrophysics. After celestial mechanics, the theory of stellar astrophysics is arguably the greatest achievement of astrophysics. Although the basics were determined by some of the most luminary names in astrophysics — Schwarzschild, Chandrasekhar, Eddington, Gamow, Hoyle — it was the development of modern computers that enabled us to really advance the field with a detailed quantitative understanding. I think of this as the first revolution in our understanding of stars.

It is because stars span all length and time-scales that they are so interesting to the physicist. Although they can be used as tools to investigate and understand the Universe, they exhibit some fascinating physics in and of themselves. To study stars requires a significant knowledge of atomic and nuclear physics, gravitation, fluid dynamics, energy transport, as well as the interactions between these processes. The diligent student will be rewarded with an understanding of the beauty and majesty of these structures.

At the present time, stellar astrophysics is in the middle of its second revolution. We are being deluged with the most exquisite stellar data coming from space missions and large-scale Earth-based surveys. The *Sloan Digital Sky Survey* has mapped the sky in three dimensions. The *Kepler* and *CoRoT* missions have fulfilled the promise of asteroseismology, where stellar oscillations of the ball of gas are used to probe the interior structure, something that Eddington thought would never be possible[1]. First *Hipparcos,* and more recently *Gaia,* have delivered beautifully accurate astrometry (hence distances and luminosities) of over a billion nearby stars. Fibre-fed spectroscopes attached to large telescopes take high resolution spectra of thousands of stars every day. We search for stellar siblings, born together in the same gas cloud, and identified by their identical chemical composition, a concept known as chemical tagging. The *ESO-Gaia, GALAH,* and *APOGEE* surveys are all examples of

---

[1] "At first sight it would seem that the deep interior of the sun and stars is less accessible to scientific investigation than any other region of the universe." A. S. Eddington in his book *The Internal Constitution of the Stars,* 1926.

research projects inspired by the field of galactic archaeology. This aims to unravel the formation and evolutionary history of our Galaxy through studying its stellar populations. We are finding planets, by analysing starlight of course, almost everywhere in the Galaxy, and we have begun to explore the connection between parent stars and their planetary system. The laboratory study of pre-solar grains, isolated from meteorites but originating in stellar winds, challenges our understanding of how the elements are produced in stars by nucleosynthesis.

If I am not careful, I will very soon get to the point, which is this: none of this is possible without a thorough understanding of stars.

When I was starting out in this field there were just a few classic texts. Eddington's (1926) book is mostly of interest historically, and well worth reading. Then there were the slightly more recent works by Chandrasekhar (1958) and Schwarzschild (1958), Clayton's (1968) book emphasising the nucleosynthesis, and the encyclopaedic two volumes by Cox and Guili (1968). But the first revolution due to modern computers was only accessible through research papers[2]. That has changed, and there are now many books available that synthesise the advances of the last 50 years. So why do we need another?

This book is thorough in dealing with the important physics, and it also describes the lives of stars including their important nucleosynthesis. There are some novel treatments here that are rarely seen, and some insights that you will not find elsewhere. A student of stellar physics will be well served by the contents of this volume. There are many interesting exercises included, which are fundamental to active learning.

The authors are always clearly evaluating the assumptions they make, and critically analysing them and their limitations. A clear distinction is made between real stars and the idealized models that we construct. The latter are used to try to understand the former.

---

[2]The reader is referred to the pioneering papers of Icko Iben Jr for inspiration, and his two volume *Stellar Evolution Physics*, 2012.

The discussion of binary stars is most welcome and serves as an excellent introduction to this enormous topic.

The explosion of stellar data that is happening now will help answer some of the research questions raised in this volume. Indeed, many of the people who will answer those questions will almost certainly have learned about stars through studying this text.

John C. Lattanzio
Monash University

June 2018

Preface

This book is based on the 24 lecture course on Structure and Evolution of Stars taught in part III of the Cambridge Mathematical Tripos from the late 1990s. A similar course has been taught since the 1960s when introduced by Professor Sir Fred Hoyle[1] and we are indebted to those who have built it up over many years as the theory has grown to maturity. Over time the course has been shared by the Natural Sciences Tripos. For a few years it was used by the Cavendish laboratory as an optional course in part III physics and since 2003 it has also been a part III astrophysics organised by the Institute of Astronomy. The level of material is designed to be accessible both to reader's who have taken either applied mathematics or theoretical physics to a level similar to a Cambridge undergraduate BA but without necessarily any specific specialisation in astrophysics.

We have throughout been influenced by those who taught us this course. Tout took part III mathematics in 1985 when Peter Eggleton was teaching this course. Over years he shared the lecturing with James Pringle, Douglas Gough and Benjamin Davies as well as Eldridge who took the course as a part III physics student in 2000.

The chapter on binary stars is necessarily terse, both mathematically and physically, because we could write an entire book on the

---

[1] A course entitled *The Structure and Evolution of Stars* was taught by Hoyle in the Lent term of 1967. In 1963 he had taught a shorter course on *Stellar Structure* that can be traced back to Lyttleton in the Lent term of 1959.

subject. However the Cambridge stars course sometimes featured a four lecture introduction to the subject. The chapter is included here to give the reader a feel for the subject and encourage further study. In recent years Tout has taught a separate sixteen lecture course on Binary Stars that goes into considerably more detail. Perhaps one day this will form the topic of another book.

The questions included at the end of each chapter are designed both to test the readers understanding of the material but also stretch the contents of the course. Many are based on example sheets and Tripos questions and their origin has been lost in time.

Where we have included the historical development of the theory, we have made every effort to examine the original scientific papers. However we do not claim anything as a definitive record and urge any readers interested in this to research the topics more carefully. In particular we make no claim to historical completeness.

J. J. Eldridge & Christopher A. Tout

Acknowledgements

We are indebted to a number of readers who have gone through various chapters to both find mistakes and make suggestions. Greatest amongst these has been Dr Ross Church of Lund Observatory and a one-time student of the course. Ross has carefully read every chapter, sometimes more than once, without frustration. His efforts are followed by those of Dr Anna Żytkow who has been the lecturer of the course for the last four years. We thank our families friends and colleagues who have supported and encouraged us in this endeavour and particularly Dr Julie Wang whose comments on the manuscript as non-physical scientist have been invaluable. We also thank our employers Churchill College and the Institute of Astronomy in the University of Cambridge and the University of Auckland for not working us so hard that we could not complete this book. We are particularly grateful to Churchill College for providing the opportunity for stimulating dinner conversations, to Professor Archie Howie for insights into the personality of George Gamow who, as a close colleague of the then master Sir John Cockcroft, had been an overseas fellow in the early days and to Dr Michael Hoskin for his expert knowledge of the Herschels. Finally we are very grateful to be a small part of the general scientific community, the members of which have, over centuries, pushed back the frontiers of our understanding of the evolution of stars, both single and interacting. Their accumulated contributions have made this book possible.

In passing we have drawn attention to a few interesting contributors but this in no way undermines the parts played by many others. Our presentation has certainly been influenced by those who taught us and with whom we have shared the teaching. In particular we give special thanks to James Pringle, Peter Eggleton and Douglas Gough for the use of their lecture notes.

Contents

Foreword   vii

Preface   xi

Acknowledgements   xiii

1. Observable Properties of Stars   1

    1.1. Modelled Quantities   8
    1.2. Time-Scales   9
    1.3. Other Stars 10
          1.3.1. The naming of stars 11
          1.3.2. Apparent and absolute magnitudes   12
          1.3.3. Distances 14
          1.3.4. Colours 17
          1.3.5. Spectral lines 19
    1.4. The Hertzsprung–Russel Diagram   22
    1.5. Stellar Masses   27
    1.6. Questions 27

2. The Equations of Stellar Structure   31

    2.1. Physical Structure 31
    2.2. Polytropes 34
          2.2.1. Solutions to the Lane–Emden equation . . .   36
          2.2.2. Mass 36
          2.2.3. Central condensation 38

2.2.4. Mass–radius relation   38
2.3.   The Virial Theorem   39
2.3.1. Estimates for stars in equilibrium   42
2.3.2. Net energy   43
2.3.3. Pulsations   44
2.4.   Questions   44

3.   The Equation of State   47
3.1.   Gas Pressure   48
3.2.   Mean Molecular Weight   49
3.3.   Radiation Pressure   50
3.4.   Ratio of Specific Heats and Virial
       Equilibrium   54
3.5.   Importance of Radiation Pressure   57
3.6.   Ionisation Equilibria   60
3.6.1. Pressure ionisation   63
3.7.   Degeneracy   64
3.7.1. Non-relativistic electrons, $p = m_e v$ 67
3.7.2. Extremely relativistic electrons, $v = c$ . . .   68
3.8.   White Dwarfs and Neutron Stars   69
3.8.1. Warm degenerate matter   71
3.9.   Choice of State Variables   73
3.10. Molecular Hydrogen   73
3.11. Coulomb Interactions   74
3.12. The Equation of State for Stellar Models   74
3.13. Questions   76

4.   Heat Transport   79
4.1.   Thermal Equilibrium   79
4.1.1. Photons   80
4.1.2. Particles   81
4.1.3. Importance of conduction   82
4.1.4. Diffusion of ions   84
4.2.   Radiative Transfer   84
4.2.1. Sources of opacity κ   90

4.2.2. Combined opacity tables 96
4.3. Conduction 98
4.4. Convection 98
  4.4.1. Stability. 99
  4.4.2. Convective energy transport 104
  4.4.3. The temperature gradient 107
  4.4.4. Other points of interest for convection . . . 109
  4.4.5. Refinements to MLT. 110
  4.4.6. Convective overshooting 111
  4.4.7. Semi-convection 111
  4.4.8. The Ledoux criterion 113
4.5. Thermohaline Mixing 116
4.6. Questions 118

5. Stellar Atmospheres   121
5.1. Specific Intensity 121
5.2. Absorption 122
5.3. Emission Coefficient 123
  5.3.1. Detailed balance and stimulated
         emission 123
5.4. Equation of Transfer 125
5.5. Scattering 126
5.6. Surface Boundary Condition 127
  5.6.1. Plane-parallel atmosphere 128
5.7. Second Surface Boundary Condition 132
5.8. Breakdown of Assumptions 134
5.9. Line Formation 135
5.10. Questions 138

6. Energy Generation   141
6.1. Gravitational Contraction 144
6.2. Nuclear Energy Generation 147
  6.2.1. Hydrogen burning 149
  6.2.2. Helium burning: The triple-α reaction . . . 155
  6.2.3. Advanced burning stages 157

6.3.   Nuclear Reaction Rates   158
       6.3.1.   The Coulomb barrier   160
       6.3.2.   Barrier energy   162
       6.3.3.   Cross-section factor   165
       6.3.4.   Gamow energy   169
       6.3.5.   Resonant reactions   172
       6.3.6.   Thermostatic control   174
       6.3.7.   Electron screening   174
6.4.   Reaction Equilibria   175
       6.4.1.   Nuclear statistical equilibrium   176
6.5.   The Origin of the Elements   179
6.6.   Neutrino Losses   184
6.7.   Questions   186

7.   Stellar Models   191

7.1.   Time Dependence and Stellar Evolution   194
7.2.   Methods of Solution   196
       7.2.1.   Shooting   197
       7.2.2.   Relaxation   198
7.3.   Homology   201
       7.3.1.   Zero-age solar-like stars   203
       7.3.2.   Higher masses   206
       7.3.3.   Stellar lifetimes   207
       7.3.4.   Fully convective stars   207
       7.3.5.   Red giants   210
7.4.   Homologous Evolution   211
7.5.   Questions   213

8.   Stellar Evolution   219

8.1.   Stellar Evolution Models   219
       8.1.1.   Convective overshooting   221
8.2.   The Evolution of a $5\,M_\odot$ Star   222
8.3.   Thermal Pulses   231
       8.3.1.   Carbon stars   236
       8.3.2.   Quantitative problems   238

8.4. The Evolution of a $1\,M_\odot$ Star 239

8.5. The Evolution of a $7\,M_\odot$ Star 244

8.6. The Evolution of Stars More Massive
Than $8\,M_\odot$ 246

8.7. Further Complications and Uncertainties 250

    8.7.1. Initial metallicity 251

    8.7.2. Mass loss and stellar winds 252

    8.7.3. Stellar rotation 256

8.8. Naked Helium Stars, White Dwarfs
and Wolf–Rayet Stars 258

8.9. The Deaths of Massive Stars 261

    8.9.1. Core collapse 262

8.10. Supernova Spectra and Light Curves 263

8.11. Evolution Summary 265

8.12. Questions 266

9. Binary Stars   271

9.1. Numbers 272

9.2. Observed Binary Stars 274

9.3. Orbits 275

    9.3.1. Newton's laws 277

    9.3.2. Angular momentum of the orbit 278

    9.3.3. Energy 279

    9.3.4. The Laplace–Runge–Lenz vector 279

    9.3.5. Orbital energy and Kepler's third law . . . 280

9.4. Orbital Elements 281

    9.4.1. Visual binary stars 281

    9.4.2. Spectroscopic binary stars 282

    9.4.3. Eclipsing binary stars 283

9.5. Tides 285

9.6. Tidal Equilibrium 286

    9.6.1. Circularisation 286

    9.6.2. Synchronization 286

    9.6.3. The tidal mechanism 288

    9.6.4. Time-scales 292

9.7. Mass Transfer 294

9.7.1. Mass transfer rate   299

9.7.2. The stream   302

9.7.3. Stability of mass transfer   302

9.8.   Period Evolution of Binary Stars   304

9.9.   The Zoo of Binary Stars   306

9.9.1. Algols   307

9.9.2. Cataclysmic variables   312

9.9.3. Common envelope evolution   315

9.9.4. Type Ia supernovae   317

9.9.5. Massive stars, neutron stars
        and black holes   319

9.10.   Questions   321

Bibliography   325

Index   333

Chapter 1

Observable Properties of Stars

"A star is basically a pretty simple structure" announced Fred Hoyle at a colloquium in the Cambridge Observatory's Library in 1954. From the audience Professor Redman retorted, "You'd look pretty simple, Fred, at ten parsecs."[1] This was at the dawn of the modern age of stellar astrophysics. The basic physics had been identified and the equations assembled. It remained to solve them in progressively more detail and so model the stars in all their glory, a process that continues apace. Today's instruments and observing techniques mean that stars do not look quite so simple at ten parsecs any more, our understanding of the details of the physical processes has improved enormously and we have identified some important complications, particularly in the fluid dynamics of convection and mixing. Nevertheless stars, in their interiors, are indeed pretty simple structures. Not far below their visible surfaces, temperatures are so hot that the complexities of chemistry are not a burden. Even in the atmospheres much of the chemistry appears to be relatively simple and biology remains out of reach. If we understand physics we can understand stars. One of the beauties of stars is that they touch upon every aspect of physics from the quantum mechanics of

---

[1] This conversation, eloquently described to us orally by John Faulkner, is reported by Fellgett (1995).

electron degeneracy and nuclear fusion, the general relativity of the neutron stars and black holes, thermodynamics and electrodynamics, fluid dynamics and magnetohydrodynamics through to solid state crystalline structures of cold white dwarfs. In the coming chapters the necessary physics is described, with some mathematical detail, to the extent that the reader can grasp the essentials and understand the details to the level required to comprehend the current literature and to proceed to push forward the frontiers of our understanding of stars.

First we shall briefly review what we can observe and measure for real stars and so learn what our models must explain. Life on Earth depends on our closest star, the Sun, but it is sometimes easy to overlook its stellar nature just because of its very proximity. It is only because the Earth orbits at a distance that allows our carbon-based life to evolve and survive, sustained by the energy the Sun radiates, that we can contemplate the stars at all. Its proximity allows us to study the Sun in quite intimate detail compared to our distant, apparently point-like, neighbours. Hence we naturally use the Sun as a standard with which we compare all other stars. From even a minimal study we can deduce three important facts about stars: they are rather luminous, hot and apparently spherical.

Early scientific study of the Sun can be traced back to observations of sunspots by Chinese astronomers as early as 206 BC. Similar observations were also undertaken by the medieval Andalusian polymath Averës in the twelfth century. However it was not until the invention of the telescope in the seventeenth century that detailed observations could be undertaken by the likes of Thomas Harriot and Galileo Galilei. By following the motion of spots on the Sun's surface Galileo deduced that it rotates with a period of about a month. However this rotation is not the uniform rotation of a solid body. The period varies from about $35\,\mathrm{d}$ at the poles to $24.5\,\mathrm{d}$ at the equator. This corresponds to an equatorial surface velocity of $1.7\,\mathrm{km\,s^{-1}}$. Today we know that this, though typical of stars like the Sun, is rather slow compared with many more massive stars.

In the same century Johannes Kepler deduced his laws of planetary motion from meticulous analysis of observations made by

Tycho Brahe. Soon after, Sir Isaac Newton's laws of motion and gravity made it possible to deduce the relative masses of the Sun and the planets from the periods of their satellites and trigonometric estimates of distances. Though such estimates of the distance to the Sun from Earth have been made since ancient times an accurate absolute scale was missing until the transits of Venus in 1761 and 1769 were observed from distant points on the surface of the Earth in a major international collaboration. This not only gives the mean distance from the Earth to the Sun, the astronomical unit ($a_{\oplus} = 1\,\mathrm{AU}$), but also the solar radius and an estimate of the Sun's absolute luminosity from measurements of its energy flux through a unit area at the Earth, the Solar irradiance. The absolute masses of the solar system bodies required the further measurement of Newton's gravitational constant which did not come until the late eighteenth century when Henry Cavendish completed John Michell's[2] proposed measurement of the gravitational force between two masses on the Earth.

These fundamental parameters of the Sun are measured more and more precisely as techniques for their measurement change and improve. In summary[3]

- the solar mass, $M_{\odot} = 1.9885 \times 10^{33}\,\mathrm{g} = 1.9885 \times 10^{30}\,\mathrm{kg}$,
- the solar radius, $R_{\odot} = 6.957 \times 10^{10}\,\mathrm{cm} = 6.957 \times 10^{8}\,\mathrm{m}$,
- the solar luminosity, $L_{\odot} = 3.828 \times 10^{33}\,\mathrm{erg\,s^{-1}} = 3.828 \times 10^{26}\,\mathrm{W}$ and
- the astronomical unit, $1\,\mathrm{AU} = 1.496 \times 10^{13}\,\mathrm{cm} = 215\,R_{\odot}$,

---

[2]The Revd John Michell (1724–1793) features prominently in the history of our study of the stars. Alongside his contributions to geophysics Michell was the first to predict the existence of black holes and to prove that binary stars exist (Chapter 9).

[3]The quoted solar properties are those chosen by the International Astronomical Union (IAU) at its general assembly in Honolulu in 2015 (Prš, 2016). These are now considered to be fixed units rather than necessarily representing precisely the properties of the Sun. The IAU actually defined the product $GM_{\odot}$, rather than $M_{\odot}$ itself, so that the deduced mass also depends on what is chosen for Newton's constant G. The astronomical unit was similarly defined to be precisely 149 597 870 700 m by the IAU in 2012.

from which we can derive

- the mean density of the Sun, $\bar{\rho}_\odot = 1.41\,\mathrm{g\,cm}^{-3}$ and
- the effective temperature, $T_\odot = 5\,772\,\mathrm{K}$,

where the effective temperature of the photosphere of the Sun is defined to be the temperature of a black body of the same radius and luminosity so that $L_\odot = 4\pi\sigma R_\odot^2 T_\odot^4$.

Accurate estimates of the age of the Sun came much later. It must be older than the solar system and the Earth itself but it was not until the geological record of the Earth was analysed carefully in the early nineteenth century that an age in excess of ten million years was even contemplated. Today the best estimates of the age of the solar system are based on the radioactive decay of long lived isotopes that were trapped in rocks on the Earth and meteorites when the solar system formed. This method was originally employed by Sir Ernest Rutherford (1929) but did not become reliable until the 1950s. The latest results compare the decays of ^{238}U and ^{235}U to ^{206}Pb and ^{207}Pb in meteoritic inclusions which we believe formed earlier in the Sun's protoplanetary disc. The measurement errors of a few million years are somewhat smaller than the systematic error of about $10^8\,\mathrm{yr}$ it probably took the Sun to form, hence we can be reasonably certain that

- the age of the Sun, $t_\odot \approx 4.6 \times 10^9\,\mathrm{yr}$.

Since Joseph Fraunhofer invented the spectroscope in the early nineteenth century, much has been learnt from analysing the solar spectrum and the spectra of other stars. Most important is the composition of the solar surface. In Chapter 5 we examine how the photosphere of a star emits light in a black-body spectrum and how this is modulated by atomic absorptions further out in the star's atmosphere (see also Sec. 1.3.5). The shape of the absorption lines gives a plethora of information including the abundances of the absorbing elements, the pressure at which the line forms, any systemic velocity, the broadening caused by the star's rotation as well as a measure of any magnetic field strong enough to induce Zeeman splitting of the lines.

In 1925 Cecilia Payne (1925) (later Payne-Gaposchkin) used Meghnad Saha (1921)'s ionisation theory (Sec. 3.6) applied to spectra to determine abundances of various elements in stellar atmospheres. Though she found many elemental abundance ratios to be consistent with terrestrial ratios she unexpectedly found the lightest elements, hydrogen and helium, to dominate. This was the first indication that the composition of stars is very different to that of the Earth. In 1926 when Eddington was postulating hydrogen fusion as the source of the Sun's luminosity he was still reluctant to admit a hydrogen abundance above the 7% he believed necessary to power the Sun for 10^{10} yr. The dominance of hydrogen and helium was only confirmed when accepted for the Sun by Henry Norris Russell (1929). Helium emission lines can be seen in the solar chromosphere during eclipse, and indeed the new element helium was first identified this way by Norman Pogson and named in 1868 by Sir Norman Lockyer. However it is not possible to make an absolute determination of the abundance of the Sun's hydrogen and helium spectroscopically because the electrons in helium are too tightly bound at the temperature of its photosphere. Measurements of composition are limited to elements more massive than helium, known collectively as metals, for which abundances relative to hydrogen can be found from spectra. Though the Sun's helium abundance can be estimated with other techniques, such as helioseismology which is able to probe how helium ionises with depth in a way that depends on its abundance, direct measurements of the solar wind or chromospheric observations, it is not well constrained and is usually fixed to ensure that solar models have the correct luminosity (Sec. 8.4). We shall use canonical abundances by mass for hydrogen $X = 0.7$, helium $Y = 0.28$ and metals, the metallicity $Z = 0.02$ so that $Z/X = 0.029$. Recent analysis of the solar spectrum, taking into account three-dimensional modelling of convective motions apparent in the solar granulation, suggests that this is too high for the solar photosphere where something as low as $Z/X = 0.018$ may be more appropriate. However helioseismological measurements remain more consistent with higher metallicity. As we write, this remains an active area of research in which a resolution is still awaited.

The relative abundances of the metals are better constrained both by the solar spectrum and by, mostly consistent, measurements on the Earth and within the solar system. However it is particularly difficult to measure abundances of all the noble gases, He, Ne, Ar, Kr,..., because not only it is hard to ionize their closed shell electrons but also they tend to escape from rocks. Among the metals the most abundant are O, C, probably Ne, N, Mg, Si and Fe. All these elements are made in stars and we discuss how in Chapter 6, where we also throw some light on why these particular metals tend to be the most abundant. As a consequence of stellar nucleosynthesis, stars that have formed more recently tend to have higher metallicities. Historically stars with metallicities similar to the Sun were labelled as population I while those of somewhat lower metallicity ($Z \approx 0.001$), found in old globular clusters, were labelled as population II. The lowest metallicity stars are found in the halo of the Galaxy which therefore formed early in its evolution while stars, such as the Sun, in the younger Galactic Disc have the highest metallicities. Recently the first stars, presumably with effectively no metals at all, have been called population III. In practice stars over a continuous range of metallicities from about $Z = 2 \times 10^{-7}$ to 0.03 have been observed within our Galaxy and its neighbours.

Another probe of the deep interior of the Sun is its flux of neutrinos. Neutrinos are created during weak nuclear reactions that take place at the Sun's very centre and escape freely through the whole body of the star. Thus they are an important indicator of the conditions, particularly the temperature, at a point in the core of the Sun that is otherwise unobservable. In Sec. 6.2.1.1 we look at the solar neutrino flux and the resolution of the old solar neutrino problem in more detail.

We have briefly alluded to helioseismology already. The Sun acts as a cavity for various acoustic, pressure and gravity standing waves of particular frequencies that depend on the variation of sound speed with depth. By measuring oscillations of the surface of the Sun we can determine how sound waves travel through the interior and we invert to investigate its structure. This process is similar to the use of seismology on the Earth to reveal its interior. Helioseismology has, rather precisely, measured that the Sun undergoes convective motions down to $0.713 \pm 0.001 \, R_{\odot}$. It is also possible to measure the rotation of the Sun as a function of both depth and latitude because rotation splits frequencies as waves travel with or against the motion. From sunspots the latitudinal differential rotation at the Sun's surface has been known for some time. Helioseismology determines that this differential rotation persists approximately radially through the convection zone to a thin layer of strong differential rotation, the tachocline, at its base. Within this the core of the Sun appears to rotate uniformly.

Any of these techniques used to probe the Sun can, in principle, be extended to other stars. Their apparent brightness and colours have been noted since ancient times. Their spectra have been examined since the middle of the nineteenth century. Asteroseismology is already beginning to reveal the interiors of other stars and will play an important role in the future. Neutrino fluxes from most nuclear reactions in other stars are too weak to be detected. Their detection does however become possible from supernovae (Sec. 8.10) when the cores of massive stars collapse at the end of their lives. Indeed neutrinos were detected from supernova 1987A in the Large Magellanic Cloud, a satellite galaxy that orbits our own. Future detectors should be able to measure neutrinos from supernovae in more distant neighbours, such as the Andromeda Galaxy, M31, and it is likely that a future Galactic supernova will first be detected by through its neutrino flux.

The proximity of the Sun allows us to see yet more detail because we can resolve the solar disc with great precision. We can similarly resolve coronal features, visible to the naked eye during total eclipses. In these images the Sun appears very active. We can see rising and falling convective cells as a solar granulation pattern across the disc. Sunspots, caused by magnetic flux tubes breaking through the surface, are seen to appear regularly and last for hours to months before disappearing. Over a chaotic eleven-year cycle they begin to appear at latitudes around 30° in both hemispheres. Their number increases as spots appear nearer and nearer to the equator and then falls off as they cease to appear at higher latitudes. Their variation points to a solar cycle over about 22 yr during which the surface magnetic field of the Sun changes sign.

The magnetic field emerging from its surface carries energy some distance away from the Sun in prominences and flares and appears to be responsible for heating a low-density corona to above 10^6 K. From this a solar wind flows out into the solar system at speeds between 300 and 800 km s^{-1} around the Sun's escape velocity of 618 km s^{-1}. At the Earth we measure an average mass-loss rate of rate of $-\dot{M}_\odot \approx 10^{-14} M_\odot$ yr^{-1}. This wind is forced to rotate with the Sun's magnetic field out to an Alfvén radius, of about thirty times that of the Sun itself, where its kinetic energy density exceeds that of the magnetic field. Because of this, the angular momentum carried off, even by its weak wind, has been sufficient to slow the Sun to its current spin rate. Though interesting in their own right none of these phenomena appear to affect the long term, nuclear time-scale, evolution of the Sun so we do not consider them further. Unlike the Sun, some unresolved stars exhibit quite drastic short term variations in luminosity which can be variously attributed to pulsation, duplicity or severe spottiness. High resolution time-dependent photometry and spectroscopy allow us to probe the physical causes in some detail. For massive and evolved stars the mass loss in stellar winds plays a very important role in their evolution. Usefully, strong stellar winds can be identified and quantified by analysis of spectral lines (Sec. 5.9) but, noting the short time-scale variability of the Sun's surface, we must always remember that such measurements don't necessarily reveal long term average mass-loss rates.

1.1.   Modelled Quantities

In Chapter 8, we examine numerical models, including one of 1 $M_\odot$, similar to the Sun, and from this we can determine typical internal conditions.

- the central density, $\rho_c = 1.48 \times 10^2$ g cm$^{-3} \approx 100\,\bar{\rho}$,
- the central temperature, $T_c = 1.56 \times 10^7$ K and
- the central pressure, $P_c = 2.29 \times 10^{17}$ dyne cm$^{-2} = 2.29 \times 10^{16}$ Pa $= 2.32 \times 10^{11}$ atm.

1.2.   Time-Scales

Experience tells us that the gross properties of the Sun do not
noticeably vary over human lifetimes. Indeed for life to have evolved
it seems necessary that the Sun has changed little since the formation
of the Earth. It is this observation that led to the need for nuclear
fusion as the energy source in the Sun (Chapter 6) so let us now
consider on what time-scales variation might occur if appropriately
driven.

The shortest time-scale is dynamical. This can be expressed as
the time for the star to collapse to a point under free fall,

$$\tau_{\rm dyn} = \sqrt{\frac{\pi^2 R_\odot^3}{8GM_\odot}} = 1.77 \times 10^3\,{\rm s} \approx \frac{1}{2}\,{\rm hr}. \tag{1.1}$$

It is similar to the period of natural acoustic oscillations of the Sun
and the Keplerian orbital period at its surface. Apart from some
notable variable stars, such as the Cepheids or Miras, most do not
change significantly on such short time-scales so we usually assume
that stars are in dynamical equilibrium.

The thermal, or Kelvin–Helmholtz, time-scale is the time it
would take the star to radiate its internal energy at its current
luminosity. By the virial theorem (Sec. 2.3), this is closely related
to the gravitational binding energy of the star and so the Kelvin–
Helmholtz time-scale

$$\tau_{\rm KH} = \frac{E_{\rm grav,\odot}}{L_\odot} = \eta\frac{GM_\odot^2}{R_\odot L_\odot} \approx 10^{15}\,{\rm s} \approx 3 \times 10^7\,{\rm yr}, \tag{1.2}$$

where η is a structural factor of order unity that can be calculated
from a stellar model. Lord Kelvin (William Thompson) and Hermann
von Helmholtz originally drew attention to its relation to the
prospective lifetime of the Sun were its energy source due to the
release of gravitational potential energy alone (Chapter 6). It is
the time-scale on which a star evolves when not supported by energy
generation. This was the case when the Sun first contracted from
a cool cloud and so $\tau_{\rm KH}$ gives a good estimate of the time taken
for the Sun to form. There are some later stages in the evolution of
single, and particularly binary stars, when a star loses its thermal

equilibrium and evolves for a while on a similar time-scale. The Sun however appears to have changed very little over the last hundred million years so we can assume that it is in both thermal and dynamical equilibrium.

Faced with the fact that the Sun produces its own energy and is supported against gravitational collapse by nuclear fusion (Chapter 6), the time-scale on which it must evolve is its nuclear time-scale. We estimate this as the time taken to fuse, or burn, all the hydrogen to helium in the Sun at its current luminosity so the nuclear time-scale

$$\tau_{\rm N} = \frac{X M_\odot Q_{\rm H \to He}}{L_\odot m_{\rm He}} = \frac{\Delta M_{\rm H \to He} c^2}{L_\odot} \approx 2 \times 10^{18} \, {\rm s} \approx 10^{11} \, {\rm yr}, \quad (1.3)$$

where X is the mass fraction of hydrogen, $Q_{\rm H \to He}$ is the energy released when four hydrogen nuclei combine to form one helium nucleus of mass $m_{\rm He}$ and $\Delta M_{\rm H \to He}$ is the total mass of the star that would be lost to nuclear binding energy were all the hydrogen converted to helium. Typically stars evolve off the main sequence when they have consumed only about 10% of their available hydrogen (Chapter 8) and so it is often said that the nuclear time-scale for the Sun is about 10^{10} yr, its core-hydrogen-burning lifetime. As stars move to later burning stages the energy Q available per reaction falls off and the nuclear time-scale decreases. Once oxygen burning has begun $\tau_{\rm N} \to \tau_{\rm dyn}$ and the star can no longer be considered to be in a dynamical, thermal or nuclear equilibrium.

1.3. Other Stars

Even given their distance and angular size, the information we can obtain from other stars is remarkably rich and rapidly increasing as new techniques and telescopes are brought into service both on the ground and in space. With the naked eye alone it is immediately apparent that not all stars are the same. They cover a wide range in brightness and they appear to have a range of colours, from blue to red, underlying their predominantly white appearance. Our eyes are attuned to the light from the Sun so, because stars emit almost as black bodies, blue stars are hotter than the Sun while orange and red stars are cooler. Surface temperatures of stars range from about 2000 to 200 000 K. Their brightness can be a result of either intrinsic luminosity or distance. The Sun is the brightest object in our sky but would be invisible to the naked eye at the distance of Betelgeuse which itself is probably some 100 000 times as luminous as the Sun. Until we can routinely detect distant neutrino emissions or gravitational radiation all the information reaching us directly from other stars is in the light they emit. Astronomers have been looking at stars since we first took an interest in the night sky. Because of this there is much that appears arcane about the way we describe them. In the next sections we aim to give the reader enough information to begin to understand what is written elsewhere.

1.3.1. *The naming of stars*

From ancient times stars have been given names and many of these, particularly those of the brightest stars, survive today. Many have their origin in the long history of Arabic and the Ancient Greek astronomy. The first systematic record to survive in Western civilisation[4] was that of Claudius Ptolemy in his Almagest of the second century AD and a product of the Roman empire. It is considered to be based on Hipparchus' catalogue dating from some 300 yr earlier. Also of ancient origin is the division of groups of stars into constellations whose, sometimes memorable, shapes allow us to navigate the heavens on a dark night. Today these have rigorously determined boundaries so that every star is a member of a single constellation that localises it on the sky. At the start of the seventeenth century Johann Bayer introduced the designation of a star by a Greek letter, α, β, ... followed by the genitive, often abbreviated to three letters, of the Latin name for its constellation. Thus Aldebaran, the brightest star in the zodiacal constellation of Taurus is also known as α Tauri or α Tau. Often the labels follow the

---

[4]Various cultures worldwide also created their own independent patterns on the sky for use in both time keeping and navigation. For instance Chinese astronomers actively named stars in the fifth century BC. Here we concentrate on the star and constellation names that make up the IAU standards used today.

order of the brightness of the stars but not always because position in the constellation also plays a role. The variable star Betelgeuse, α Ori, is never as bright as Rigel, β Ori, as seen from the Earth. To complicate matters further stars have been catalogued many times over in more recent years. For instance Aldebaran is also 87 Tau in Flamsteed's catalogue of 1725 and HD 29139 in the Henry Draper catalogue from the early twentieth century to add but two examples. Today the naming of stars is strictly controlled by the International Astronomical Union. A precise position, in terms of coordinates such as declination and right ascension, on the celestial sphere at a particular time provides an equally useful but less memorable identification of any star.

1.3.2.   *Apparent and absolute magnitudes*

To begin the story of classifying the brightness of a star, we again go back to Hipparchus. He sorted stars into six groups by what he called size. Stars are point-like so this apparent size is a measure of brightness. The relation between the two is easily seen on a photographic plate where brighter stars cast larger images than fainter stars. Hipparchus called the brightest group first magnitude and the faintest visible to the naked eye sixth magnitude. In the nineteenth century comparisons of starlight entering a telescope to artificial light sources revealed that the human eye perceives brightness logarithmically and that stars of first magnitude are about 100 times as bright as those of sixth magnitude. So Norman Pogson (1856a) proposed the system of magnitudes, which persists today, in which one magnitude corresponds precisely to a factor $\sqrt[5]{100} \approx 2.512$ in brightness. In terms of the radiation flux F reaching the Earth the apparent magnitude of a star can then be written as

$$m_* = -2.512 \log_{10} F + m_0, \qquad (1.4)$$

for some constant m_0.

   In practice we do not see the full spectrum of light that reaches us from any given star, partly because of the response of our eyes, or whatever the detector we happen to be using, and partly because of absorption by the Earth's atmosphere. The latter can be eliminated by observations from space or significantly reduced at high altitude. The former is quantified by the use of specific filters that transmit a known amount of light over a range of wavelengths. For instance a visual filter, V-band, covers most of what our eyes can detect while an infrared filter, I-band, transmits only light redward of what we can see. Various conventions are used to fix the constant m_0. The simplest fixes $m_* = 0$ for all filters for the star Vega, α Lyr, but like all stars Vega has a small variability so this is not ideal. In practice we must take great care, when using precise magnitudes, to avoid problems with ambiguities in definitions. Comparison between two stars in the same filter circumvents the need for m_0. We may write the difference in magnitudes

$$m_A - m_B = -2.512 \log_{10}\left(\frac{F_A}{F_B}\right) \tag{1.5}$$

for two stars, A and B, with fluxes F_A and F_B.

In the Vega system the Sun has an apparent visual magnitude $m_V = -27$ while Sirius, the brightest star in the night sky has $m_V = -1.4$. There are four other stars with $m_V < 0$, eleven with $0 < m_V < 1$, thirty-two with $1 < m_V < 2$ and about 5000 in total, visible to the naked eye, with $m_V < 6$. The current detection limit with the largest telescopes is $m_V \approx 30$ which is about that of a candle on the Moon or a star like the Sun in the Andromeda Galaxy.

To understand more about the stars themselves we are interested in their absolute luminosity rather than the radiation flux that reaches the Earth. Without absorption, the total luminosity through any sphere centred on a star remains constant so that the flux per unit area falls of as the square of the distance. We define an absolute magnitude M_x for each filter x to be the apparent magnitude m_x the star would have if at a distance of $10\,\mathrm{pc}$, a unit of which the definition and significance will become apparent in the next section. For a star at a distance d from the Earth the absolute magnitude is given by

$$M_x = m_x - 5.024 \log_{10}\left(\frac{d}{10\,\mathrm{pc}}\right). \tag{1.6}$$

We also want to know the total rate L at which energy is emitted by the star over all wavelengths rather than just what passes through a particular filter. This is encapsulated in the bolometric magnitude M_{bol}. The absolute bolometric magnitude of the Sun is $M_{bol} = 4.755$ so that the bolometric magnitude of a star of luminosity L is

$$M_{bol} = -2.512 \log_{10} \left(\frac{L}{L_\odot} \right) + 4.755, \qquad (1.7)$$

though this is to some extent a definition itself. For any given star observed through a given filter x its bolometric magnitude M_{bol} is related to its observed absolute magnitude M_x by an associated bolometric correction BC_x, which depends on the temperature and more weakly on other properties of the star's spectrum, such that

$$M_{bol} = M_x - BC_x. \qquad (1.8)$$

Unfortunately, to complicate matters further, the bolometric magnitude of Vega is not zero in the Vega system. Because any filter cuts out some of a star's luminosity it is reasonable to define bolometric corrections so that they are always positive. Thus the usual definition is to fix $BC_x = 0$ at its minimum, where the greatest fraction of the total light passes through the particular filter, for main-sequence stars.[5] In this case the visual bolometric correction for the Sun, with an absolute visual magnitude $M_V = 4.83$ is 0.07, while that of Vega turns out to be about 0.03. At the extremes of both hot and cold stars $BC_V \approx 4$. Once again care must be taken to understand exactly what has been defined how when particular measurements are considered.

1.3.3.   *Distances*

Because stars are so very far away accurate measurement of their distance is not easy. A simple geometric method is to measure the angular displacement of a nearby object against a background of distant, fixed stars, as the Earth moves around its orbit (Fig. 1.1).

---

[5]Quite often the sign convention is changed in Eq. (1.8) and the corrections are defined to be always negative.

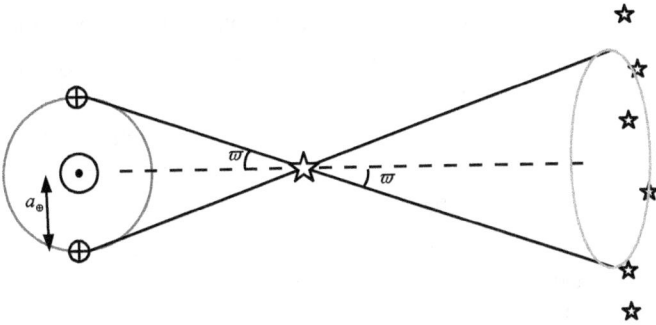

Fig. 1.1. Distances from parallax. A nearby star appears to move against a background of more distant, fixed, stars as the Earth orbits the Sun. The parallax ϖ is half the angle subtended by the star at the limits of the Earth's orbit of radius $a_\oplus = 1\,\mathrm{AU}$.

The maximum angular displacement of the star, when the Earth is at extremes of its orbit, is twice the parallax ϖ. Though the concept is simple and known from our everyday experience of judging distances by binocular vision, its practice remained elusive for centuries. No star has a parallax as large as 1 arcsec $= 1'' = 1°/3600$ and some idea of the distance to stars and the difficulties of parallax measurement can be realised by noting that a typical set of human eyes, set 8 cm apart subtend an angle of $1''$ at a distance of 16.5 km. A convenient unit of distance is the parsec, the distance to a star that would have a parallax of $1''$. The small angle means that $\tan \varpi \approx \varpi$ and

$$\frac{d}{\mathrm{pc}} = \frac{1''}{\varpi} \tag{1.9}$$

so that

$$1\,\mathrm{pc} = \frac{1\mathrm{AU}}{1''} = \frac{360°}{(2\pi)''}\mathrm{AU} = 2.0626 \times 10^5 \,\mathrm{AU}$$

$$= 4.437 \times 10^7 \, R_\odot = 3.086 \times 10^{16}\,\mathrm{m} = 3.262\,\mathrm{ly}, \tag{1.10}$$

where 1 ly, a light year, is the distance light travels in 365.25 d.

Measurements are complicated by the fact that the nearest stars also tend to have a large angular velocity, known as proper motion, owing to their orbital motion about our Galaxy. Indeed it was the indication of proximity suggested by a large proper motion that led astronomers to select particular stars for parallax determination. The elliptical parallax generated by the Earth's orbit is superimposed on the linear proper motion and the two must be deconvolved. The first stellar parallax, that of 61 Cyg by Freidrich Bessel, was not measured until 1838. Bessel's measurement of $\varpi = 0.3136''$ though not able to live up to its implied accuracy was a major and long awaited breakthrough. The annual proper motion of 61 Cyg is more than ten times its parallax. In fact 61 Cyg is a relatively wide binary star with a period of some 678 yr and the orbital motion, though annually slight, adds further complexity to the measurement. As telescope engineering has improved so has the number and accuracy of measured parallaxes. Our nearest star is the triple system α Cen, prominent in the Southern sky, with a parallax of $0.7471'' \pm 0.0012''$ and so at a distance of $1.339 \pm 0.002\,\mathrm{pc} = 4.366 \pm 0.007\,\mathrm{ly}$. The system consists of two relatively bright stars in an orbit with period of 79.9 yr together with a faint third companion in a very wide orbit with a period of some hundreds of thousands of years. Presently this third component, known as Proxima Centauri is some 0.04 pc closer to the Earth than its companions.

The smallest parallax reasonably determined from the Earth's surface is about $0.01''$ which corresponds to a distance of only 100 pc. A significant improvement was made by the space mission *Hipparcos* which operated from 1989 to 1996 and measured parallaxes down to about $0.001'' = 1\,\mathrm{mas}$, a milliarcsecond, so distances to 1000 pc. As we write the *Gaia* satellite, launched in December 2013, is measuring parallaxes accurate to between $7\,\mu\mathrm{as}$ and $200\,\mu\mathrm{as}$ for stars ranging from less than 10th magnitude to as faint as 20th magnitude. With this astrometric distances out to 150 kpc become measurable. For comparison the Sun lies about 8 kpc from the Galactic Centre in a disc of stars about 300 pc in thickness and extending to a radius of about 20 kpc. The distance to the Large Magellanic Cloud, a satellite galaxy visible in southern hemisphere, is about 50 kpc. With *Gaia* we shall not only gain unprecedented knowledge of the structure of our Galaxy but accurate measurements of the distances of some one thousand million stars that will revolutionise our understanding of the stars themselves.

1.3.4.　*Colours*

The spectrum of a star resembles that of a black body of the same effective temperature as the star. Knowing the transmission of a particular filter we can calculate the fraction of the total energy that passes through and for two filters we can calculate the ratio of energies transmitted for a black body of a given temperature. This difference in magnitude defines a colour. To obtain a precise relation between effective temperature and colour we must know the detailed shape of the spectrum of a star over the range of the filters. Figure 1.2 shows a particular set of filters, the Johnson–Cousins filters, below the spectra of three stars of different effective temperatures. The deviation from a smooth black bodies is quite apparent but so is the underlying similarity. As an example, with blue and red filters we can define a colour

$$B - R = M_B - M_R = m_B - m_R. \qquad (1.11)$$

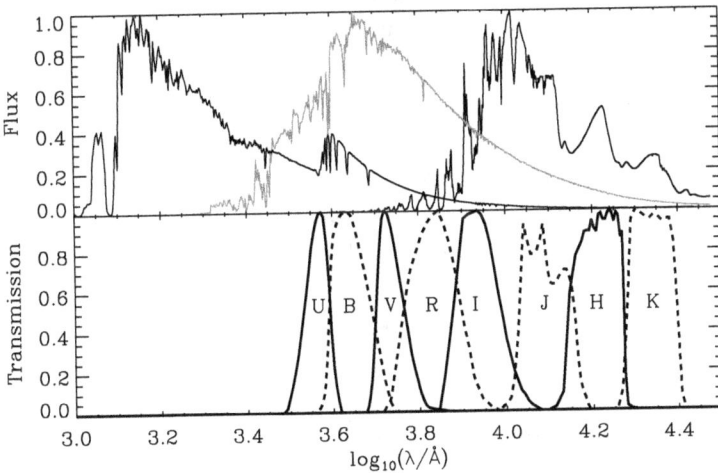

Fig. 1.2. Lower panel: the transmission functions of the broad-band Johnson–Cousins (1975) UBVRIJHK filters, arranged left to right with alternate filters identified by solid and dashed lines. Upper panel: normalised spectral-energy distributions (Westera *et al.*, 2002) of three stars with surface temperatures of 12 000 K (left peak), 6000 K (central peak) and 2500 K (right peak). The stars of different temperatures overlap the broad-band filters by different amounts. The differences between the measured magnitudes of a star in any two filters is a colour.

Hot stars emit relatively more energy in the blue than the red while cool stars emit more in the red so the $B-R$ colour of hot stars is less than that of cool stars. From Fig. 1.2 it is apparent that different pairs of filters are better suited to different effective temperatures. A reliable determination of T_{eff} requires that a significant fraction of the star's emission passes through both filters. Thus U, B and V filters are not suitable for the star of 2500 K which is better observed with filters in the infrared. As temperature rises the redward filters move into the Rayleigh–Jeans tail of the black-body spectrum where colour converges to a limit independent of temperature. So for the hottest stars it is necessary to observe in the ultraviolet.

A complication to the determination of temperature by colour is introduced by absorption of starlight en route from the star to the telescope. Significant absorption, particularly outside the visual wavelengths, takes place in the Earth's atmosphere but this can be carefully measured because it is the same for all stars. More problematic is interstellar absorption by any intervening gas or, most significantly, dust. When any dust particles are small compared with the wavelength, the Rayleigh regime, photon scattering rises as the fourth power of frequency so that bluer light is scattered much more readily than the red. This is the same effect that makes the sky blue away from the Sun and red when sunlight is seen directly through a haze of particles. It also introduces the concept of reddening of stars. All magnitudes are increased but magnitudes in blue filters increase more than red so that colours generally become redder. Sometimes the absorption around a star is associated with its own circumstellar material but more often it is due to intervening dust in our Galaxy so that all stars in a particular direction are affected by the same absorption. By comprehensive observation it has been possible to construct a three-dimensional reddening map according to Galactic coordinates that facilitates colour and magnitude corrections. Reddening increases with distance, particularly when looking towards the Galactic Centre, so once we look deep enough, infrared telescopes become necessary to see stars at all.

1.3.5. *Spectral lines*

A second and in many ways much more reliable, though more complex both technologically and physically, method to determine the temperature of a star is by the detailed structure of its spectral lines. The temperatures of the atmospheres of stars, particularly that of the Sun, are just those temperatures at which atomic transitions, electrons moving from one orbital to another, are easily excited, both by collisions between atoms and by absorption of photons. The spectrum of the 6000 K star, illustrated in Fig. 1.2, is typical. The underlying shape is a black body, superimposed on which are a set of dips that manifest themselves as dark absorption lines. These come about as the black-body radiation emitted by the stellar photosphere passes through the slightly cooler stellar atmosphere. Photons of just the right energy to excite electrons from one orbital to a more excited or ionised state are absorbed. The atmosphere is in local thermodynamic equilibrium so these photons are eventually reemitted but isotropically and at a lower temperature so that the intensity of the radiation reaching us is reduced.

Steeped by historical epithets, the labelling (Table 1.1) of stellar spectra can appear quite arcane to the uninitiated. Our modern Morgan–Keenan (MK) system (Morgan and Kennan, 1973) has its

Table 1.1. The standard Harvard spectral classification scheme.

Stellar Type	Temperature/K	Apparent Colour	Hydrogen Lines	Prominent Lines
O	≥30000	Blue	Weak	Ionised helium
B	10000–30000	Blue-white	Medium	Neutral helium, hydrogen
A	7500–10000	White	Strong	Hydrogen
F	6000–7500	Yellow	Medium	Neutral hydrogen, ionised calcium
G	5200–6000	Yellow	Weak	Neutral hydrogen, strong ionised calcium
K	3700–5200	Orange	Very weak	Neutral metals, ionised calcium
M	≤3700	Red	Very weak	Molecules and neutral metals

roots in the Harvard Observatory System that developed around the turn of the 19th to 20th centuries after Edward Pickering began a comprehensive survey of stellar spectra in the 1880s. Much of the classification was completed by Willamina Fleming and the Draper Catalogue of Stellar Spectra, named in memory of Henry Draper, who had first photographed the spectrum of Vega in 1872, was published in 1890 (Pickering, 1890). The classification used was loosely based on the strength of the hydrogen lines. Those stars with the strongest hydrogen lines were designated type A while those with the weakest type M. By 1901 the classification had been reorganised into fewer types and in decreasing stellar temperature by Annie Jump Cannon. Hydrogen ionises at about $10\,000\,\mathrm{K}$ and it is stars of this temperature that have the most prominent hydrogen lines. As the temperature rises fewer atoms have bound electrons and the lines disappear from the spectra. As the temperature falls the electrons around the hydrogen nuclei become more energetically confined to the ground state orbits. This in turn leads to fewer weaker hydrogen lines in the spectra. However lines from the more weakly bound electrons of other atoms and molecular rotation and vibration bands become more prominent. So it is straightforward to distinguish the very hot O stars from the relatively very cool M stars. The sequence of spectral types from the hottest to the coolest normal stars follows

$$\mathrm{O} \quad \mathrm{B} \quad \mathrm{A} \quad \mathrm{F} \quad \mathrm{G} \quad \mathrm{K} \quad \mathrm{M}$$

and Table 1.1 indicates the appropriate temperature range, colour and spectral characteristics of each type. By 1912 Cannon had further subdivided each spectral type into ten, adding an Arabic numeral so that the sequence in decreasing temperature becomes $\ldots, \mathrm{O8}, \mathrm{O9}, \mathrm{B0}, \mathrm{B1}, \ldots$ (Cannon and Pickering, 1912). Three other spectral types R, N and S, not associated directly with temperature designate cool carbon stars, cooler carbon stars and stars with zirconium oxide bands.

Spectral lines also contain information about the pressure in the atmosphere of a star because as the collision rate of atoms increases so the time for the absorption or emission process is shortened below the natural lifetime of an excited state. By Heisenberg's uncertainty relation, this increases the width of the line $\Delta \nu$ because $\Delta \nu \propto \Delta E$

the spread in energy across the state and $\Delta E \Delta t = \hbar/2$, where Δt is the lifetime of the state. In Chapter 5 we shall see that the pressure at a stellar photosphere is directly proportional to its surface gravity. So pressure broadening gives a measure of surface gravity which has proved quantitatively highly accurate in the presence of the very high surface gravities of white dwarfs. The information is included in the Yerkes MK system by a luminosity class designated by a capital Roman numeral after the spectral type. Class I stars are supergiants, II intermediate giants, III normal giants, IV subgiants and V main-sequence dwarfs. The Sun is a G2 V star. Philip Keenan continued to revise the MK classification system up until his death in 2000. In addition to the basic picture painted above it contains many more spectral types and a plethora of qualifiers to indicate peculiarities.

As well as these basic properties of effective temperature and surface gravity stellar spectra, when fully analysed, contain a wealth of other information. We have already drawn attention to the fact that strength of various spectral lines for a given spectral type reveals details of the composition of a star. This information is most easily extracted for nearby giant stars that are cool enough to show many atomic lines, large enough to not suffer excessive pressure broadening and bright enough for us to easily obtain very high resolution spectra. By examining molecular bands it is even possible to separate abundances of different isotopes of lighter elements such as carbon and nitrogen. Strong surface magnetic fields can be measured when spectral lines are split by the Zeeman effect. Doppler line broadening reveals vertical motions in the stellar atmosphere as well as surface rotation, because one edge of a star approaches us while the opposite recedes. The Doppler shift of an entire spectral line measures the radial velocity of a star relative to the Earth. There is a component of the radial velocity owing to the star's motion in the Galaxy. Superimposed on this can be a cyclic motion owing to a binary orbit. In Chapter 9 we shall examine in some detail the information that can be gleaned from stars in binary systems, with the most important being an absolute measure of mass. It is principally the measurement of radial velocities to accuracies of the order of $1\,\mathrm{m\,s^{-1}}$ that allows us to detect planets around other stars. Stellar spectra may also contain emission lines which usually arise from either circumstellar material that is somehow heated or from excited atoms in an expanding shell or stellar wind. The Sun's chromosphere, visible during eclipse of the stellar disc, shows strong emission lines associated with a temperature in excess of 25 000 K, rising to over 10^6 K in the X-ray emitting corona. Both are probably magnetically heated. Other stars show emission associated with magnetic activity and yet others with outflowing material. In extreme cases the shape of such spectral lines can be used to estimate a mass-loss rate from the star Sec. 5.9).

1.4. The Hertzsprung–Russel Diagram

The most important tool of the stellar astronomer is the Hertzsprung–Russell (H–R) diagram. In its most theoretical form it is a plot of the logarithm of luminosity as a function of the decreasing logarithm of effective temperature for a group of stars. Observationally similar diagrams are obtained by plotting an absolute magnitude against a colour, such as decreasing M_V against $B - V$. Different diagrams are related by knowledge of bolometric corrections that depend on model atmospheres for the stars. The origin of the diagram goes back to the early twentieth century when Henry Norris Russell, an American who worked partly in Cambridge at the time, and the Danish astronomer and chemist Ejnar Hertzsprung constructed the first examples. Figure 1.3 is a H–R diagram for a set of well-measured, detached binary stars. These stars are observationally biased to be both relatively nearby and bright so do not reflect the distribution of all stars. Different appearances are found when the apparently brightest or the nearest stars are plotted. The number of luminous stars, both hot and blue and cool and red, increases because, though they are rarer these can be seen, to greater distances. Interestingly stars do not uniformly populate the whole diagram. Most lie on a band extending from hot bright blue stars to cool faint red stars. This is the main sequence and the Sun lies close to its most populated region.

Because the effective temperature T_{eff} of the star plotted on the x-axis is defined to be the temperature of a black body of the same total luminosity L the two are related to the radius of the star by

$$L = 4\pi\sigma R^2 T_{\text{eff}}^4, \tag{1.12}$$

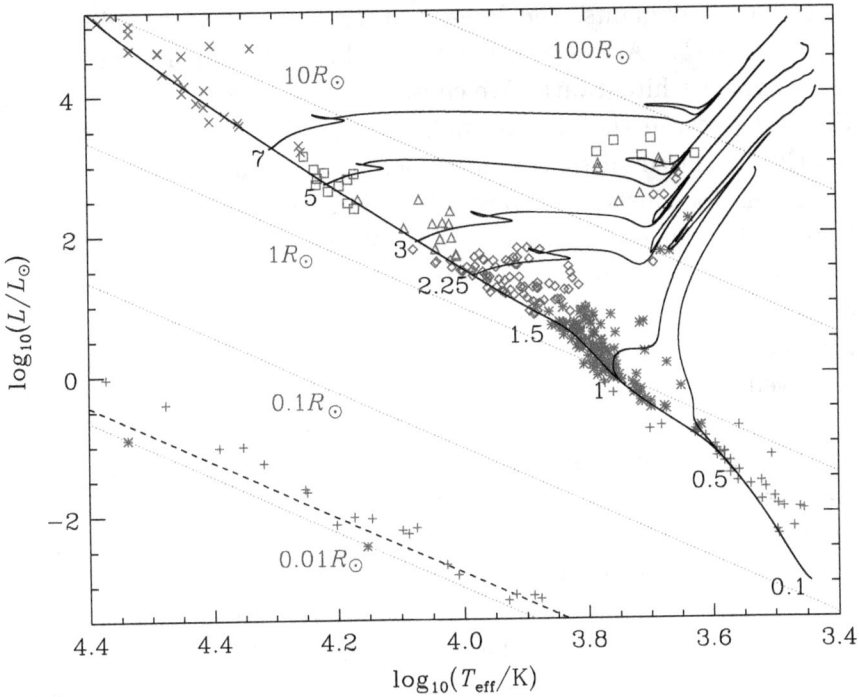

Fig. 1.3. A Hertzsprung–Russell diagram for both components of relatively nearby and bright, well-measured double-lined eclipsing but detached binary stars (see Sec. 9.4.3) in DEBCat (Southworth, 2015) and eclipsing white dwarf binary stars (Parsons *et al.*, 2017). The stars are identified by mass such that upright crosses (+) have $M < 0.75\,M_\odot$, asterisks (*) have $0.75 < M/M_\odot < 1.5$, diamonds (◊) have $1.5 < M/M_\odot < 2.5$, triangles (△) have $2.5 < M/M_\odot < 4$, squares (□) have $4 < M/M_\odot < 6$ and St. Andrew's crosses (×) have $M > 6\,M_\odot$. Superimposed solid lines are theoretical zero-age main-sequence and stellar evolution tracks, labelled in solar masses, described in Chapter 8 and the dashed line is the track of a cooling white dwarf of mass $0.6\,M_\odot$. Dotted lines are of constant radius as labelled.

where σ is the Stefan–Boltzmann constant, and the loci of stars of the same radius are straight lines of slope -4 in a H–R diagram. So stars at the top left of the main sequence are blue giants or supergiants and those at the bottom right are red dwarfs. A second prominent group of stars is seen to form a near vertical strip rising from the redder end of the main sequence. It becomes more populated and extended to higher luminosities in a diagram of the brightest stars.

These are red giants, the largest observed stars with radii up to about $1000 R_\odot$. At the other extreme, below the main sequence lie the compact white dwarfs. We chose to use detached binary stars for Fig. 1.3 because they are at well determined distances. Otherwise poorly measured distances introduce an artificial dispersion in L. Diagrams for clusters of stars, all at almost the same distance, or indeed stars in other galaxies, do not suffer from such dispersion even when the absolute distance cannot be well determined. Typical globular clusters are made up of around a million stars, often of the same age and metallicity, many of which can be resolved when observed by the likes of the *Hubble Space Telescope*. In these high precision cluster H–R diagrams the main sequence is truncated to the blue and stars are seen to turn off towards the giants. A few stars, known as blue stragglers are seen to lie blueward of the main sequence. The giant branch itself splits into two distinct parts, the normal red giants (RGB) and the asymptotic giant branch (AGB), and a red clump, often with a prominent horizontal branch extending to the blue, appears at the base of the AGB. These features can be seen in the H–R diagram of the bright globular cluster ω Cen shown in Fig. 1.4. In H–R diagrams of nearby stars of any magnitude, the fainter but relatively common white dwarfs appear in a strip, scattered around a line of constant radius that is below the main sequence. Other groups of stars appear in separate but distinct regions. The supergiants extend from blue to red across the very top of the diagram. Subgiants lie between the main sequence and the true red giants. Horizontal branch stars extend bluewards from the red clump in many, but not all, globular clusters. Figure 1.5 schematically identifies these various groupings. A good theory of stellar evolution must explain how each of these regions is populated.

Comparison of the H–R diagrams of associations or clusters of stars reveals further clues to stellar evolution. Often all the stars in the cluster formed together, or at least within a relatively short time compared with stellar evolution time-scales, so that today they have the same age. The H–R diagram of the very young association NGC 604 is shown in Fig. 1.6. Most of the stars are on the main sequence or near to it. Such associations tend to be rather young. As time passes, they lose mass because stars are ejected, drive strong

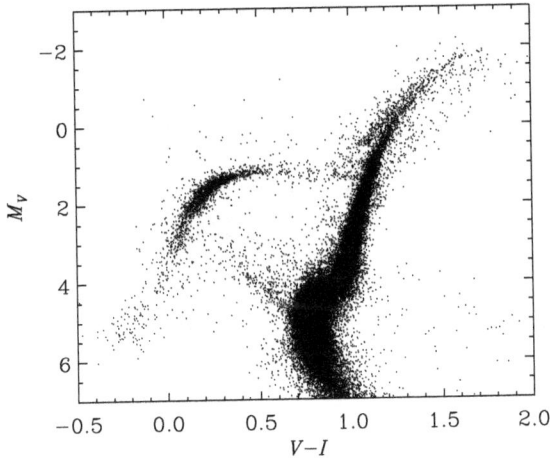

Fig. 1.4. A Hertzsprung–Russell diagram, plotted as a colour–magnitude diagram, for stars in the old populations of the globular cluster ω Centauri. The stellar data are from the compilation by Bellini *et al.* (2009) and we have plotted only those stars which they are 90% confident are members of the cluster. The main-sequence turn off, blue stragglers, horizontal branch and the distinction between the red giant and asymptotic giant branches are all resolved.

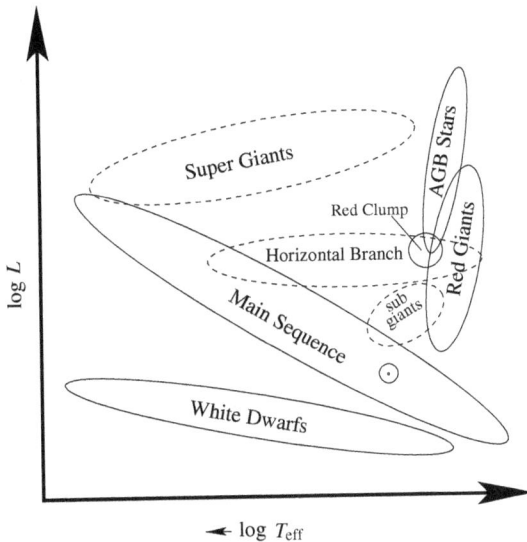

Fig. 1.5. A composite schematic Hertzsprung–Russell diagram illustrating the various groups of stars to be explained by a theory of stellar evolution. Groups enclosed in solid ellipses show up in almost all H–R diagrams. Rarer groups, that appear in some cases are enclosed by dashed ellipses.

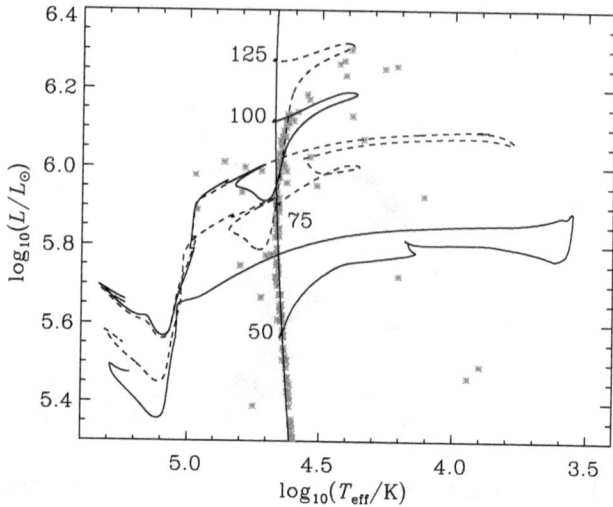

Fig. 1.6. A Hertzsprung–Russell diagram for the stellar population in the young star-forming association NGC 604 in the galaxy M33. Data are from Eldridge and Relaño (2011). Evolutionary tracks for very massive stars are overlayed and we see that the main sequence in this cluster extends to around 100 $M_\odot$.

winds or explode as supernovae and eventually these stellar nurseries disperse their stars through their host galaxies. Open clusters, such as the Pleiades and Hyades, easily found in the northern sky, are relatively loosely bound groups of typically thousands of stars. The particularly young Pleiades appears to still contain some of the gas from which the stars formed. The more tightly bound globular clusters, such as 47 Tucanae and the more unusual ω Centauri easily found in the Southern hemisphere, or M13 in the constellation of Hercules in the northern, are much denser and older. So analysing the H–R diagrams of clusters we see evidence of stellar evolution with age. The youngest clusters have a full main sequence extending from blue supergiants to red dwarfs while the oldest have only the red end of the main sequence, together with rather more populated giant branches as well as an increasing number of white dwarfs, hinting immediately that their blue main-sequence stars have evolved to giants and beyond. Much of the morphology of cluster H–R diagrams is now explained well by the application of the physics discussed in this book. In Chapter 8 we describe how a typical star such as the

Sun evolves from the main sequence to the red giant branch, back to the horizontal branch or the red clump at its cool end, then on to the asymptotic giant branch and finally to a white dwarf. A white dwarf remnant is left whenever the star has lost enough mass to avoid a supernova explosion. More massive stars end their lives in these spectacular explosive deaths to leave even more exotic remnants, neutron stars or black holes.

1.5. Stellar Masses

Just as the Sun can be weighed by application of Newton's laws and measurement of G so can other stars when they have binary companions. We shall discuss how this is done in Sec. 9.4.2. Unfortunately, absolute masses cannot be found for many stars and very few of these are evolved giants. So most comparisons between observation and theory of stellar evolution must be made through the H–R diagram alone. For stars on the main sequence there exists a well calibrated mass to luminosity relation (see Sec. 7.3) and this is often used to estimate mass. For more massive white dwarfs (Sec. 3.8), for which the surface gravity can be well measured by pressure broadening of spectral lines, an accurate mass can be inferred because the radius is a function of the total mass only.

1.6. Questions

1. The flux of energy emitted by a perfect black body in photons with wavelengths between λ and $\lambda + d\lambda$ is proportional to

$$B_\lambda d\lambda \propto \frac{1}{\lambda^5} \frac{1}{\exp\left(\frac{hc}{k_B T \lambda}\right) - 1} d\lambda,$$

where h is Planck's constant, c is the speed of light and k_B is Boltzmann's constant. Define apparent stellar magnitudes m_U, m_B and m_V by

$$m_U = -2.5\log_{10}(f_{\lambda U}/\mathrm{erg\,cm^{-2}\,s^{-1}\,\mathring{A}^{-1}}) + C_1,$$

$$m_B = -2.5\log_{10}(f_{\lambda B}/\mathrm{erg\,cm^{-2}\,s^{-1}\,\mathring{A}^{-1}}) + C_2 \quad \text{and}$$

$$m_V = -2.5\log_{10}(f_{\lambda V}/\mathrm{erg\,cm^{-2}\,s^{-1}\,\mathring{A}^{-1}}) + C_3,$$

Table 1.2. Main-sequence stars have approximately.

T/K	3000	10 000	40 000
$U - B$	1.2	0.0	-1.15
$B - V$	1.6	0.0	-0.35

where f_λ is the flux received at Earth at wavelengths of $\lambda_U = 3600\,\text{Å}$, $\lambda_B = 4400\,\text{Å}$ and $\lambda_V = 5500\,\text{Å}$.

Normalise the constants C_i so that for a star of temperature $10^4\,\mathrm{K}$, $m_U = m_B = m_V$. Show that the colours $U - B = m_U - m_B$ and $B - V = m_B - m_V$ are independent of stellar distance.

Calculate and plot $U - B$ versus $B - V$ for $T = 3 \times 10^3$, 10^4 and $4 \times 10^4\,\mathrm{K}$. What is the limit as $T \to \infty$? Comment on the effect of measurement error on estimates of stellar temperatures.

How do your calculated colours compare with measured colours listed in Table 1.2?

Interstellar dust absorbs radiation preferentially at ultraviolet wavelengths. A star which is dimmed by E mag at λ_V, is dimmed by $1.3E$ mag at λ_B and by $1.5E$ mag at λ_U. How does interstellar extinction affect position in a colour–colour diagram?

Show that it is possible to define a parameter

$$Q = U - B - K(B - V)$$

for some constant K, to be determined, such that Q is independent of the effect of interstellar extinction. What might such a parameter be used for?

2. Sirius A has an apparent visual magnitude $m_V = -1.5$ and a spectral type A0 V. Its distance is 2.7 pc. Calculate its luminosity and radius in solar units.

Sirius B is a binary companion to Sirius A. It has the same spectral type and $m_V = 8.5$. Calculate its luminosity and radius.

Measurements of the orbits of Sirius A and B give a binary period of 50 yr, mass ratio $M_A/M_B = 3$ and an angular semi-major axis of $8''$.

Calculate the masses and mean densities of both components. Which star is the main-sequence star and what is the other?

[*For an A0 V spectrum you may assume* $T_{\text{eff}} = 10\,000\,K$ *and Bolometric Correction,* $m_V - m_{\text{bol}} = 0.4$. *An absolute bolometric magnitude* M_{bol} *of zero corresponds to a luminosity of* $3.0 \times 10^{35}\,\text{erg s}^{-1}$. *For a binary orbit you may use the fact that the semi-major axis a, total mass M and period P are related by Kepler's law*

$$\frac{a}{\text{AU}} = \left(\frac{M}{M_\odot}\right)^{1/3}\left(\frac{P}{\text{yr}}\right)^{2/3}.]$$

3. The star γ^2 Velorum is a binary system comprising the nearest Wolf-Rayet star to the Sun and an O star. It is an astrometric and spectroscopic binary. Observations reveal a parallax of 0.00292", apparent visual magnitudes of $m_V = 1.78$ for the O star and $m_V = 4.27$ for the WR star. The angular separation of the stars is 0.00357", the inclination of the binary orbit to the line of sight is $i = 67.4°$ and the orbital period is 78.53 d. The radial velocity semi-amplitude of the WR star is $K_{\text{WR}} = 122\,\text{km s}^{-1}$ and of the O star is $K_O = 38.4\,\text{km s}^{-1}$. Bolometric corrections for the WR and O star are 4.85 and 2.91 and their radii are $6\,R_\odot$ and $17\,R_\odot$ respectively.

Determine the distance d to the binary system, its orbital separation a, the absolute magnitudes M_V of the two stars, their luminosities and surface temperatures.

Use Kepler's third law and the relation between mass ratio and radial velocities (see Sec. 9.4.2) to determine the masses of the two stars.

[*You may wish to refer to Chapter 9 or simply use Newton's laws for a circular orbit.*]

4. A meteorite, believed to have formed soon after the birth of the solar system, contains minerals that incorporate both lead and uranium. Uranium has two long-lived isotopes, ^{238}U with a half-life of 4.683×10^9 yr and ^{235}U with a half-life of 7.038×10^8 yr. The decay of ^{238}U proceeds through a series of short-lived nuclides that

terminates at the stable ^{206}Pb. Similarly the ^{235}U series terminates at ^{207}Pb. Let the number density of each of these isotopes at time t be $n_i(t)$ where i is the atomic weight of the isotope. The meteorite formed at $t = 0$ and neither uranium nor its daughters, including lead, can escape from the mineral.

Given that no new uranium can be created naturally after the formation of the solar system, show that the ratio of number densities of the uranium isotopes in any mineral at time t is independent of the amount of uranium present at $t = 0$. This ratio can be measured accurately in terrestrial uranium ores to be $n_{238}/n_{235} = 137.8$.

Show further that, in a particular mineral,

$$n_{207}(t) - n_{207}(0) = [n_{206}(t) - n_{206}(0)] \frac{n_{235} \left(e^{\lambda_{235} t} - 1\right)}{n_{238} \left(e^{\lambda_{238} t} - 1\right)},$$

where $\lambda_i n_i$ is the rate of decay of the number density of isotope i.

The isotope ^{204}Pb is not the product of any radioactive decay and so its number density in any inert mineral has remained unchanged since it formed. In the Canyon Diablo iron meteorite $n_{206}/n_{204} = 9.46$ and $n_{207}/n_{204} = 10.34$ with measurement errors of about 1% while the Nuevo Laredo stone meteorite has $n_{206}/n_{204} = 50.28$ and $n_{207}/n_{204} = 34.86$ with measurement errors of about 1%. Both are believed to have formed at the same time early during the formation of the solar system. Deduce that

$$\frac{n_{235} \left(e^{\lambda_{235} t} - 1\right)}{n_{238} \left(e^{\lambda_{238} t} - 1\right)} \approx 0.6,$$

within about 3% and show that this is consistent with an age for the solar system of $t > 4.55 \pm 0.07 \times 10^9$ yr.

5. A type Ia supernova is typically 2.5 magnitudes brighter than a core-collapse supernova. In a volume limited survey, that includes all supernovae within a fixed distance from us, approximately 30% of supernovae are type Ia. Estimate the fraction of type Ia supernovae seen in a magnitude limited survey, that includes all supernovae brighter than a fixed apparent magnitude.

Chapter 2

The Equations of Stellar Structure

Four differential equations are required to model the structure of a star. These are actually partial differential equations in space, or mass, and time but we shall often treat a star as quasi-static and solve the equations of stellar structure in space at a given time and then consider changes, including in composition, with time as evolution from one such quasi-static model to another. Two equations describe the physical structure of the star, its distribution of mass and pressure. These form the body of this chapter. The other two equations for the transport and generation of energy determine the luminosity and temperature through the star. We consider them in Chapters 4 and 6.

2.1. Physical Structure

We begin our analysis with the assumption that a star is non-rotating, spherically symmetric and without internal motion. Though none of these assumptions is truly valid for any star we shall justify them for the case of the Sun but also note that all are violated by most stars at some stage in their lives.

First the hydrodynamic equation of mass conservation is

$$\frac{\partial \rho}{\partial t} = -\mathbf{\nabla} \cdot (\rho \mathbf{v}).$$ (2.1)

In a spherically symmetric quasi-static star the velocity $\boldsymbol{v} = \boldsymbol{0}$ and so the density $\rho(r) = \text{const}$. We may therefore write the first equation of stellar structure as

$$\frac{dm}{dr} = 4\pi r^2 \rho, \tag{2.2}$$

where $m(r)$ is the mass within a radius r measured from the centre of the star where $m = 0$ and $r = 0$. At the surface $m = M$, the total mass of the star, and $r = R$, its radius. For a given mass m the appropriate radius r changes quasi-statically with time.

The second equation is derived from the equation of motion for fluid in the star, the Navier–Stokes equation,

$$\rho \left(\frac{\partial \boldsymbol{v}}{\partial t} + (\boldsymbol{v} \cdot \boldsymbol{\nabla}) \boldsymbol{v} \right) = -\boldsymbol{\nabla} \cdot \mathbf{P} + \boldsymbol{F}, \tag{2.3}$$

where $\mathbf{P}$ is the general symmetric stress tensor, with components P_{ij} for the fluid. Its divergence has components $(\boldsymbol{\nabla} \cdot \mathbf{P})_i = \partial P_{ij}/\partial x_j$, where we use the Einstein summation convention that repeated suffices implicitly indicate a sum of the form $a_i b_i \equiv \sum_{i=1}^{3} a_i b_i$. For a perfect non-viscous gas $P_{ij} = P\delta_{ij}$, where P is the pressure so that Eq. (2.3) reduces to the Euler momentum equation,

$$\rho \left(\frac{\partial \boldsymbol{v}}{\partial t} + (\boldsymbol{v} \cdot \boldsymbol{\nabla}) \boldsymbol{v} \right) = -\boldsymbol{\nabla} P + \boldsymbol{F}. \tag{2.4}$$

Usually microscopic, or atomic-scale, viscosity is negligible in stellar material but it is important in degenerate material (Sec. 3.7) in which case we would add a Stokes' stress

$$-\mu \left(\frac{\partial v_i}{\partial x_j} + \frac{\partial v_j}{\partial x_i} \right), \tag{2.5}$$

to P_{ij}. Otherwise the Reynolds number is generally high so that any flow is turbulent and can be usefully written as

$$\boldsymbol{v} = \bar{\boldsymbol{v}} + \boldsymbol{v}', \tag{2.6}$$

where $\boldsymbol{v}'$ varies on short time- and length-scales compared with the mean flow $\bar{\boldsymbol{v}}$ but can transport momentum leading to a Reynolds stress

$$-\overline{\rho v_i' v_j'} \tag{2.7}$$

to be added to P_{ij}. A similar approach can be used to describe convective motions. Near the surface of the Sun the velocity of rising and falling convective cells is $v' \approx 0.1\,\mathrm{km\,s^{-1}}$ and the scale of the granulation is about $1000\,\mathrm{km}$ so that, for a density $\rho \approx 1\,\mathrm{g\,cm^{-3}}$, the Reynolds stress contribution is $|\nabla \cdot \mathbf{P}|/\rho \approx 0.01\,\mathrm{m\,s^{-2}} \ll g \approx GM_\odot/R_\odot^2 \approx 300\,\mathrm{m\,s^{-2}}$. In the presence of a significant magnetic field of strength $\mathbf{B}$ we would also need to add the Maxwell stress,

$$P_{ij} = \frac{1}{\mu_0}\left(\frac{1}{2}B^2\delta_{ij} - B_i B_j\right), \tag{2.8}$$

where μ_0 is the permeability of free space and $B = |\mathbf{B}|$. The typical magnetic field strength at the surface of the Sun is $B \approx 2\,\mathrm{G}$ and varies on scales of $1,000\,\mathrm{km}$ so the magnetic contribution to $|\nabla \cdot \mathbf{P}|/\rho \approx 3 \times 10^{-3}\,\mathrm{m\,s^{-2}} \ll g$. However the magnetic field is known to be much stronger deep inside the Sun, rising by a factor of about 10^4 at the tachocline just below the base of the convective envelope. However the thickness of the tachocline is about $0.04\,R_\odot$, much larger than the size of the surface granulation, and $|\nabla \cdot \mathbf{P}|/\rho \approx 0.1\,\mathrm{m\,s^{-2}} \ll g$. So even though the magnetic field of the Sun is not well known, we can assume its contribution to be negligible. Any body forces are included in $\mathbf{F}$ but only gravity is important and $\mathbf{F}/\rho = \mathbf{g}$ the gravitational acceleration.

We are interested in the bulk properties of the star. Assuming the star grows or shrinks on a time-scale τ the contribution from the first term on the left-hand side of Eq. (2.4) can be estimated with

$$\frac{\partial v}{\partial t} \approx \frac{R}{\tau^2}. \tag{2.9}$$

This is only important if $\tau \approx \tau_{\mathrm{dyn}} \approx \frac{1}{2}\mathrm{hr}$ so it is certainly negligible for the Sun, the radius of which has not noticeably changed in several thousands of years of eclipse observations. From models we know that the Sun is expanding at about one inch every year and this will not change much before the end of its main-sequence life and $\tau \approx 10^{10}$ yr. So, on the evolutionary time-scales in which we are interested, the first term in Eq. (2.4) can be neglected. The largest bulk contribution to the second term comes from rotation and is largest at the solar equator. The Sun rotates slowly with a spin period $2\pi/\Omega \approx 30$ d so, in its rotating frame, the centrifugal acceleration $R\Omega^2 \approx 0.6\,\mathrm{cm\,s^{-2}} \ll g$ and can also be neglected. Other short time-scale but large length-scale variations in the Sun are measured by helioseismology. Oscillations of the Sun drive surface velocity variations of the order of $0.5\,\mathrm{km\,s^{-1}}$ varying on a time-scale of about 5 min and a length-scale of about the solar radius. For these the first term in Eq. (2.4) is of order $4\,\mathrm{m\,s^{-2}}$ and the second $4 \times 10^{-4}\,\mathrm{m\,s^{-2}}$ both much smaller than g. Though small and unimportant in the Sun, similar oscillations become somewhat more significant in potently pulsating stars such as Cepheid, RR Lyrae and Mira variables.

Overall our assumptions of static non-rotating stars without significant internal flows are quite generally valid and conservation of momentum reduces to hydrostatic equilibrium, the second equation of stellar structure,

$$\frac{\mathrm{d}P}{\mathrm{d}r} = -\rho g = -\frac{Gm\rho}{r^2}. \tag{2.10}$$

A simple boundary condition at the stellar surface is that $\rho = 0$ when $m = M$ but we shall look at this more carefully in Chapter 5. We also require $\mathrm{d}P/\mathrm{d}r \to 0$ at the centre because, physically, there cannot be a cusp in the acceleration there.

2.2. Polytropes

With only these two equations we are now able to begin to create very simple models of the interior structure of stars. We can achieve this if the pressure is a function of density only, $P = P(\rho)$. With this closure Eqs. (2.2) and (2.10) can be solved. In practice the equation of state of stellar material is somewhat more complicated (see Chapter 3) but certain stars such as cool white dwarfs and fully convective stars can be well approximated in this way.

Simplifying further suppose that $P \propto \rho^\gamma$ so we may write

$$P = K\rho^{1+\frac{1}{n}}, \tag{2.11}$$

where K is a constant and n is known as the polytropic index. Differentiating Eq. (2.2) with respect to r and substituting in Eq. (2.10) we find

$$\frac{1}{r^2}\frac{d}{dr}\left(\frac{r^2}{\rho}\frac{dP}{dr}\right) = -4\pi G\rho. \tag{2.12}$$

We remove dimensions by scaling the density to the central density ρ_c, as $\rho = \rho_c\theta^n$, with $\theta = 1$ at $r = 0$ and $\theta = 0$ at the stellar surface $r = R$. Then

$$P = K\rho_c^{1+\frac{1}{n}}\theta^{n+1} = P_c\theta^{n+1} \tag{2.13}$$

and

$$\left[\frac{(n+1)K}{4\pi G}\rho_c^{\frac{1}{n}-1}\right]\frac{1}{r^2}\frac{d}{dr}\left(r^2\frac{d\theta}{dr}\right) = -\theta^n. \tag{2.14}$$

Removing the radial dimension by setting $r = \alpha\xi$ with

$$\alpha^2 = \frac{(n+1)}{4\pi G}K\rho_c^{\frac{1}{n}-1} \tag{2.15}$$

we obtain the Lane–Emden equation,

$$\frac{1}{\xi^2}\frac{d}{d\xi}\left(\xi^2\frac{d\theta}{d\xi}\right) = -\theta^n. \tag{2.16}$$

Solving this we are able to create simple models of the interior structure of stars of a particular polytropic index n. As an example, an isentropic monatomic gas has $\gamma = 5/3$, so a star composed entirely of such material is a polytrope of index $n = 3/2$.

The central boundary conditions are $\theta = 1$ and, because $dP/dr = 0$, $d\theta/d\xi = 0$ when $\xi = 0$. The surface of the star is at $\xi = X_n$, where θ first vanishes for index n.

2.2.1. *Solutions to the Lane–Emden equation*

There are three analytic solutions to the equations with the required boundary conditions,

$$n = 0, \qquad \theta = 1 - \frac{\xi^2}{6}, \qquad\qquad X_0 = \sqrt{6} = 2.45, \qquad (2.17)$$

$$n = 1, \qquad \theta = \frac{\sin \xi}{\xi}, \qquad\qquad X_1 = \pi, \qquad (2.18)$$

$$n = 5, \qquad \theta = \left(1 + \frac{\xi^2}{3}\right)^{-1/2}, \qquad X_5 \to \infty. \qquad (2.19)$$

Other solutions can be found by relatively straightforward numerical integration of the initial value problem from $\xi = 0$ until $\theta = 0$. Example integrations are shown in Fig. 2.1 and their scaled radii are listed in Table 2.1 where X_n is the first zero of θ for the given n. The radius of the star is then

$$R = \alpha X_n. \qquad (2.20)$$

2.2.2. *Mass*

With solutions to the Lane–Emden equation we can also calculate the mass of a star of a given radius and polytropic index. The internal mass $m(r)$ is given by the integral

$$m = \int_0^r 4\pi \rho r'^2 \, dr' = 4\pi \rho_c \alpha^3 \int_0^\xi \theta^n \xi'^2 \, d\xi'. \qquad (2.21)$$

Into this we substitute θ^n from the Lane–Emden equation so that

$$m = -4\pi \rho_c \alpha^3 \int_0^\xi \frac{d}{d\xi'} \left(\xi'^2 \frac{d\theta}{d\xi'}\right) d\xi' = -4\pi \rho_c \alpha^3 \xi^2 \frac{d\theta}{d\xi}. \qquad (2.22)$$

and the total mass

$$M = 4\pi \left[\frac{(n+1)K}{4\pi G}\right]^{3/2} \rho_c^{\frac{3-n}{2n}} \left(-\xi^2 \frac{d\theta}{d\xi}\right)_{\xi=X_n}. \qquad (2.23)$$

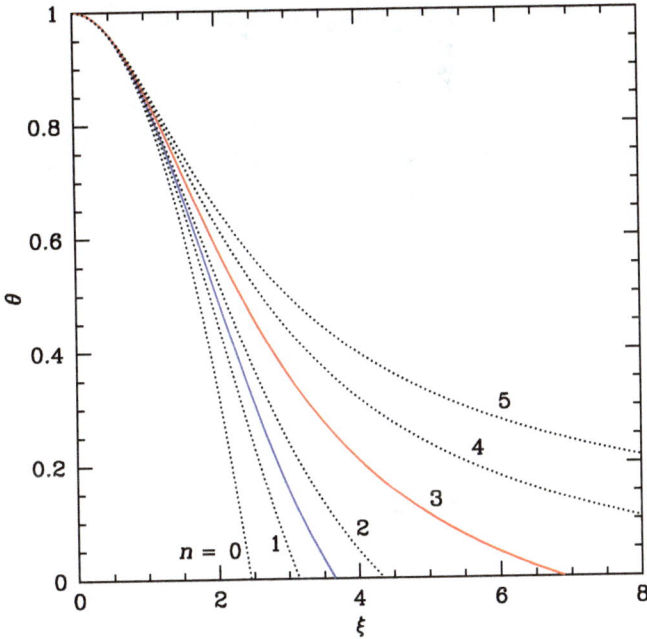

Fig. 2.1. Numerical solutions to the Lane–Emden equations for $n = 0$, 1, $3/2$ (solid blue), 2, 3 (solid red), 4 and 5. The higher the polytropic index the more diffuse the star and so the larger $\xi = X_n$ at the first zero of θ for index n. The $n = 4$ solution has its first zero at $\xi = 15.0$ while the $n = 5$ solution has no zeros. The solution for $n = 3/2$ is a good model for non-relativistic degenerate stars and convective stars. The $n = 3$ solution is a special case in which the dimensionless mass Y_n is independent of radius. This is the case for extremely relativistic degenerate matter and was used by Eddington to investigate radiation pressure support in stars. These special cases are discussed further in Chapter 3.

Table 2.1. Dimensionless radii X_n and masses Y_n and central condensations of polytropes.

n	0	1	3/2	2	3	4	5
X_n	2.45	3.14	3.65	4.35	6.90	15.0	∞
Y_n	4.90	3.14	2.71	2.41	2.02	1.80	1.73
$\rho_c/\bar{\rho}$	1.00	3.29	5.99	11.4	54.2	622	∞

Noting that it depends only on n we define

$$Y_n = \left(-\xi^2 \frac{d\theta}{d\xi}\right)_{\xi=X_n}. \tag{2.24}$$

For the analytical solutions

$$n = 0, \quad Y_0 = 2\sqrt{6} = 4.90, \tag{2.25}$$
$$n = 1, \quad Y_1 = \pi, \tag{2.26}$$
$$n = 5, \quad Y_5 = \sqrt{3} = 1.73 \tag{2.27}$$

and we combine these with numerical calculations in Table 2.1. For $n = 3$, M is independent of ρ_c. When $n = 5$, M is finite but $R \to \infty$ and for $n > 5$ there are no solutions of finite M or R.

2.2.3.   *Central condensation*

To better understand the physical significance of n we consider the ratio of the central density ρ_c to the overall mean density $\bar{\rho}$ of the object,

$$\bar{\rho} = \frac{M}{\frac{4}{3}\pi R^3} = \left(-\frac{3}{\xi}\frac{d\theta}{d\xi}\right)_{\xi=X_n} \rho_c. \tag{2.28}$$

The central concentration $\rho_c/\bar{\rho}$ is also a function of n only and we list it for various indices in Table 2.1. Recall for the Sun $\rho_c/\bar{\rho} \approx 100$. In fact an $n = 3.25$ polytrope fits the structure of the Sun quite well. Its mass $M_\odot$ and radius $R_\odot$ then determine the appropriate ρ_c and K.

2.2.4.   *Mass–radius relation*

It is interesting to consider how the radius of a polytrope relates to its total mass. We know that

$$R = \alpha X_n \propto K^{1/2} \rho_c^{(1-n)/2n}, \tag{2.29}$$

while

$$M \propto K^{3/2} \rho_c^{(3-n)/2n} \tag{2.30}$$

so, eliminating ρ_c, we find

$$M^{n-1} \propto R^{n-3} K^n. \tag{2.31}$$

For fixed K,

$$R \propto M^{(n-1)/(n-3)}. \tag{2.32}$$

In particular, when $n = 3/2$, $R \propto M^{-1/3}$. This is important for low-mass white dwarfs and convective stars for which $\gamma = \frac{5}{3}$ because it shows that they shrink in radius as they grow in mass. More generally,

- if $n < 1$, R grows as M increases,
- if $1 < n < 3$, R shrinks as M increases and
- if $n > 3$, R grows as M increases,

all at fixed K. We note that $n = 3$ corresponds to $P \propto \rho^{4/3}$. We shall see in Chapter 3 that this limit is approached by stars for which support is dominated by either radiation pressure or relativistic degeneracy pressure.

2.3.   The Virial Theorem

Integrating the equation of motion of stellar material we obtain the virial theorem for stellar objects. This is a general integral theorem for stellar structure with which we can make useful and widely applicable deductions about stars and their evolution.

Consider constructing a star by bringing together fluid elements of masses δm at position vectors $\boldsymbol{r}(t)$ relative to some origin (Fig. 2.2).

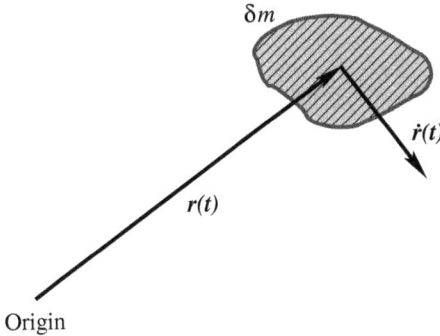

Fig. 2.2.   The position $\boldsymbol{r}(t)$ and velocity $\dot{\boldsymbol{r}}(t)$ of a small packet of material of mass δm relative to an arbitrary origin.

The equation of motion of each fluid element can be written as

$$\ddot{\boldsymbol{r}} = -\frac{1}{\rho}\boldsymbol{\nabla}\cdot\mathbf{P} + \boldsymbol{F}, \tag{2.33}$$

where $\mathbf{P}$ is again the general stress tensor, with components P_{ij}. Any other body forces are included in $\boldsymbol{F}$. Once again only gravity is important and $\boldsymbol{F} = \boldsymbol{g}$.

We take the dot product with $\boldsymbol{r}$ and integrate over the whole star of mass M and volume V, noting that a volume element $\delta V = (1/\rho)\delta m$, to obtain

$$\int_M \boldsymbol{r}\cdot\ddot{\boldsymbol{r}}\,\mathrm{d}m = -\int_V \boldsymbol{r}\cdot(\boldsymbol{\nabla}\cdot\mathbf{P})\,\mathrm{d}V + \int_M \boldsymbol{r}\cdot\boldsymbol{F}\,\mathrm{d}m. \tag{2.34}$$

The left-hand side of this equation can be expanded with

$$\boldsymbol{r}\cdot\ddot{\boldsymbol{r}} = \frac{\mathrm{d}}{\mathrm{d}t}(\boldsymbol{r}\cdot\dot{\boldsymbol{r}}) - \dot{\boldsymbol{r}}\cdot\dot{\boldsymbol{r}} = \frac{1}{2}\frac{\mathrm{d}^2}{\mathrm{d}t^2}(\boldsymbol{r}\cdot\boldsymbol{r}) - \dot{\boldsymbol{r}}\cdot\dot{\boldsymbol{r}} \tag{2.35}$$

to obtain

$$\int_M \boldsymbol{r}\cdot\ddot{\boldsymbol{r}}\,\mathrm{d}m = \frac{1}{2}\ddot{I} - 2T, \tag{2.36}$$

where $I = \int_M \boldsymbol{r}\cdot\boldsymbol{r}\,\mathrm{d}m$ is a moment of inertia and $T = \frac{1}{2}\int_M(\dot{\boldsymbol{r}}\cdot\dot{\boldsymbol{r}})\,\mathrm{d}m$ is the internal kinetic energy which includes such motions as rotation and turbulence.

For the first term on the right-hand side of Eq. (2.34), we note that

$$\boldsymbol{r}\cdot(\boldsymbol{\nabla}\cdot\mathbf{P}) = x_i\frac{\partial}{\partial x_j}P_{ij} = \frac{\partial}{\partial x_j}(x_i P_{ij}) - P_{ii}, \tag{2.37}$$

and write $P_{ii} = \mathrm{Trace}\,\mathbf{P} = 3P$. So, applying the divergence theorem, this term becomes

$$3\int_V P\,\mathrm{d}V - \int_V \boldsymbol{\nabla}\cdot(\boldsymbol{r}\cdot\mathbf{P})\,\mathrm{d}V = 3\int_V P\,\mathrm{d}V - \int_S (\boldsymbol{r}\cdot\mathbf{P})\cdot\mathrm{d}\boldsymbol{S}, \tag{2.38}$$

where the surface S encloses the volume V that defines the star. If on the surface of the star $\mathbf{P}$ is isotropic and $P_{ij} = P_s\delta_{ij}$ is a constant then at the surface we have

$$(\boldsymbol{r} \cdot \mathbf{P})_i = r_j P_{\mathrm{s}} \delta_{ij} = P_{\mathrm{s}} r_i \tag{2.39}$$

so, again by the divergence theorem,

$$\int_S (\boldsymbol{r} \cdot \mathbf{P}) \cdot \mathrm{d}\boldsymbol{S} = P_{\mathrm{s}} \int_S \boldsymbol{r} \cdot \mathrm{d}\boldsymbol{S} = P_{\mathrm{s}} \int_V \boldsymbol{\nabla} \cdot \boldsymbol{r} \, \mathrm{d}V = 3 P_{\mathrm{s}} V. \tag{2.40}$$

The second term on the right-hand side of Eq. (2.34), often referred to as *the virial of Clausius*, is the internal gravitational energy of the star. To see this consider gravity acting alone so that $\boldsymbol{F} = -\boldsymbol{\nabla}\Phi$ with

$$\Phi = -\int_M \frac{G \mathrm{d}m'}{|\boldsymbol{r} - \boldsymbol{r}'|} \tag{2.41}$$

for an isolated star. We write

$$\int_M \boldsymbol{r} \cdot \boldsymbol{F} \, \mathrm{d}m$$

$$= -\int_M \boldsymbol{r} \cdot \boldsymbol{\nabla}\Phi \, \mathrm{d}m$$

$$= G \int_M \int_M \boldsymbol{r} \cdot \boldsymbol{\nabla} \left(\frac{\mathrm{d}m'}{|\boldsymbol{r} - \boldsymbol{r}'|} \right) \mathrm{d}m$$

$$= G \int_M \int_M \boldsymbol{r}' \cdot \boldsymbol{\nabla}' \left(\frac{\mathrm{d}m}{|\boldsymbol{r}' - \boldsymbol{r}|} \right) \mathrm{d}m'$$

$$= \frac{1}{2} G \int_M \int_M \boldsymbol{r} \cdot \boldsymbol{\nabla} \left(\frac{1}{|\boldsymbol{r} - \boldsymbol{r}'|} \right) + \boldsymbol{r}' \cdot \boldsymbol{\nabla}' \left(\frac{1}{|\boldsymbol{r} - \boldsymbol{r}'|} \right) \mathrm{d}m \, \mathrm{d}m'$$

$$= -\frac{1}{2} G \int_M \int_M \frac{\boldsymbol{r} \cdot (\boldsymbol{r} - \boldsymbol{r}')}{|\boldsymbol{r} - \boldsymbol{r}'|^3} - \frac{\boldsymbol{r}' \cdot (\boldsymbol{r} - \boldsymbol{r}')}{|\boldsymbol{r} - \boldsymbol{r}'|^3} \, \mathrm{d}m \, \mathrm{d}m'$$

$$= -\frac{1}{2} G \int_M \int_M \frac{\mathrm{d}m \, \mathrm{d}m'}{|\boldsymbol{r} - \boldsymbol{r}'|} = \Omega, \tag{2.42}$$

the gravitational energy required to assemble the star from infinity.

Combining the terms we arrive at a useful form for the virial theorem,

$$\frac{1}{2}\ddot{I} = 2T + 3\int_V P\,dV - 3P_sV + \Omega. \tag{2.43}$$

For an ideal gas this simplifies somewhat. Let the ratio of specific heat capacities at constant pressure and constant volume $c_P/c_V = \gamma$ so that $P = (\gamma-1)u$, where u is the internal energy per unit volume (see Sec. 3.4). For a monatomic gas with $\gamma = \frac{5}{3}$, $P = nkT$ and $u = \frac{3}{2}nkT$, where n is the particle number density and k is Boltzmann's constant. Generally, for constant γ through the star, we have

$$\int_V P\,dV = (\gamma - 1)U, \tag{2.44}$$

where U is the total internal energy. For a star in a vacuum the surface pressure $P_s = 0$ and so we arrive at the form

$$\frac{1}{2}\ddot{I} = 2T + 3(\gamma - 1)U + \Omega, \tag{2.45}$$

applicable to typical stars.

2.3.1.   *Estimates for stars in equilibrium*

Consider the case of a non-rotating dynamically stable star for which $T = 0$ and I is constant so that

$$U = -\frac{1}{3(\gamma - 1)}\Omega. \tag{2.46}$$

Generally $\gamma > 1$ for any real gas. In particular for a monatomic ideal gas, with $\gamma = 5/3$,

$$U = -\frac{1}{2}\Omega = \frac{3}{2}\overline{P}V, \tag{2.47}$$

where $\overline{P}$ is the pressure averaged over the volume of the star. The gravitational potential can be written in the form

$$\Omega = -\eta\frac{GM^2}{R}, \tag{2.48}$$

where η is a dimensionless structural factor of order unity, while the volume of the star

$$V = \frac{4}{3}\pi R^3.$$ (2.49)

Thus

$$\bar{P} = \frac{\eta GM^2}{4\pi R^4} \approx 10^{15}\left(\frac{M}{M_\odot}\right)^2\left(\frac{R_\odot}{R}\right)^4 \text{ dyn cm}^{-3}.$$ (2.50)

Models tell us that the pressure at the centre of a star $P_c \approx \bar{P}$ and, again for a perfect gas $P \approx \rho \mathcal{R} T$, where $\mathcal{R}$ is the gas constants so

$$\bar{T} \approx \frac{\bar{P}}{\bar{\rho}\mathcal{R}} \approx 10^7\left(\frac{M}{M_\odot}\right)\left(\frac{R_\odot}{R}\right) \text{ K},$$ (2.51)

which is close to the temperature at which hydrogen fusion can begin (see Chapter 6). Conversely we could reverse the argument and argue that stars must attain a structure in which their central temperature is high enough for nuclear reactions and so estimate their typical radii on the main sequence.

2.3.2. *Net energy*

The total or net energy E of a star is the sum of its internal and gravitational energies

$$E = U + \Omega = U - 3(\gamma - 1)U = -(3\gamma - 4)U = \frac{3\gamma - 4}{3(\gamma - 1)}\Omega.$$ (2.52)

The assembly of a star from matter at infinity is, by the attractive nature of gravity, always energetically favourable. Therefore a star's gravitational energy $\Omega < 0$ always. The star can remain bound if and only if its total energy $E < 0$. That is, if and only if $\gamma > \frac{4}{3}$. In Chapter 3 we shall see that objects supported entirely by either radiation pressure or degeneracy pressure of extremely relativistic electrons behave as if $\gamma = \frac{4}{3}$. In these extreme limiting cases stars behave as $n = 3$ polytropes, which cannot remain in a bound equilibrium state.

A further consequence of Eq. (2.52) is that a bound star in equilibrium must shrink if its only source of energy is gravity. As a star radiates $\dot{E} < 0$. Its luminosity $L = -\dot{E}$. So $\dot{\Omega} < 0$ and $\dot{U} > 0$. Without any other source of energy, as a star radiates, it contracts and heats up. This is the essence of star formation and evolution. A gas cloud contracts until its central temperature becomes high enough for nuclear fusion to begin and halt the collapse. Whenever a nuclear energy source is exhausted in the core of a star the first response is for the ashen core to contract and heat up.

2.3.3.   *Pulsations*

Like any confined fluid stars wobble if they are perturbed. The Sun oscillates at many frequencies with periods of the order of minutes. For small perturbations we can treat their kinetic energy as negligible and so set $T = 0$ and also assume that the total energy E is conserved. Question 4 explores how we can then use the virial theorem in the form of Eq. (2.45) to make a good estimate of the typical oscillation time-scale. The fundamental oscillation period of the Sun is about half an hour. The first oscillations to be confirmed had a period of about five minutes. In the Cepheids and RR Lyrae stars pulsations are driven by an opacity instability when hydrogen ionises. Such pulsations cannot be considered small but the longer periods of days reflect the smaller mean densities of these stars. Similarly the Mira variables that are some of the largest giant stars have pulsation periods of years.

2.4.   Questions

1. In any equilibrium configuration prove that

$$\frac{\mathrm{d}}{\mathrm{d}r}\left(P + \frac{Gm^2}{8\pi r^4}\right) < 0,$$

where

$$m(r) = \int_0^r 4\pi \rho r^2\,\mathrm{d}r.$$

Deduce a lower limit for the central pressure P_c in terms of the stellar mass and radius. Evaluate the limit for the Sun.

2. When ρ decreases outwards, show that

$$m < \frac{4}{3}\pi\rho_c r^3.$$

Thence prove that

$$P_c < \frac{1}{2}G\left(\frac{4}{3}\pi\right)^{\frac{1}{3}}\rho_c^{\frac{4}{3}}M^{\frac{2}{3}}.$$

Evaluate a lower limit to the density at the centre of the Sun.

3. Show that the gravitational binding energy

$$\Omega = -4\pi\int_0^R \frac{Gm}{r}\rho r^2\,dr.$$

Hence show that the gravitational binding energy of a polytrope of index n is

$$\Omega = -\frac{3}{5-n}\frac{GM^2}{R}.$$

The recombination energy of stellar matter per unit mass is I. Show that, when the internal structure of a star corresponds to an $n = 3/2$ polytrope, total disruption of the star is energetically feasible if

$$R > \frac{3}{7}\frac{GM}{I}.$$

Given that the ionisation energy of a hydrogen atom is $13.6\,\text{eV}$, estimate the maximum radius for a star of $1\,M_\odot$ and explain the existence of red giants.

4. Using the virial theorem, or otherwise, show that the period Π of spherically symmetric pulsations of a star satisfying

$$I = 4\pi\int_0^R \rho r^4\,dr = kMR^2$$

and

$$\Omega = -\eta GM^2/R,$$

where k and η are constants, is given by

$$\Pi = \left[\frac{3\pi k}{(3\gamma - 4)\eta G \bar{\rho}} \right]^{\frac{1}{2}},$$

where $\bar{\rho}$ is the mean density of the star.

Comment on the relation of Π to the free-fall time and the acoustic travel time from surface to centre.

Chapter 3

The Equation of State

In most stars, the equation of state is not everywhere sufficiently simple that we may write pressure $P = P(\rho)$ as a function of density ρ only. Generally we need two state variables, such as density and temperature T, and a description of the composition, such as the set of mass fractions X_i of nuclide i. An equation of state might then be written in the form $P = P(\rho, T, \{X_i\})$. The first two equations of stellar structure (2.2) and (2.10) are no longer closed and we shall need to include two more, the equation of heat transport and the equation of energy generation that form the subjects of Chapters 4 and 6.

It is often convenient to divide the pressure into contributions from the ions, electrons and radiation,

$$P = P_i + P_e + P_{rad} \tag{3.1}$$

but in a non-degenerate gas the electron contribution can be combined with the ion contribution in a single gas pressure,

$$P_g = P_i + P_e. \tag{3.2}$$

We begin our discussion with this case before examining the contribution of radiation, the importance of which increases with temperature. We then examine the quantum mechanical contribution of electron degeneracy, important in the dense material of the cores of giants and white dwarfs.

3.1.   Gas Pressure

Cool but sufficiently tenuous stellar material is usually well described simply as an ideal gas. This is a reasonably good approximation in the Sun and a very good one outside its relatively dense core and we can write

$$P = P_\mathrm{g} = \rho \frac{\mathcal{R}T}{\mu},$$ (3.3)

where $\mathcal{R}$ is the gas constant and μ is the mean molecular weight of the ions and electrons combined. The gas constant

$$\mathcal{R} = \frac{k}{m_\mathrm{H}} = 8.314 \times 10^7 \,\mathrm{erg\,mol^{-1}\,K^{-1}}$$ (3.4)

is the ratio of Boltzmann's constant k to the atomic mass unit m_H. One mole of hydrogen, or protons, has a mass very close to $1\,\mathrm{g}$ so in the form commonly used for stellar evolution

$$\mathcal{R} \approx 8.314 \times 10^7 \mathrm{erg\,g^{-1}\,K^{-1}}$$ (3.5)

too. Formally m_H is defined to be one-twelfth of the mass of a carbon-12 atom so that one mole of protons actually has a mass of $1.007\,\mathrm{g}$. In our discussions in this chapter we shall ignore small differences from integer atomic masses to make the discussion simpler. They do not make very significant changes to the equation of state but should be taken properly into account in quantitative calculations. On the other hand the differences are essential when we consider nuclear fusion in Sec. 6.2 which involves the conversion of mass to energy.

With SI units a complication arises because we must write the mass of $1\,\mathrm{mol}$ of protons as $10^{-3}\,\mathrm{kg}$ and then

$$P_\mathrm{gas} = \rho \frac{\mathcal{R}^*T}{\mu},$$ (3.6)

where $\mathcal{R}^* = 8.314 \times 10^3 \,\mathrm{J\,kg^{-1}\,K^{-1}}$ while $\mathcal{R}$ is still $8.314\,\mathrm{J\,mol^{-1}\,K^{-1}}$. Though this is easily dealt with it remains a compelling reason to use cgs units for stellar interiors.

3.2. Mean Molecular Weight

The mean molecular weight μ is the mean mass of the particles that contribute to the gas pressure. Consider the ideal-gas law in the form

$$P = \frac{n_{\text{mol}} \mathcal{R} T}{V},$$ (3.7)

for n_{mol} moles of particles in a volume V to see that

$$\frac{1}{\mu} = \text{number of moles per gramme},$$ (3.8)

which is equivalent to the number of particles per nucleon, as long as we make the approximations that the mass of an electron and the difference in the mass of a proton, a neutron and atomic mass unit are negligible. For gases composed purely of a single species we can then just count the number of particles per nucleon to find $1/\mu$.

- Neutral hydrogen atoms have one particle, the atom, per nucleon, the proton, so $\mu = 1$.
- Ionised hydrogen has two particles, a proton and an electron, so $\mu = 1/2$.
- Ionised helium has three, an α-particle and two electrons, per four nucleons, two protons and two neutrons, so $\mu = 4/3$.
- Ionised metals, of atomic weight A and atomic number Z, for which typically $A = 2Z$, have $1 + Z$ particles, the nucleus and Z electrons per A nucleons so $\mu = A/(1 + Z) \approx 2$.

In stars we commonly call any element other than hydrogen or helium a metal so that the most common metals in the composition of a star are usually carbon, oxygen and nitrogen, as well as elements, such as iron, that behave as metals on the Earth. For a mixture that has fractions X_i by mass of each component i, with a pure mean molecular weight μ_i, the number density of particles

$$n = \frac{\rho}{m_{\text{H}}} \left(\sum_i \frac{X_i}{\mu_i} \right) = \frac{\rho}{\mu m_{\text{H}}}$$ (3.9)

so

$$\frac{1}{\mu} = \sum_i \frac{X_i}{\mu_i}.$$

(3.10)

A mixture of fully ionised atoms of hydrogen, with mass fraction $X_H = X$, helium, predominantly ^4He, with mass fraction $X_{He} = Y$, and metals with combined mass fraction[1] Z has

$$\frac{1}{\mu} \approx 2X + \frac{3}{4}Y + \frac{1}{2}Z.$$

(3.11)

Mass fractions must sum to unity so

$$X + Y + Z = 1$$

(3.12)

and then, for a solar-like composition in which $Z = 0.02$, we have

$$\mu = \frac{4}{2.98 + 5X}.$$

(3.13)

For the Sun $X = 0.7$ initially throughout, and still at the surface now, so that $\mu = \mu_{\text{initial}} = 0.617$. As hydrogen fuses to helium X decreases so μ increases. Eventually, when $X = 0$, $\mu = 1.34 > 2\mu_{\text{initial}}$ and hydrogen fusion will have doubled the mean molecular weight at its centre. It is the gradual change in the central μ that currently drives the slow evolution of the Sun.

3.3. Radiation Pressure

Stellar interiors are extremely close to a state of local thermodynamic equilibrium of both particles and radiation (see Chapter 4). The radiation field is therefore that of an almost perfect black body at the local temperature T. To calculate the internal energy per unit volume u and associated pressure P_r, we proceed by recalling Planck's

---

[1]Though there is risk of confusion of the metal mass fraction with atomic number the context always makes a distinction and we prefer to stick with standard notation used elsewhere. We are careful to redefine variables whenever their meaning changes.

derivation of the black-body spectrum. In any cavity individual photons have a wave function $\psi(\mathbf{r}, t)$ that satisfies the wave equation

$$\frac{1}{c^2}\frac{\partial^2 \psi}{\partial t^2} = \nabla^2 \psi, \qquad (3.14)$$

where c is the speed of light. The radiation in a cavity connected by a small hole to a large cubical box of side L and volume $V = L^3$ at the same temperature T must have the same distribution of photons to be in equilibrium. Let us describe such a box with Cartesian coordinates $0 < x < L$, $0 < y < L$ and $0 < z < L$. Photons must reflect from the walls so ψ must vanish on all sides and, for photons of frequency ν, the solutions to Eq. (3.14) are standing waves

$$\psi(x, y, z, t) = \sin\frac{n_x \pi x}{L} \sin\frac{n_y \pi y}{L} \sin\frac{n_z \pi z}{L} \cos 2\pi\nu t, \qquad (3.15)$$

with

$$\frac{\pi^2}{L^2}\left(n_x^2 + n_y^2 + n_z^2\right) = \frac{4\pi^2 \nu^2}{c^2}, \qquad (3.16)$$

with n_x, n_y and n_z positive integers. The number of distinct states with frequencies between ν and $\nu + \delta\nu$ is found by counting the states in the positive octant of sphere of radius $2\nu L/c$ in (n_x, n_y, n_z) space and noting that there are two distinct polarisations for each combination of n_x, n_y and n_z giving a total of

$$N = 2\frac{1}{8}\frac{4}{3}\pi \left(\frac{2\nu L}{c}\right)^3 \qquad (3.17)$$

states with frequency less than ν. Differentiating we find the number density of states with frequency between ν and $\nu + d\nu$

$$n(\nu)\,d\nu = \frac{1}{V}\,dN = \frac{1}{L^3}\frac{dN}{d\nu}\,d\nu = \frac{8\pi\nu^2}{c^3}\,d\nu. \qquad (3.18)$$

Photons are bosons so any number, including zero can occupy each state. Each photon has energy $E_\nu = h\nu$, where h is Planck's constant. In thermal equilibrium the states are occupied according to a Boltzmann distribution so the mean energy per state

$$\langle E_\nu \rangle = \frac{\sum_{l=0}^{\infty} l h \nu e^{-l h \nu / k_{\rm B} T}}{\sum_{l=0}^{\infty} e^{-l h \nu / k_{\rm B} T}}. \tag{3.19}$$

We put $x = e^{-h\nu/k_{\rm B}T} < 1$ and note that

$$\sum_{l=0}^{\infty} x^l = 1/(1-x) \quad \text{and} \quad \sum_{l=0}^{\infty} l x^l = x \frac{\rm d}{{\rm d}x} \left(\frac{1}{(1-x)} \right) = \frac{x}{(1-x)^2} \tag{3.20}$$

so that

$$\langle E_\nu \rangle = -k_{\rm B} T \log_e x \frac{x}{1-x} = \frac{h\nu}{e^{h\nu/k_{\rm B}T} - 1}. \tag{3.21}$$

Multiplying by Eq. (3.18) we find the energy density of photons with frequency between ν and $\nu + {\rm d}\nu$ to be

$$u_\nu \, {\rm d}\nu = \frac{8\pi \nu^3 h}{c^3 \left(e^{h\nu/kT} - 1 \right)} \, {\rm d}\nu \tag{3.22}$$

and integrating over all states the specific energy density by volume is

$$
\begin{aligned}
u &= \frac{8\pi h}{c^3} \int_0^\infty \frac{\nu^3}{e^{h\nu/k_{\rm B}T} - 1} \, {\rm d}\nu \\
&= \frac{k_{\rm B}^4 T^4}{h^3 c^3} 8\pi \int_0^\infty \frac{x^3}{e^x - 1} \, {\rm d}x = \frac{8\pi^5 k^4 T^4}{15 h^3 c^3} = a T^4,
\end{aligned}
\tag{3.23}
$$

where the integral $\int_0^\infty x^3/(e^x - 1) \, {\rm d}x = \pi^4/15$ can be evaluated in various ways including contour integration, the radiation constant

$$a = \frac{8\pi^5 k^4}{15 h^3 c^3} = \frac{4\sigma}{c} = 7.5657 \times 10^{-15} \, {\rm erg \, cm^{-3} \, K^{-4}}, \tag{3.24}$$

and σ is the Stefan–Boltzmann constant for black-body radiation. At any point these photons are moving isotropically at c so the intensity B_ν, defined such that $B_\nu \, {\rm d}A \, {\rm d}\Omega \, {\rm d}\nu$ is the energy flux through an area ${\rm d}A$ into solid angle ${\rm d}\Omega$ perpendicular to ${\rm d}A$, is

$$B_\nu = \frac{c}{4\pi} u_\nu = \frac{2h\nu^3}{c^2} \frac{1}{e^{h\nu/k_{\rm B}T} - 1}. \tag{3.25}$$

To find the radiation pressure, we need the momentum passing through an elemental area. Because the walls of any cavity filled with black-body radiation must reflect all photons, this momentum flux equals the pressure on the wall. This we can relate to the energy flux because a photon of energy $h\nu$ carries momentum $h\nu/c$ so we can calculate the rate momentum is carried through an elemental surface with vector area $d\boldsymbol{\sigma}$ carried by photons with frequency between ν and $\nu + d\nu$ as

$$P_{r,\nu}\, d\nu = \int_{\text{sphere}} \frac{B_\nu}{c} c \cos^2 \theta\, d\Omega\, d\nu, \qquad (3.26)$$

where θ is the angle to $d\boldsymbol{\sigma}$, one factor of $\cos\theta$ is from the projection of dA on to $d\Sigma$ and the other because a photon moving through the area at θ moves at $c\cos\theta$ parallel to $d\boldsymbol{\sigma}$ and the integral is over all solid angles $d\Omega$ (Fig. 3.1). By symmetry we may write $d\Omega = 2\pi \sin\theta\, d\theta$ so

$$P_{r,\nu}\, d\nu = \frac{2\pi B_\nu}{c} \int_0^\pi \cos^2 \theta \sin\theta\, d\theta\, d\nu = \frac{4\pi}{3c}\, d\nu = \frac{u_\nu}{3}\, d\nu. \qquad (3.27)$$

Integrating over frequency we find a total radiation pressure of

$$P_{\text{r}} = \frac{1}{3}u = \frac{1}{3}aT^4. \qquad (3.28)$$

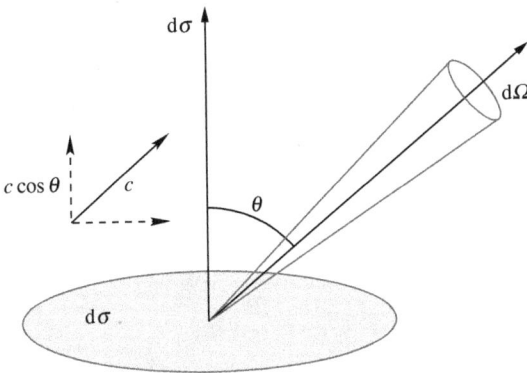

Fig. 3.1. Photon momentum flux. An elemental surface has vector area $d\boldsymbol{\sigma}$. Photons of momentum $h\nu/c$ move at a projected speed $c\cos\theta$ in the direction of $d\boldsymbol{\sigma}$. They carry momentum $h\nu\cos\theta/c$ into a solid angle $d\Omega$ at an angle θ to $d\boldsymbol{\sigma}$.

3.4.   Ratio of Specific Heats and Virial Equilibrium

Before examining the radiation pressure contribution in stars we shall look at the relation between energy density and pressure through the ratio of specific heat capacities to which we referred when looking at the virial theorem in Sec. 2.3. The first law of thermodynamics states that the heat dQ absorbed by a fluid is

$$dQ = T\,dS = dU + P\,dV, \tag{3.29}$$

where S is the entropy and U is the internal energy of fluid occupying a volume V at pressure P. The heat capacity at constant volume is therefore

$$C_V = \left(\frac{\partial Q}{\partial T}\right)_V = \left(\frac{\partial U}{\partial T}\right)_V. \tag{3.30}$$

Defining the enthalpy

$$H = U + PV, \tag{3.31}$$

we may also write

$$dQ = T\,dS = dH - V\,dP \tag{3.32}$$

from which the heat capacity at constant pressure

$$C_P = \left(\frac{\partial Q}{\partial T}\right)_P = \left(\frac{\partial H}{\partial T}\right)_P. \tag{3.33}$$

Now consider a volume V of ideal gas consisting of N monatomic particles. Kinetic theory of gases tells us that the mean energy per particle is $k_{\mathrm{B}}T/2$ for each degree of freedom, of which we have three translational, so that the internal energy

$$U = \frac{3}{2}Nk_{\mathrm{B}}T \tag{3.34}$$

and

$$C_V = \frac{3}{2}Nk_{\mathrm{B}}. \tag{3.35}$$

By the ideal gas equation of state the pressure

$$P = \frac{Nk_{\mathrm{B}}T}{V} \tag{3.36}$$

so the enthalpy

$$H = \frac{3}{2}Nk_\mathrm{B}T + Nk_\mathrm{B}T \tag{3.37}$$

and

$$C_P = \frac{5}{2}Nk_\mathrm{B}. \tag{3.38}$$

Thence the ratio of the heat capacities

$$\gamma = \frac{C_P}{C_V} = \frac{5}{3} = \frac{c_P}{c_V}, \tag{3.39}$$

where

$$c_V = \frac{C_V}{\mu N m_\mathrm{H}} \quad \text{and} \quad c_P = \frac{C_P}{\mu N m_\mathrm{H}} \tag{3.40}$$

are the specific heat capacities by mass. Each particle in ionised gases and neutral atoms has only the three translational degrees of freedom and so $\gamma = 5/3$. Molecules have additional degrees of freedom that contribute to U but not P, which is due only to translation. Below about 10^3 K hydrogen molecules H_2 have two additional rotational degrees of freedom and $\gamma = 7/5$. At higher temperatures vibrational modes become important too and γ falls further. In general for ideal gases

$$PV = (\gamma - 1)U. \tag{3.41}$$

For pure black-body radiation

$$U = aT^4V \tag{3.42}$$

so

$$C_V = 4aT^3V \tag{3.43}$$

but C_P is undefined because at constant pressure P temperature T is also constant. However

$$P = \left(\frac{4}{3} - 1\right)U \tag{3.44}$$

and we can define an effective $\gamma = 4/3$ to satisfy Eq. (3.41) in the case of solely radiation pressure support. Recall from the virial theorem of

Sec. 2.3 that the total energy of a star $E < 0$ if and only if $\gamma > 4/3$ so that a star cannot be stably supported by radiation pressure. However, while there is any gas pressure contribution, there is an effective $\gamma > 4/3$ and a stable configuration is possible.

Associated with the ratio of heat capacities and of importance for stars, particularly when we come to consider convective energy transport in Chapter 4, are the thermodynamic derivatives at constant entropy S,

$$\Gamma_1 = \left(\frac{\partial \log P}{\partial \log \rho}\right)_S,$$

$$\frac{\Gamma_2}{\Gamma_2 - 1} = \left(\frac{\partial \log P}{\partial \log T}\right)_S \quad \text{and} \quad \Gamma_3 - 1 = \left(\frac{\partial \log T}{\partial \log \rho}\right)_S, \tag{3.45}$$

with

$$\left(\frac{\partial \log P}{\partial \log \rho}\right)_S \Big/ \left(\frac{\partial \log T}{\partial \log \rho}\right)_S = \frac{\Gamma_1}{\Gamma_3 - 1} = \frac{\Gamma_2}{\Gamma_2 - 1} = \left(\frac{\partial \log P}{\partial \log T}\right)_S. \tag{3.46}$$

For a fixed mass of fluid

$$\frac{d\rho}{\rho} = -\frac{dV}{V} \tag{3.47}$$

so that

$$\Gamma_1 = -\left(\frac{\partial \log P}{\partial \log V}\right)_S \quad \text{and} \quad \Gamma_3 - 1 = -\left(\frac{\partial \log T}{\partial \log V}\right)_S, \tag{3.48}$$

and we can continue the definitions to black-body radiation even though it is massless. Whenever $PV = (\gamma - 1)U$ with constant γ we may write

$$P \, dV + V \, dP = (\gamma - 1) \, dU. \tag{3.49}$$

The first law of thermodynamics (3.29) with $dQ = T \, dS = 0$ leads us to

$$\frac{dP}{P} = -\gamma \frac{dV}{V} \tag{3.50}$$

and thence, for this effective γ,

$$\Gamma_1 = \gamma \tag{3.51}$$

for both ideal gases and black-body radiation. To find Γ_2 we need the equation of state. For radiation $P_r = aT^4/3$ is independent of S and

$$\Gamma_2 = \frac{4}{3} \tag{3.52}$$

directly. For the ideal gas $PV = Nk_BT$ so we may write the enthalpy

$$H = U + PV = U + (\gamma - 1)U = \gamma U = \frac{\gamma}{\gamma - 1}Nk_BT. \tag{3.53}$$

Again with $đQ = 0$ in Eq. (3.32) we arrive at

$$\frac{dP}{P} = \frac{\gamma}{\gamma - 1}\frac{dV}{V} \tag{3.54}$$

and $\Gamma_2 = \gamma$. It follows that

$$\Gamma_1 = \Gamma_2 = \Gamma_3 = \gamma, \tag{3.55}$$

for an ideal gas while for pure black-body radiation

$$\Gamma_1 = \Gamma_2 = \Gamma_3 = \frac{4}{3}. \tag{3.56}$$

For a mixture of gas and radiation (see Question 3)

$$\frac{4}{3} < \Gamma_1 < \gamma, \tag{3.57}$$

where $\gamma = C_P/C_V$ for the appropriate ideal gas.

3.5. Importance of Radiation Pressure

The contribution of radiation pressure to stellar structure is often exaggerated based on Eddington's early work before the importance of electron degeneracy was realised. Because stars adapt to their equation of state radiation pressure is relatively unimportant for much of their lifetime other than in a few short phases of evolution for very massive stars. A $100\,M_\odot$ main-sequence star typically has half its pressure contributed by radiation and only at $1000\,M_\odot$ can we claim almost full radiation-pressure support. Even then there must still be some gas pressure contribution because without it a star would be unstable.

To estimate importance of radiation pressure recall from the virial theorem that

$$\overline{P} \approx 10^{15} \left(\frac{M}{M_\odot}\right)^2 \left(\frac{R_\odot}{R}\right)^4 \text{dyn cm}^{-2} \tag{3.58}$$

and

$$\overline{\rho} \approx 1.4 \left(\frac{M}{M_\odot}\right) \left(\frac{R_\odot}{R}\right)^3 \text{g cm}^{-3}. \tag{3.59}$$

For an ideal gas we can then estimate the temperature

$$\overline{T} = \frac{\mu \overline{P}}{\mathcal{R} \overline{\rho}} = 10^7 \mu \left(\frac{M}{M_\odot}\right) \left(\frac{R_\odot}{R}\right) \text{K} \tag{3.60}$$

and thence the radiation pressure

$$\overline{P_\text{r}} \approx 2.5 \times 10^{13} \mu^4 \left(\frac{M}{M_\odot}\right)^4 \left(\frac{R_\odot}{R}\right)^4 \text{dyn cm}^{-2}. \tag{3.61}$$

For the Sun this gives us $\overline{P_\text{r}} = 0.02 \overline{P_\text{g}}$. From detailed models of the centre of the Sun the ratio is actually $P_\text{r} = 0.00065P$ and our estimate is rather poor because of the contribution from electron degeneracy. It is apparent that $\overline{P_\text{r}}/\overline{P_g}$ increases with mean molecular weight μ and the mass of the star M but decreases with its radius R. This latter dependence means that a star can nearly always expand to reduce the contribution of radiation pressure. It also counteracts the evolutionary effect of increasing μ because stars also tend to grow as they evolve.

In a more careful analytic analysis, due to Eddington, we write $P_\text{g} = \beta P$, so that $P_\text{r} = (1 - \beta)P$. Then

$$\frac{1}{3}aT^4 = \frac{1 - \beta}{\beta} \frac{\rho \mathcal{R} T}{\mu} \tag{3.62}$$

so

$$T = \left(\frac{3\mathcal{R}}{a\mu} \frac{1 - \beta}{\beta}\right)^{1/3} \rho^{1/3} \tag{3.63}$$

and

$$P = \left[\left(\frac{\mathcal{R}}{\mu} \right)^4 \frac{3}{a} \frac{1-\beta}{\beta^4} \right]^{1/3} \rho^{4/3}. \tag{3.64}$$

Hence given P and ρ we can find β by rearranging to find Eddington's quartic,

$$\frac{P^3}{\rho^4} \beta^4 = \left(\frac{\mathcal{R}}{\mu} \right)^4 \frac{3}{a} (1-\beta). \tag{3.65}$$

Taking $\beta = $ const and using our estimates from the virial theorem we expect that

$$\frac{P^3}{\rho^4} \propto \frac{M^6 R^{-12}}{M^4 R^{-12}} \propto M^2. \tag{3.66}$$

This kind of analysis assumes the structure of the star to be homologous. We take a closer look at such models in Sec. 7.3 but here we are implicitly assuming that stars are $n = 3$ polytropes so that they remain homologous even as $\beta \to 0$. We find that

$$\frac{1-\beta}{\beta^4} \propto M^2 \mu^4 \tag{3.67}$$

so $P_{\rm r}$ is most important in massive and evolved stars. Note that the homologous assumption removes any dependence on R in this case of a polytrope with $n = 3$.

In Table 3.1 we list the radiation pressure contribution at the centre of zero-age main-sequence models of various masses. As predicted the importance of radiation pressure increases with mass. More massive stars are consequently more extended than they otherwise would be but not sufficiently to make them unstable. The contribution from chemical evolution is weaker. In Table 3.2 for a

Table 3.1. Contribution of radiation to total pressure at the centres of stars of various masses.

$M/M_\odot$	1	10	20	40	100
$1 - \beta_{\rm c}$	0.00053	0.0073	0.15	0.29	0.50

Table 3.2. Contribution of radiation to total pressure at the centre of the Sun as it evolves across the main sequence.

X_c	0.7	0.4	0.1
$1 - \beta_c$	0.00053	0.00064	0.00082

model of the Sun we list the radiation pressure at the centre as the central hydrogen abundance X_c falls and mean molecular weight μ_c rises with evolution.

3.6.   Ionisation Equilibria

So far we have thought of our gas as either ionised or neutral and of a fixed number of particles per unit mass. However the state of ionisation changes through the star and so impacts on the equation of state. While deep in the interior we can assume all elements are completely ionised, near the surface of stars this is not true. Most important are the ionisation of hydrogen and helium which are usually the most abundant species in stars at temperatures and densities where ionisation matters. For hydrogen we consider the reaction

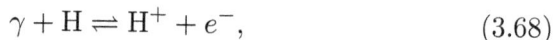

$$\gamma + \mathrm{H} \rightleftharpoons \mathrm{H}^+ + e^-, \tag{3.68}$$

where γ represents an ionising photon and e^- an electron. More generally for a species S between stages of ionisation in which it has lost i and $i+1$ electrons we have

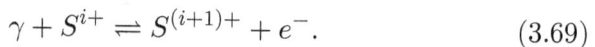

$$\gamma + S^{i+} \rightleftharpoons S^{(i+1)+} + e^-. \tag{3.69}$$

Spectroscopically neutral hydrogen H is often referred to as HI, ionised hydrogen H^+ as HII and, for example, triply ionised carbon as CIV.

In thermal equilibrium the relative abundances of different ionisation states are related by Saha's equation

$$\frac{n_{i+1} n_e}{n_i} = \left(2 \frac{(2\pi m_e kT)^{\frac{3}{2}}}{h^3} \right) \frac{u_{i+1}}{u_i} \exp\left(-\frac{\chi_i}{k_B T} \right), \tag{3.70}$$

where n_i, n_{i+1} and n_e are the number densities of ions in stage i, ions in stage $i + 1$ and free electrons. The first term in parentheses is the density of free electron states, the initial factor of two accounts for the spin degeneracy of the free electrons, χ_i is the ionisation energy and u_i is the partition function for ionisation stage i given by

$$u_i = \sum_{n=0}^{\infty} g_n \exp\left(-\frac{E_n}{k_B T}\right), \tag{3.71}$$

where the sum is over all atomic states n, with degeneracy g_n, of the particular ionisation stage. The ground state energy $E_0 = 0$ and $E_n \rightarrow \chi_i$ as ionisation is approached (Fig. 3.2). Mathematically a problem appears to arise because the degeneracy of energy levels grows too quickly with n for isolated atoms. For example for a single hydrogen atom $g_n = 2(n + 1)$ and so the sum in Eq. (3.71) diverges. In practice the proximity of other atoms means that only the lowest states are available to electrons bound to a single atom, χ_i is effectively lowered and the sum is truncated. Usually it is sufficient to use $u_i \approx g_{i,0}$ for the equation of state of stellar material. This is not however so when calculating opacities (Sec. 4.2.1.2).

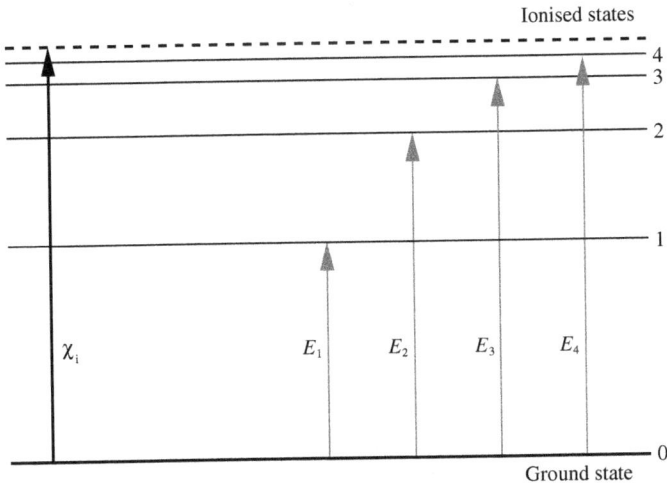

Fig. 3.2. Energy levels E_i and ionisation energy χ_i of a typical atom.

In hydrogen the ground state electron has two spin states and $u_0 = 2$ while the ionised proton has only one state and $u_1 = 1$ so

$$\log_e \left(\frac{n_{H+}}{n_H} \right) = \log_e \left(\frac{(2\pi m_e kT)^{\frac{3}{2}}}{n_e h^3} \right) - \frac{\chi_i}{k_B T}. \qquad (3.72)$$

We write the first term of the right-hand side as $\Theta(n_e, T)$. In general n_e depends on the ionisation of all species and the full set of ionisation equilibria must be solved together. However during a particular ionisation phase Θ is a slowly varying function when compared with the second term and it is temperature that tends to determine whether a species is ionised or not.

In the Sun, near the surface, $n_e \approx 10^{13}\,\mathrm{cm}^{-3}$ and $T \approx 6 \times 10^3\,\mathrm{K}$ so that $\Theta \approx 18.5$. Consequently, ionisation occurs quickly when $\chi_i/k_B T \approx 18.5$. For hydrogen $\chi = 13.6\,\mathrm{eV}$ and so it ionises at about $8500\,\mathrm{K}$. The variations of the relative fractions of ionised and atomic hydrogen with temperature are illustrated in Fig. 3.3. For single electron atoms χ is proportional to the square of the number of protons so helium fully ionises above $34\,000\,\mathrm{K}$ while iron does not fully ionise until between 10^5 and $10^6\,\mathrm{K}$ in the Sun.

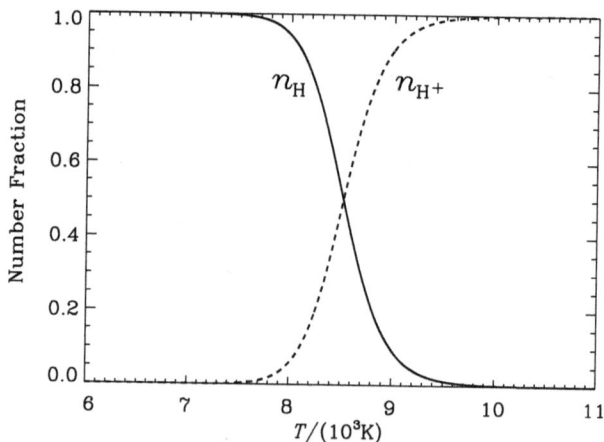

Fig. 3.3. The fraction of ionised and atomic hydrogen as a function of temperature in the Sun's ionisation zone. The solid line is the number fraction of neutral hydrogen n_H or HI while the dashed line is the number fraction of ionised hydrogen n_{H+} or HII.

3.6.1. *Pressure ionisation*

In very dense environments ionisation is complicated because the separation of atoms can become smaller than the ground state electron orbitals. At high enough densities, even in cold material, all atoms behave as if they are fully ionised. At high pressures nuclei are very close and electrons are not bound to a particular atom. The gas behaves like a terrestrial metal in which electrons are able to move freely among the nuclei (Fig. 3.4). Pressure ionisation is very important in the cores of planets. The mostly hydrogen core of Jupiter can be considered as metallic in this sense. In the Sun the effect is small while in the degenerate cores of evolved stars, the subject of the next section, the effect is extreme and the material can simply be treated as fully ionised. For low-mass stars and planets the situation is somewhat more complicated. Their interiors pass through the region in the equation of state in which pressure ionisation contributes a large fraction of the pressure support at cool

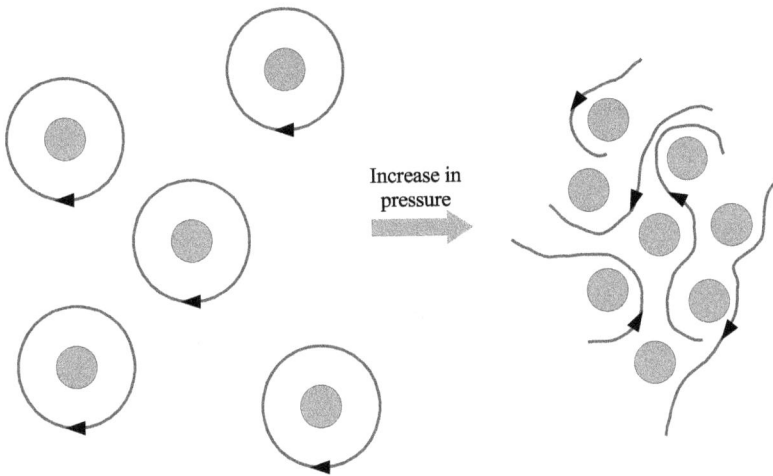

Fig. 3.4. A schematic diagram of electrons in normal atoms (left-hand side) and at high pressure (right-hand side). At low pressure, and low enough temperature, electrons occupy bound states around nuclei. At high pressure nuclei are sufficiently close that electrons cease to be bound to a single nucleus even at very low temperatures. The fluid is pressure ionised and behaves in a similar way to a terrestrial metal.

temperatures, at which the gas is neutral at low densities but at high densities is fully pressure ionised and degenerate. Calculation of the correct equation of state under such circumstances is still an active area of research.

3.7.   Degeneracy

Thermal pressure support requires a suitable temperature or density gradient to maintain a sufficient pressure gradient. While nuclear fusion proceeds at the centre of a star this is naturally provided. However once fusion ceases the core becomes isothermal. While burning continues in a shell around it, the burnt-out core remains hot and gas pressure support is possible as long as the growing core has a small enough mass relative to the surrounding envelope. Otherwise the core must contract until electrons are forced close enough together that quantum mechanical exclusion becomes important. As electrons are forced into a smaller and smaller volume they must occupy higher and higher momentum states. Raising them to these states requires energy and work, $đQ = PdV$, must be done to compress the volume, so a pressure is exerted.

To understand the quantitative details we must turn to quantum mechanics. Individual electrons obey the Heisenberg uncertainty relation for momentum p and position x in each dimension,

$$\Delta p \Delta x \geq \frac{1}{2}\hbar, \qquad (3.73)$$

where $\hbar = h/(2\pi)$ and h is Planck's constant. This corresponds to a density of states in a six-dimensional position and momentum phase space for the electrons at $\boldsymbol{x}$ with momentum $\boldsymbol{p}$ of the form

$$f(\boldsymbol{x},\boldsymbol{p})\,\mathrm{d}^3x\,\mathrm{d}^3p = \frac{2}{h^3}\,\mathrm{d}^3x\,\mathrm{d}^3p. \qquad (3.74)$$

Mathematically this distribution function can be derived by considering discrete particle states in a cubical box of side L with periodic boundary conditions at the walls. The states of the momentum operator are then also discrete energy states of the Hamiltonian operator. They have spatial wave functions of the form

$$\psi_k(x) = \frac{1}{\sqrt{V}} e^{ik \cdot x}, \tag{3.75}$$

where $i^2 = -1$, $V = L^3$ is the volume of the box and k, the wave vector of the state, is related to momentum and quantised by

$$k = \frac{p}{\hbar} = \frac{2\pi}{L} n, \tag{3.76}$$

where $n = (n_1, n_2, n_3)^T$ for any integers, positive, negative and zero, n_1, n_2 and n_3. We then allow $L \to \infty$ because our box, the space in which our electrons are free to move is very large compared to $\hbar/p$. The number of states with momentum between p and $p + dp$ in each of the three orthogonal directions is then L/h. Electrons are fermions with spin one half so that there are two spin states that can be occupied for each spatial state. The use of periodic boundary conditions is a mathematical device that allows us to choose particle momentum states that are also energy states. Identical results can be obtained by confining particles to an infinite cubical well with $\psi = 0$ on the boundaries. In this model the momentum states are not states of the Hamiltonian except in the limit as $L \to \infty$ and we must work with a distribution of states in position and energy space instead of the more symmetric position and momentum space until the limit is taken.

In a cold electron gas energy states are filled from the lowest up to the Fermi energy E_f with momentum of magnitude p_f. Consider a volume V containing N_e electrons. The number density of electrons is

$$n_e = \frac{N_e}{V} = \frac{\rho}{\mu_e m_H}, \tag{3.77}$$

where μ_e is the mean molecular weight of electrons and m_H is again the atomic mass unit. Once again applying our integer atomic weights for clarity, we find that $1/\mu_e$ is the number of electrons per nucleon. So for ionised hydrogen, appropriate for brown dwarfs or planets, $\mu_e = 1$ while for helium and metals, as in normal stellar cores, $\mu_e = 2$.

Integrating over the filled states, with $d^3p = 4\pi p^2 dp$ for spherical shells of constant p, we find that

$$N_e = \int_V \int_0^{p_f} \frac{2}{h^3} 4\pi p^2 \, dp \, d^3x = \frac{8\pi p_f^3 V}{3h^3}, \qquad (3.78)$$

a relation between density and Fermi momentum.

Just as for radiation pressure and photons (Sec. 3.3), the pressure P_e is the momentum flux per unit area carried by electrons (Fig. 3.5). The number of electrons with momentum between p and $p + dp$ passing through area $d\sigma$ into solid angle $d\Omega$ around direction S at an angle θ to the vector area $d\sigma$ in time dt is

$$f \frac{d\Omega}{4\pi} v \, dt \, d\sigma \, \cos\theta \, d^3p. \qquad (3.79)$$

These electrons carry a momentum $p\cos\theta$ along $d\sigma$. So, integrating over spherical shells up to the Fermi momentum, we obtain the electron pressure

$$P_e = \int_0^{p_f} \int_{\text{sphere}} \frac{2}{h^3} 4\pi p v \cos^2\theta \frac{d\Omega}{4\pi} 4\pi p^2 dp. \qquad (3.80)$$

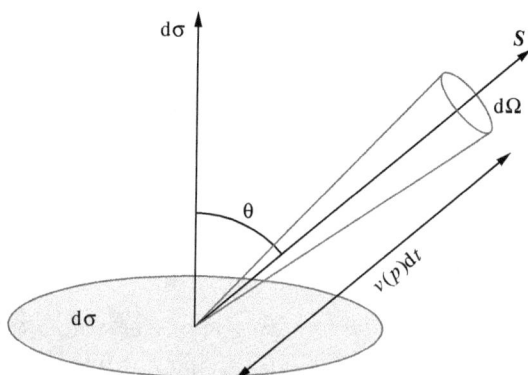

Fig. 3.5. Electron momentum flux. An elemental surface has vector area $d\sigma$. Electrons of momentum p move at speed v parallel to p. They carry momentum $p\cos\theta$ into a solid angle $d\Omega$ around S at an angle θ to $d\sigma$ at a rate $v(p)\cos\theta \, dt$.

Now

$$\int_{\text{sphere}} \cos^2 \theta \, d\Omega = \int_0^{\pi} 2\pi \sin \theta \cos^2 \theta \, d\theta = \frac{4}{3}\pi \qquad (3.81)$$

so

$$P_e = \frac{8\pi}{3h^3} \int_0^{p_f} p^3 v \, dp. \qquad (3.82)$$

3.7.1. *Non-relativistic electrons, $p = m_e v$*

For non-relativistic electrons $p = m_e v$, where m_e is the mass of the electron, and

$$P_e = \frac{8\pi}{3h^3} \int_0^{p_f} \frac{p^4}{m_e} \, dp = \frac{8\pi p_f^5}{15 m_e h^3}. \qquad (3.83)$$

Eliminating p_f between Eqs. (3.78) and (3.82), with (3.77) we arrive at

$$P_e = \left(\frac{3}{\pi}\right)^{2/3} \frac{h^2}{20 m_e} \frac{\rho^{5/3}}{(\mu_e m_H)^{5/3}} = K_{nr} \rho^{5/3}. \qquad (3.84)$$

This is an equation of state of a polytrope with $n = \frac{3}{2}$ and fixed

$$K_{nr} = \frac{1.004 \times 10^{13}}{\mu_e^{5/3}} \, \text{dyn cm}^{-2} (\text{g cm}^{-3})^{-5/3}. \qquad (3.85)$$

We are now able to construct our first complete stellar model, that of a cold white dwarf supported entirely by non-relativistic degeneracy pressure. From Eqs. (2.15), (2.20) and (2.23), the radius and mass M are related to the central density ρ_c by

$$R = \sqrt{\frac{5 K_{nr}}{8\pi G}} \rho_c^{-1/6} X_{3/2} \qquad (3.86)$$

and

$$M = 4\pi \left(\frac{5 K_{nr}}{8\pi G}\right)^{3/2} \rho_c^{1/2} Y_{3/2}, \qquad (3.87)$$

where $X_{3/2} = 3.65$ and $Y_{3/2} = 2.71$. Equivalently, the central density

$$\rho_c = \left(\frac{M}{4\pi Y_{3/2}}\right)^2 \left(\frac{5K_{nr}}{8\pi G}\right)^{-3} = 1.322 \times 10^5 \mu_e^5 \left(\frac{M}{M_\odot}\right)^2 \text{g cm}^{-3}.$$

$$(3.88)$$

Eliminating ρ_c, we find

$$R \propto M^{-1/3}, \qquad (3.89)$$

as expected from Eq. (2.32). For a given K_{nr}, which depends only on μ_e and fundamental constants, the radius is fixed by the mass. We have

$$R = 0.04\mu_e^{-5/3}\left(\frac{M}{M_\odot}\right)^{-\frac{1}{3}} R_\odot = 0.0126 \left(\frac{M}{M_\odot}\right)^{-\frac{1}{3}} R_\odot \qquad (3.90)$$

for helium and carbon–oxygen white dwarfs for which $\mu_e = 2$.

Non-relativistic models are valid until the central pressure, needed to support the star, requires that electrons at the centre become relativistic. This is when $p_f \approx m_e c$ or

$$\rho_c \approx \mu_e m_H \frac{8\pi}{3h^3}(m_e c)^3 \approx 2 \times 10^6 \text{g cm}^{-3}, \qquad (3.91)$$

for $\mu_e = 2$. Such densities are reached when $M \approx 0.7\,M_\odot$.

3.7.2.   *Extremely relativistic electrons, $v = c$*

As the density increases so more and more electrons are forced to relativistic energies and move at the speed of light. Ultimately we can write $v = c$ in Eq. (3.82) so that

$$P_e = \frac{8\pi c}{3h^3} \int_0^{p_f} p^3 \, \mathrm{d}p = \frac{2\pi c}{3h^3} p_f^4. \qquad (3.92)$$

Again we eliminate p_f between this and Eq. (3.78) to arrive at

$$P_e = \left(\frac{3}{\pi}\right)^{\frac{1}{3}} \frac{hc}{8} \left(\frac{\rho}{\mu_e m_H}\right)^{\frac{4}{3}} = K_{er}\,\rho^{4/3}. \qquad (3.93)$$

In this case

$$K_{er} = 1.244 \times 10^{15} \, \text{dyn cm}^{-2} (\text{g cm}^{-3})^{-4/3}. \qquad (3.94)$$

is independent of the electron mass m_e because particles moving at c are effectively massless. We now have the equation of state for a polytrope of $n = 3$ and this, as we have already remarked, cannot be bound. Its mass

$$M = 4\pi \left(\frac{K_{er}}{\pi G} \right) Y_3 \qquad (3.95)$$

is independent of ρ_c and R. Substituting the fundamental constants

$$M = \frac{5.75}{\mu_e^2} M_\odot = M_{Ch}. \qquad (3.96)$$

This is the Chandrasekhar mass, the maximum mass for an object supported by electron degeneracy pressure. When $\mu_e = 2$

$$M_{Ch} = 1.438 \, M_\odot, \qquad (3.97)$$

the maximum degenerate core mass that can survive in a giant or as a white dwarf remnant. Further increase in mass must either ignite more nuclear fusion or result in catastrophic gravitational collapse.

3.8. White Dwarfs and Neutron Stars

Actual white dwarfs are mostly degenerate and quickly cool to temperatures at which their interiors behave very much as the cool fluids we have just described. It is hard for an electron to scatter from one momentum state to another when nearly all states are already occupied. So electrons have very long mean free paths and conduct heat very well. The interior of a white dwarf is therefore quite isothermal. Similarly electrons transport momentum so that degenerate matter is also rather viscous. However actual white dwarfs do differ from polytropes near their surface because pressure and density drop sufficiently that gas pressure becomes more important than degeneracy pressure. This non-degenerate envelope tends to have very little mass and, in the case of massive white dwarfs, is very thin so that our derived mass radius relation is still quite valid. At

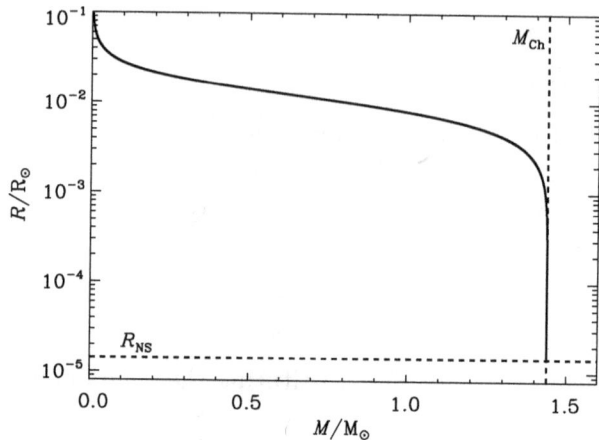

Fig. 3.6. The radii of white dwarfs as a function of their mass. At low mass $R \propto M^{-1/3}$. Above about $0.7\,M_\odot$ electrons begin to move relativistically in the centre. As $M \to M_{\mathrm{Ch}}$ electron degeneracy pressure can no longer support the white dwarf which collapses until degenerate neutrons take over.

low masses the gas pressure supported envelope can become thick. The actual radius of the white dwarf then exceeds the polytropic prediction and also depends on the composition of the envelope material.

Figure 3.6 illustrates how the radius of a white dwarf varies with mass. Once the Chandrasekhar mass is reached electron, degeneracy pressure can no longer provide sufficient support and the star collapses until neutrons take over. The mass of a neutron $m_{\mathrm{n}} \approx m_{\mathrm{H}} \approx 2000 m_{\mathrm{e}}$ so at $M = M_{\mathrm{Ch}}$ neutrons are not moving relativistically. Assuming all electrons combine with protons so that the gas is entirely composed of degenerate neutrons we can naively compute the neutron degeneracy pressure by replacing m_{e} and μ_{e} with the mass and mean molecular weight of the neutrons, m_{n} and $\mu_{\mathrm{n}} = 1$ in K_{nr} (Eq. 3.84). Thence the ratio of the radius of a neutron star to a white dwarf of the same mass but with effectively non-relativistic electron degeneracy support is

$$\frac{R_{\mathrm{NS}}}{R_{\mathrm{WD}}} \approx \left(\frac{m_{\mathrm{e}}}{m_{\mathrm{n}}}\right)\left(\frac{\mu_{\mathrm{e}}}{\mu_{\mathrm{n}}}\right)^{5/3} \approx 1.5 \times 10^{-3}. \tag{3.98}$$

At M_{Ch} we have $R_{\text{NS}} \approx 11\,\text{km}$ and the mass of the Sun in a sphere of radius no bigger than an average city. The Schwarzschild radius of the event horizon of a black hole of mass M is

$$R_{\text{GR}} = \frac{2GM}{c^2} = 4.24 \left(\frac{M}{1.438\,M_{\odot}} \right) \text{km}. \qquad (3.99)$$

So it is necessary to model neutron stars with general relativistic corrections to both space time and their equation of state. As a consequence, a significant fraction of the baryonic mass that collapses to a neutron star is absorbed by its binding energy and so its effective gravitational mass is smaller than that of its progenitor. In reality it is unlikely that the material exists as free neutrons through the entire star. Deep inside densities and pressures lie beyond experimental verification. At the surface a crust of super-heavy elements condenses. This crust can be probed by examination of X-ray bursts generated if material falls on to the surface of a neutron star while the equation of state of the interior is gradually being elucidated by mass and radius measurements of neutron stars in binary systems and gravitational wave signatures of merging neutron stars. As the mass increases so the neutrons reach relativistic velocities and direct application of Eq. (3.96) gives a maximum mass of $M_{\text{max}} = 5.75\,M_{\odot}$. This is reduced in a relativistic calculation and current estimates actually place

$$2.5 < \frac{M_{\text{max}}}{M_{\odot}} < 3, \qquad (3.100)$$

where the low limit is from the measured mass of a binary pulsar and the upper limit is theoretical.

3.8.1. *Warm degenerate matter*

Even in old white dwarfs the material is not completely cold and in a full model we must allow particles to have energies that exceed the Fermi energy. In warm degenerate matter not every particle fills the first available energy state. So, at the high energy end, there is a spread of energies around p_{f}. Figure 3.7 illustrates how the cold degenerate matter distribution, with number density $n(p)\,\mathrm{d}^3 p$

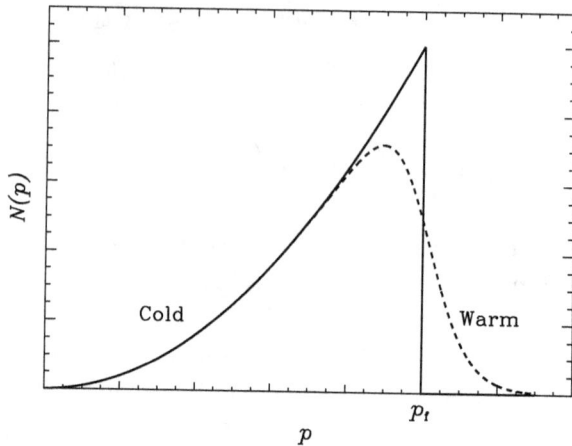

Fig. 3.7. The distribution of electron momenta in cold and warm degenerate matter. In the cold gas all momentum levels up to p_f are filled. At non-zero temperatures some states above the Fermi level are occupied at the expense of those below.

of momentum states between p and $p + dp$,

$$n(p)\, d^3 p = \begin{cases} \dfrac{2}{h^3} 4\pi p^2\, dp, & 0 < p \leq p_f, \\ 0 & \text{otherwise,} \end{cases} \quad (3.101)$$

is replaced by that derived by application of Fermi–Dirac statistics to the electrons,

$$n(p)\, d^3 p = \frac{2}{h^3} \frac{4\pi p^2\, dp}{1 + \exp(\frac{E'}{k_B T} - \psi)}, \quad (3.102)$$

where $E' = E - m_e c^2$, with $E^2 = p^2 c^2 + m_e^2 c^4$, is the kinetic energy of an electron with momentum p and ψ is our degeneracy parameter, related to the electron chemical potential minus the electron rest mass energy all divided by $k_B T$. So

$$\frac{\rho}{\mu_e m_H} = \frac{8\pi}{h^3} \int_0^\infty \frac{p^2}{1 + \exp(\frac{E'}{k_B T} - \psi)}\, dp, \quad (3.103)$$

determines $\psi = \psi(\rho, T)$. As $T \to 0$, $\psi \to (E_f - m_e c^2)/k_B T$. The equation of state is non-degenerate when ψ is large and negative and extremely degenerate when ψ is large and positive. To obtain the electron pressure we use Eq. (3.82) with $p = \gamma m_e v$ and $\gamma = (1 - v^2/c^2)^{-1/2}$. Thence we have $P_e = P_e(\psi, T)$ so $P_e = P_e(\rho, T)$ and ψ can be treated as another state variable. To find the total gas pressure $P_g = P_e + P_i$ it is usually correct to treat the ions as a mixture of ideal gases so that

$$P_i = \rho \frac{\mathcal{R}T}{\mu_i}, \tag{3.104}$$

where $1/\mu_i$ is the number of ions per m_H.

3.9. Choice of State Variables

When making stellar models we need to choose two state variables as independent and then derive all others from these together with the composition of the material. For example we can use $P = P(\rho, T, \{X_i\})$ for pressure, $s = s(\rho, T, \{X_i\})$ for specific entropy and so on. Alternatively pressure P or specific internal energy u are sometimes used in place of ρ. In practice a particularly good choice is ψ, T and $\{X_i\}$ because then n_e is known before we calculate the ionisation equilibria and this avoids unnecessary iteration to produce a solution.

3.10. Molecular Hydrogen

In the outer parts of low-mass stars the temperature can drop enough for molecular hydrogen H_2 to form and impact on the equation of state. We must then solve for the further equilibrium

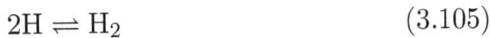

$$2H \rightleftharpoons H_2 \tag{3.105}$$

with the Saha equation

$$\frac{n_H^2}{n_{H_2}} = \frac{(2\pi m_H k_B T)^{\frac{3}{2}}}{h^3} \frac{4}{U_{H_2}} \exp\left(-\frac{\chi_D}{k_B T}\right), \tag{3.106}$$

where the molecular dissociation energy $\chi_D = 4.48\,\text{eV}$ and U_{H_2} is the partition function for H_2. This includes its rotational and vibration states and can be calculated theoretically. Other molecules such as H_2O, CO and CN also form in stellar atmospheres but their abundance is sufficiently low as to not greatly affect the equation of state. They are however very important for opacity (Sec. 4.2.1.5). The same is true of H^- formed by adding a second electron to a neutral hydrogen atom.

3.11.   Coulomb Interactions

At low T and high ρ electrostatic forces between ions become important too. Forces similar to those experienced by water molecules in their transition from steam to water to ice must be included alongside atomic ionisation and any molecular dissociation. In the weakest limit of Coulomb interactions we include the effects of plasma physics and electron screening. In the strong limit liquefaction and solidification, in particular crystallisation in cold white dwarfs and the associated solid state physics must be added too. As we move from stars through brown dwarfs to planets these extra interactions play an increasingly important part in the equation of state.

3.12.   The Equation of State for Stellar Models

Some stars can be modelled with simple equations of state but in general we must cover a wide range of temperatures and densities if we are to make general models. In practice interpolation in tables is often used but then all the thermodynamic derivatives required by the equations of stellar evolution must also be tabulated. We illustrate the analytical equation of state used in the stellar models presented in this book in Fig. 3.8. There we identify regions where each of the major contributions of radiation, gas and electron degeneracy pressure are important as well as the contribution of pressure ionisation and coulomb corrections. Shaded regions identify where ionisation and molecular dissociation take place and solid lines show where detailed stellar models of various masses lie on this plane. We note that this particular equation of state has a phase transition in the sense that pressure is not a single-valued function of temperature and density in the region of pressure ionisation for

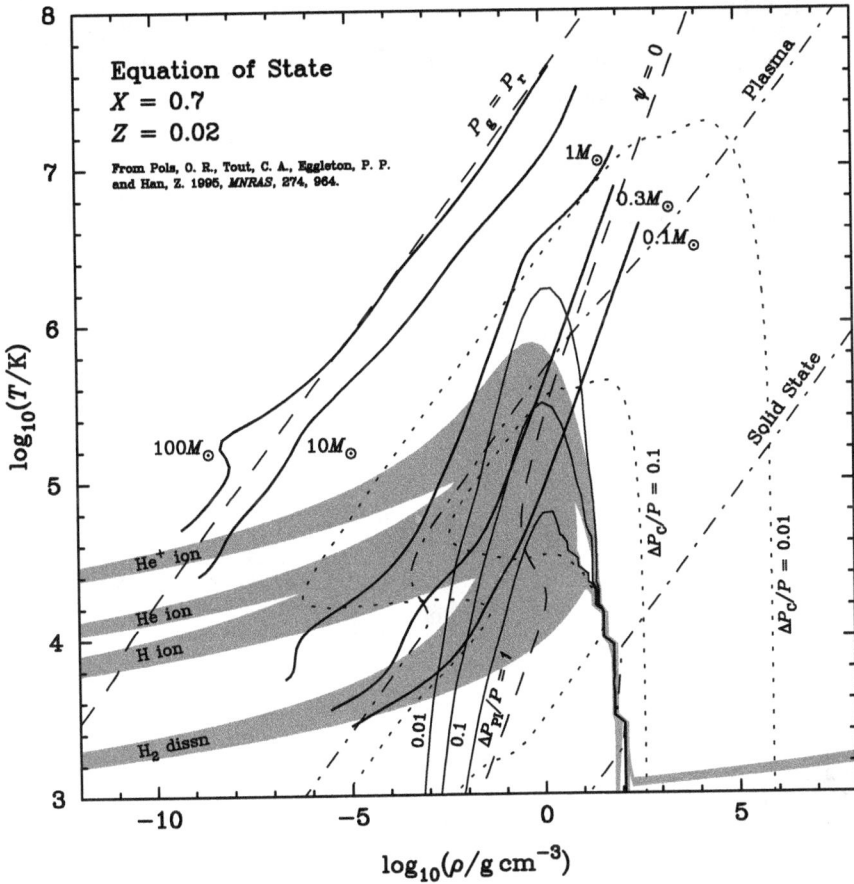

Fig. 3.8. The equation of state is used to make all the detailed stellar models we present in this book. Regions where the various contributions discussed in this chapter become important are delineated. The contours ΔP_C indicate the contribution of Coulomb corrections to the pressure and ΔP_{PI} that of pressure ionisation. Shaded regions show where ionisation of hydrogen and helium and molecular dissociation of H_2 are important and solid lines show where detailed models of various masses lie on the plane.

cool gas, below about 10^4 K and between about 10 and 100 g cm^{-3}. Fortunately, this does not impinge on the structure of the lowest mass stars that can burn hydrogen in their cores but it does need a more careful treatment to model brown dwarfs or planets.

3.13. Questions

1. Rough estimates of stellar fluxes can be obtained under the assumption that stars radiate as black bodies. The Planck function for black-body radiation at temperature T is

$$B_\nu(T)\, d\nu = \frac{2h\nu^3}{c^2} \frac{1}{\exp(h\nu/k_B T) - 1}\, d\nu,$$

where $B_\nu\, d\nu$ is the energy per unit area emitted into one steradian per unit time in radiation of frequency between ν and $\nu + d\nu$. Show that the corresponding function for wavelengths between λ and $\lambda + d\lambda$ is

$$B_\lambda(T)\, d\lambda = \frac{2hc^2}{\lambda^5} \frac{1}{\exp(hc/k_B T\lambda) - 1}\, d\lambda.$$

Show that the flux F emitted per unit area integrated over outward solid angle and all frequencies is

$$F = \pi \int_0^\infty B_\nu\, d\nu = \sigma T^4,$$

where σ is the Stefan–Boltzmann constant.

[*You may use the relation* $\int_0^\infty \frac{u^3}{e^u - 1}\, du = \frac{\pi}{15}$ *without proof.*]

2. Deduce that when $\beta = P_{\text{gas}}/P$,

$$1 - \beta_c \leq 1 - \beta^*,$$

where $\beta = \beta_c$ at the centre of the star and β^* satisfies Eddington's quartic

$$M = \left(\frac{6}{\pi}\right)^{\frac{1}{2}} \left[\left(\frac{R}{\mu_c}\right)^4 \frac{3}{a}\left(\frac{1 - \beta^*}{\beta^{*4}}\right)\right]^{\frac{1}{2}} G^{-\frac{3}{2}}.$$

Evaluate M for $1 - \beta^* = 0.2, 0.5, 0.8$ and comment on what you find.

3. An enclosure containing an ideal gas (specific heat ratio γ) and radiation undergoes an adiabatic change. Show that

$$\frac{\mathrm{d}P}{P} + \Gamma_1 \frac{\mathrm{d}V}{V} = 0,$$

where

$$\Gamma_1 = \beta + \frac{(4 - 3\beta)^2(\gamma - 1)}{\beta + 12(\gamma - 1)(1 - \beta)}$$

and $P_\mathrm{g} = \beta P$.

Sketch how Γ_1 varies as β grows from 0 to 1.

4. By use of the Fermi–Dirac distribution

$$N(p)\,\mathrm{d}p = \frac{8\pi}{h^3} \frac{p^2\,\mathrm{d}p}{\exp\left[(p^2/2mk_\mathrm{B}T) - \psi\right] + 1},$$

where ψ is defined in terms of the number density n by

$$n = \int_0^\infty N(p)\,\mathrm{d}p,$$

show that the electron pressure can be written as

$$P_\mathrm{e} = \frac{4\pi}{h^3}(2mk_\mathrm{B}T)^{3/2}k_\mathrm{B}T\left[\frac{2}{3}F_{\frac{3}{2}}(\psi)\right]$$

and the density as

$$\rho/\mu_\mathrm{e}m_\mathrm{H} = \frac{4\pi}{h^3}(2mk_\mathrm{B}T)^{3/2}F_{\frac{1}{2}}(\psi),$$

where

$$F_n(\psi) = \int_0^\infty \frac{u^n}{e^{u-\psi} + 1}\,\mathrm{d}u.$$

Express the equation of state for a partially degenerate electron gas in the form $P = P(\rho, \psi)$. Show that when $\psi \ll 0$ the ideal gas law holds. What is the equation of state when $\psi \gg 0$?

5. For cold degenerate matter, whether relativistic or not, show that the electron pressure

$$P_e = \frac{\pi m_e^4 c^5}{3h^3} \left[y\sqrt{1+y^2}\,(2y^2 - 3) + 3\sinh^{-1} y \right],$$

where $y = p_f/m_e c$ and p_f is the Fermi momentum. Demonstrate that we recover $P_e \propto \rho^{4/3}$ in the extremely relativistic case when $p_f \gg m_e c$ and $P_e \propto \rho^{5/3}$ in the non-relativistic case when $p_f \ll m_e c$.

Chapter 4

Heat Transport

We now turn to the transport of energy within a star. Typically the energy flow is from the stellar interior, where nuclear fusion is taking place, towards the stellar surface where it can be radiated into space. We define the local luminosity L_r at radius r from the centre of the star to be the total energy flux outwards through the sphere of radius $4\pi r^2$. This luminosity depends on the temperature gradient $\mathrm{d}T/\mathrm{d}r$ through the star. Three major processes contribute,

(1) radiation — the diffusion of photons,
(2) conduction — the diffusion of particles and
(3) convection — bulk transport within the fluid.

Usually one of these dominates at any one radius at any time during a star's evolution. Before discussing each in more detail we shall first consider how reasonable it is to assign a specific temperature, and indeed the same temperature, to all particles and the radiation within the fluid at any point in a star.

4.1. Thermal Equilibrium

A star cannot be in complete thermal equilibrium with its surroundings. If it were it would not shine. However material in stars is somewhat closer to thermal equilibrium than almost anything encountered on Earth. At the centre of the Sun the temperature $T_c \approx 2 \times 10^7$ K. Its radius $R_\odot \approx 7 \times 10^{10}$ cm so we estimate the size of the temperature gradient through the Sun to be

$$\left|\frac{dT}{dr}\right| \approx 3 \times 10^{-4}\,\mathrm{K\,cm^{-1}} = 0.3\,\mathrm{K}\,(10\,\mathrm{m})^{-1}. \qquad (4.1)$$

This is certainly smaller than the temperature gradient around you as you read this book.

What matters for the physics of stars is whether or not everything is in local thermodynamic equilibrium, LTE, so that it is meaningful to write $T = T(r)$ and $P = P(r)$ with the radiation behaving as a black body in equilibrium with the particles, the velocities of which obey a Maxwellian distribution. To test this assertion we consider the mean free paths of the relevant particles that transport energy in a stellar interior. For LTE we require that the mean free path λ of such particles is smaller than the length scale for changes in T. That is

$$\lambda \ll T \left/ \frac{dT}{dr} \right. \approx R_\odot. \qquad (4.2)$$

4.1.1.   *Photons*

We write the mean free path of photons as λ_γ so that the number of mean free path lengths to the surface

$$N_\gamma \approx \frac{R_\odot}{\lambda_\gamma}. \qquad (4.3)$$

In radiative regions photons diffuse to the surface in a random walk which requires N_γ^2 steps each taking a typical time λ_γ/c. So the diffusion time

$$\tau_{\mathrm{diff}} \approx \frac{N_\gamma^2 \lambda_\gamma}{c} \approx \frac{N_\gamma R_\odot}{c}. \qquad (4.4)$$

If no new photons were generated this would be equivalent to the time for all photons to leak out of the star

$$\tau_{\mathrm{leak}} \approx \frac{\overline{P_r}}{\overline{P}}\tau_{\mathrm{KH}} \approx 2 \times 10^{13}\,\mathrm{s} \approx 6 \times 10^5\,\mathrm{yr}, \qquad (4.5)$$

where $\overline{P_r}/\overline{P}$, small for the Sun, estimates the ratio of the photon to thermal energy density in the interior and τ_{KH}, the Kelvin–Helmholtz time-scale, is the time that it would take to radiate the thermal energy from the star. Thus

$$N_\gamma \approx 8 \times 10^{12} \gg 1 \qquad (4.6)$$

so that

$$\lambda_\gamma = \frac{R_\odot}{N_\gamma} \approx \frac{R_\odot^2}{c\eta_{\text{leak}}} \approx \frac{R_\odot^2}{c} \frac{\overline{P_r}}{\overline{P}} \frac{1}{\tau_{\text{KH}}} \approx 10^{-2} \, \text{cm} \ll R_\odot. \qquad (4.7)$$

So we expect LTE to be an extremely good approximation for photons in the solar interior and indeed in all stars. We note here that at its surface, where a star radiates into space, $N_\gamma \to 0$ and a more careful analysis is required for stellar atmospheres (see Chapter 5).

4.1.2. *Particles*

The mean free path of particles, atoms, ions or electrons, can be estimated with their scattering cross-section. Electrons are the least massive, move fastest and travel furthest. They are preferentially scattered by the slowly moving ions so we consider electrons scattering off positively charged ions. Define an impact parameter b as the closest approach between the paths of two charged particles were they not to affect each other (Fig. 4.1). Scattering is strong when an electron's kinetic energy is approximately equal to its electrostatic potential energy, so when

$$\frac{1}{2} m_e v_e^2 \approx \frac{e^2}{4\pi\epsilon_0 b}, \qquad (4.8)$$

where e is the electronic charge and ϵ_0 is the permittivity of free space. In LTE we have

$$\frac{1}{2} m_e v_e^2 \approx \frac{3}{2} k_B T, \qquad (4.9)$$

where k_B is Boltzmann's constant. At $T = 10^7$ K, typical of the Sun's interior, $v_e \approx 2 \times 10^9$ cm s^{-1}. Notice that $\gamma = (1 - v_e^2/c^2)^{-1} \approx 1.0045$ so we do not need to consider relativistic effects until somewhat

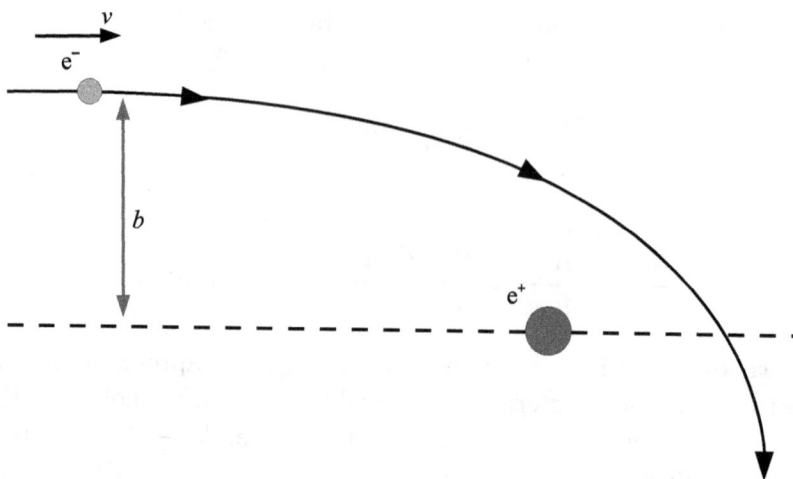

Fig. 4.1. Two particles, here an electron and a positively charged ion, approach one another from infinity where their relative speed is v and pass each other in a parabolic orbit with impact parameter b.

higher temperatures are reached. Thus we have

$$b \approx \frac{e^2}{6\pi\epsilon_0 k_{\mathrm{B}} T} \approx 10^{-10}\,\mathrm{cm}. \tag{4.10}$$

To find the mean free path consider the moving electron sweeping out a cylinder of radius b as it moves through the scattering particles. The length of the cylinder is equal to the mean free path λ_{e} when its volume contains, on average, one scattering particle (Fig. 4.2). So

$$\lambda_{\mathrm{e}} n \pi b^2 \approx 1, \tag{4.11}$$

where $n \approx \rho/m_{\mathrm{H}}$ is the number density of the scattering particles. In the Sun $\bar{\rho} = 1.4\,\mathrm{g\,cm}^{-3}$ so that $\bar{n} \approx 10^{24}\,\mathrm{cm}^{-3}$ and $\lambda_{\mathrm{e}} \approx 3 \times 10^{-5}\,\mathrm{cm} \ll \lambda_\gamma \ll R_\odot$. Consequently, the number of steps an electron must take to diffuse through the Sun is given by, $N_{\mathrm{e}} \approx R_\odot/\lambda_{\mathrm{e}} \approx 2 \times 10^{15} \gg N_\gamma \gg 1$. Particles are even closer to LTE than photons.

4.1.3. *Importance of conduction*

It is the much longer mean free path of photons that makes radiation much more efficient than conduction for heat transport in the Sun. To estimate how much more efficient we must consider the relative

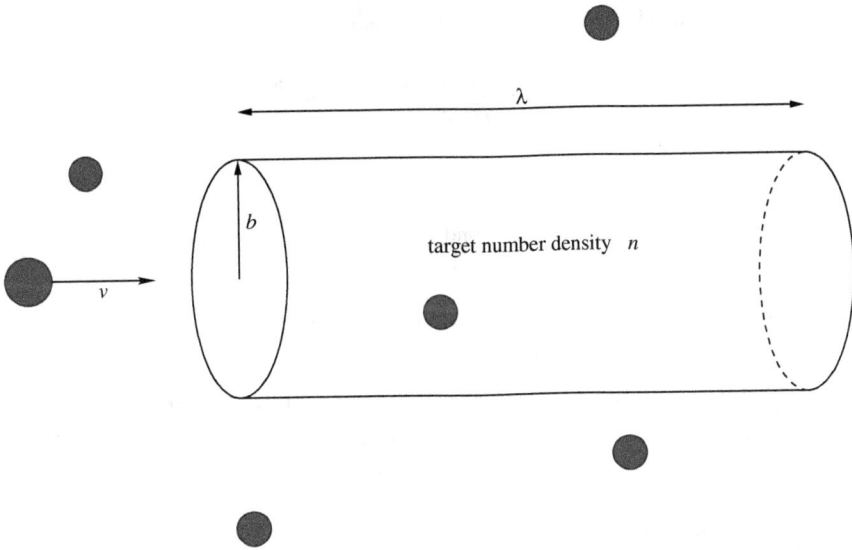

Fig. 4.2. An incoming particle moving at speed v sweeps out a cylinder of potential interaction with a radius equal to its critical impact parameter b. Potential targets are distributed with number density n. The cylinder has a length equal to its mean free path λ when there is, on average, one target within the cylinder.

energies of photons and electrons. From the diffusion times $\tau_{\text{diff}} \approx N R_\odot / v_{\text{d}}$ for carriers requiring N steps at a typical speed v_{d} to diffuse through a star we can estimate a diffusion velocity. For photons this is

$$v_{\text{d},\gamma} = \frac{R_\odot}{t_{\text{diff},\gamma}} \approx \frac{c}{N_\gamma} \approx \frac{c \lambda_\gamma}{R_\odot}. \tag{4.12}$$

Similarly for electrons

$$v_{\text{d,e}} = \frac{v_{\text{e}} \lambda_{\text{e}}}{R_\odot}. \tag{4.13}$$

To estimate energy fluxes we multiply these diffusive velocities by the energy density of the carriers. The ratio of the energy densities is just the ratio of the contributions of the carriers to the pressure so, again for $P_{\text{r}} \ll P_{\text{g}}$, the ratio of the energy flux of conduction by electrons F_{cond} to that of radiation F_{rad} is

$$\frac{F_{\text{cond}}}{F_{\text{rad}}} \approx \frac{\overline{P_g}}{\overline{P_r}} \frac{v_e}{c} \frac{\lambda_e}{\lambda_\gamma} \approx \lambda_e v_e \frac{\tau_{\text{KH}}}{R_\odot^2}. \tag{4.14}$$

In the Sun this amounts to

$$\frac{F_{\text{cond}}}{F_{\text{rad}}} \approx 0.01, \tag{4.15}$$

so conduction is mildly important even though $\lambda_e \ll \lambda_\gamma$ because electrons carry much more energy than photons. Radiation usually dominates over conduction in stars except when the matter becomes degenerate and λ_e becomes large because the density of free states into which an electron can scatter falls.

4.1.4.   *Diffusion of ions*

In LTE ions of mass m_{ion} preferentially scatter off one another with a similar mean free path to electrons. However they move somewhat more slowly at $v_{\text{ion}} \approx \sqrt{m_e/m_{\text{ion}}}/v_e < 0.023 v_e$ so their typical diffusion time-scale

$$\tau_{\text{diff}} \approx \frac{R_\odot^2}{v_{\text{ion}} \lambda_e} \approx 3 \times 10^{18}\,\text{s} \approx 10^{11}\,\text{yr.} \tag{4.16}$$

While this is long, it is only ten times the nuclear burning time-scale τ_N of the Sun and some small effect should be expected. Indeed settling of the heavier helium ions relative to hydrogen ions can be measured by helioseismology. In white dwarfs the effect of settling is strong enough that only hydrogen lines are seen in their spectra (DA WDs) unless they are convective and helium is mixed back to the surface (DB WDs) or some recent accretion event has polluted them with metals (DZ WDs).

4.2.   **Radiative Transfer**

In LTE the radiation field is that of a black body so that, according to Planck's law (Sec. 3.3), the total energy flux through an area $\text{d}A$ entering a solid angle $\text{d}\Omega$ perpendicular to that area (see Fig. 4.3)

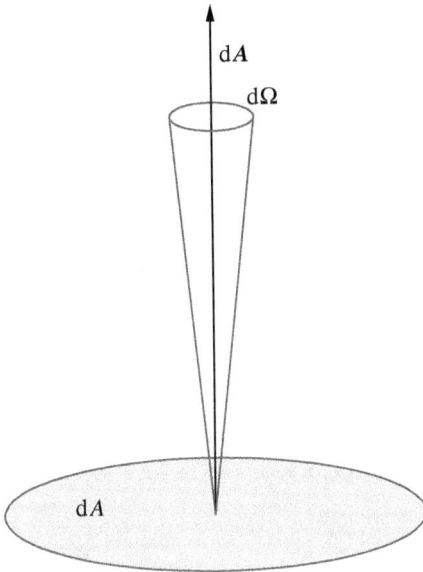

Fig. 4.3. Flux through an elemental area dA enters a solid angle $d\Omega$ parallel to the vector area $d\boldsymbol{A}$.

carried by photons with frequency between ν and $\nu + d\nu$ is

$$B_\nu(T)\, d\nu\, d\Omega\, dA = \frac{2h\nu^3}{c^2} \frac{1}{\exp(h\nu/k_{\mathrm{B}}T) - 1}\, d\nu\, d\Omega\, dA. \qquad (4.17)$$

The photons, each carrying energy $h\nu$, that make up the radiation field interact with matter by the three processes of absorption, emission and scattering. In all cases the probability of an interaction is independent of distance travelled by a photon so these are Poisson processes.

The Stefan–Boltzmann constant σ is defined so that σT^4 is the energy flux per unit area emitted into the space outside a locally flat surface of a black body of temperature T. For such a flat surface of area dA let $B_\nu^*\, d\nu\, dA$ be the flux, of energy in photons with frequency between ν and $\nu + d\nu$ emitted to all space, rather than into unit solid angle so that

$$\int_0^\infty B_\nu^*(T)\, d\nu = \sigma T^4. \qquad (4.18)$$

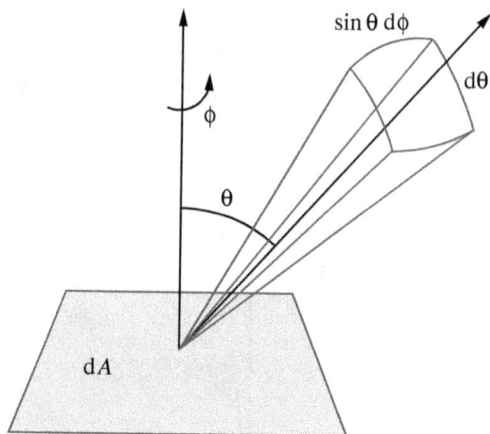

Fig. 4.4. Flux through area dA enters an element of solid angle $d\Omega = \sin\theta\, d\phi\, d\theta$ at an angle θ to dA.

To relate B_ν^* to B_ν consider the flux through a solid angle $d\Omega$ along a direction that is not perpendicular to dA. We can set up local spherical polar coordinates (Fig. 4.4) such that θ is the angle measured from the perpendicular to dA and ϕ is the angle measured around that perpendicular. The solid angle $d\Omega$ then sees a projected area $\cos\theta\, dA$ on the surface of the black body and we may write $d\Omega = \sin\theta\, d\theta\, d\phi$. We integrate this over a hemisphere to account for all the space outside the black body so that

$$B_\nu^* = \int_{\phi=0}^{2\pi} \int_{\theta=0}^{\frac{\pi}{2}} B_\nu \cos\theta\, \sin\theta\, d\theta\, d\phi = \pi B_\nu. \qquad (4.19)$$

We are interested in the total flux L_r through a sphere of radius r in the star. Let $d\sigma$ be an elemental area on the surface of such a sphere. Radiation passing through $d\sigma$ can originate from any point in the star but preferentially from a small sphere of radius equal to the mean free path of photons around $d\sigma$. So consider a small sphere S of general radius l centred on $d\sigma$ (Fig. 4.5). We want to calculate the net outward flux F through $d\sigma$ owing to photons originating on S. The combined radial flux is

$$F = \frac{L_r}{4\pi r^2} = \int_0^\infty F_\nu\, d\nu, \qquad (4.20)$$

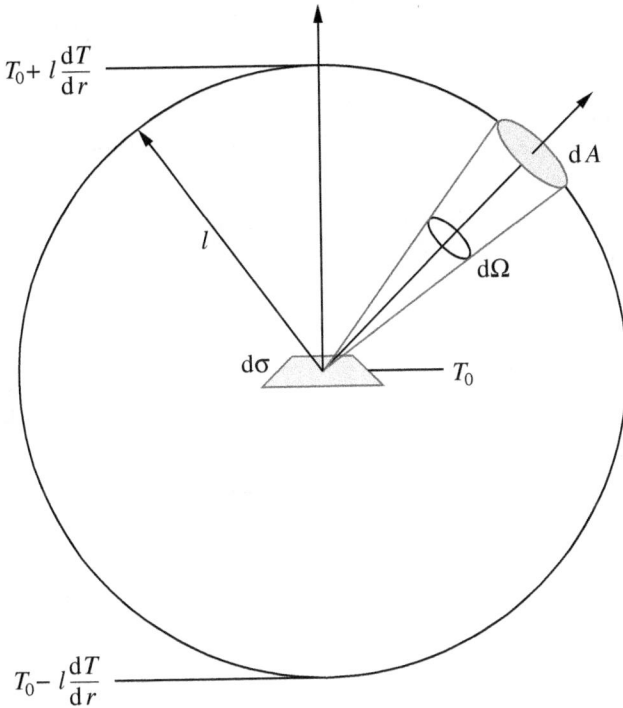

Fig. 4.5. The net flux of radiation through a small surface area $d\sigma$ perpendicular to a radius in the star can be found by integrating the flux originating from elemental areas dA on a surrounding sphere of radius l.

where F_ν is the radial flux through $d\sigma$ carried by photons with frequencies between ν and $d\nu$. In local spherical polar coordinates centred on $d\sigma$ at an angle θ measured from a radius of the star and ϕ measured around that radius consider an emitting area $dA = l^2 \sin\theta \, d\theta \, d\phi$ on S. The temperature at dA is

$$T = T_0 + \frac{dT}{dr} l \cos\theta = T_0 + \Delta T, \qquad (4.21)$$

where $T_0(r)$ is the temperature of $d\sigma$ at radius r.

Seen from dA the central area $d\sigma$ subtends a solid angle of $d\Omega' = (\cos\theta/l^2) \, d\sigma$. Were it not impeded by opacity, the flux emitted from dA, perpendicular to dA that would pass through $d\sigma$ deeper into the star would be

$$dF_{\nu-} = -B_\nu(T_0 + \Delta T)\, d\Omega'\, dA$$

$$= -B_\nu(T_0 + \Delta T)\left(\frac{\cos\theta}{l^2}\, d\sigma\right)(l^2\sin\theta\, d\theta\, d\phi). \quad (4.22)$$

For each area at θ and $T_0 + \Delta T$ there is an equal area at $\pi - \theta$ and $T_0 - \Delta T$, from which the flux

$$dF_{\nu+} = B_\nu(T_0 - \Delta T)\, d\Omega'\, dA \quad (4.23)$$

passes outwards through $d\sigma$. So the net outward flux through $d\sigma$ owing to photons originating from these two areas would be

$$dF_\nu = [B_\nu(T - \Delta T) - B_\nu(T + \Delta T)]\sin\theta\cos\theta\, d\theta\, d\phi\, d\sigma. \quad (4.24)$$

Recall that $\Delta T = l(dT/dr)\cos\theta$ so that, in the limit $\Delta T \to 0$,

$$\Delta T\frac{B_\nu(T - \Delta T) - B_\nu(T + \Delta T)}{\Delta T} \to -2l\cos\theta\frac{dB_\nu}{dT}\frac{dT}{dr}. \quad (4.25)$$

Suppose that actually only a fraction f_l of the photons emitted by dA actually travel far enough to reach $d\sigma$, with the remainder absorbed or scattered on the way, then the total outward flux through $d\sigma$ originating from photons emitted from surfaces S with radii between l and $l + dl$ with frequencies between ν and $\nu + d\nu$ is

$$F_\nu\, dl = -2f_l l\, dl\frac{dT}{dr}\frac{dB_\nu}{dT}\int_0^{2\pi}\int_0^{\frac{\pi}{2}}\sin\theta\cos^2\theta\, d\theta\, d\phi$$

$$= -f_l\frac{4\pi}{3}\frac{dB_\nu}{dT}\frac{dT}{dr}l\, dl, \quad (4.26)$$

Integrating over all spheres S we calculate that the mean free path of photons with frequency ν is

$$\int_0^\infty f_l l\, dl = \lambda_\nu, \quad (4.27)$$

so that

$$F_\nu\, d\nu = -\frac{4}{3}\pi\frac{dB_\nu}{dT}\frac{dT}{dr}\lambda_\nu\, d\nu. \quad (4.28)$$

Now define an opacity κ_ν as the effective cross-section per unit mass of stellar material to absorption or scattering of photons of frequency ν so that

$$\lambda_\nu = \frac{1}{\rho\kappa_\nu} \qquad (4.29)$$

and

$$F_\nu = -\frac{4\pi}{3}\frac{\mathrm{d}B_\nu}{\mathrm{d}T}\frac{\mathrm{d}T}{\mathrm{d}r}\frac{1}{\rho\kappa_\nu}. \qquad (4.30)$$

Integrating over all frequencies ν the total energy flux through $\mathrm{d}\sigma$ is

$$F = -\frac{4\pi}{3}\frac{\mathrm{d}T}{\mathrm{d}r}\frac{1}{\rho}\int_0^\infty \frac{1}{\kappa_\nu}\frac{\mathrm{d}B_\nu}{\mathrm{d}T}\,\mathrm{d}\nu. \qquad (4.31)$$

Integrating

$$\int_0^\infty \frac{\mathrm{d}B_\nu}{\mathrm{d}T}\,\mathrm{d}\nu = \frac{\mathrm{d}}{\mathrm{d}T}\left(\frac{\sigma T^4}{\pi}\right) = 4\frac{\sigma T^3}{\pi}, \qquad (4.32)$$

we define a Rosseland mean opacity κ by

$$\frac{1}{\kappa} = \frac{\int_0^\infty \frac{1}{\kappa_\nu}\frac{\mathrm{d}B_\nu}{\mathrm{d}T}\,\mathrm{d}\nu}{\int_0^\infty \frac{\mathrm{d}B_\nu}{\mathrm{d}T}\,\mathrm{d}\nu} \qquad (4.33)$$

and write the total flux per unit area through a sphere of radius r as

$$F = \frac{L_r}{4\pi r^2} = -\frac{16\sigma T^3}{3\kappa\rho}\frac{\mathrm{d}T}{\mathrm{d}r}. \qquad (4.34)$$

Writing the Stefan–Boltzmann constant $\sigma = ac/4$, in terms of the radiation constant a and speed of light c, we arrive at the conventional form for our third equation of stellar structure, the equation of radiative transfer,

$$\frac{\mathrm{d}T}{\mathrm{d}r} = -\frac{3\kappa\rho L_r}{16\pi acr^2 T^3}. \qquad (4.35)$$

Note that the weighting of opacity κ_ν in the Rosseland mean κ is by $\mathrm{d}B_\nu/\mathrm{d}T$ and not the intensity B_ν or the fraction of photons $B_\nu/h\nu$. There is a maximum in $\mathrm{d}B_\nu/\mathrm{d}T$ when $h\nu = 3.8k_\mathrm{B}T$. Typically κ_ν also decreases with ν and the main contribution is often from photons with energies $h\nu \approx 5k_\mathrm{B}T$.

4.2.1.   *Sources of opacity* κ

Opacity is the effective photon cross-section per unit mass of stellar material. This we can write as

$$\kappa = \frac{\sigma n_t}{\rho}, \qquad (4.36)$$

where n_t is the number density of targets and σ is the cross-section of each target. We can also determine σ as the rate of events divided by the incident photon flux. A large κ means that photons are significantly impeded in their travel through the star while a low κ indicates that they can move relatively freely with long mean free paths. There are several contributions to the opacity of stellar material and discussion of these makes up the remainder of this section. The calculation of stellar opacities is a small industry in itself and so, rather than attempting to explain every detail, we give a qualitative discussion of the processes that must be included.

4.2.1.1.   *Electron scattering*

Photons incident on free particles scatter directly. Because they are the least massive, electrons present the largest effective cross-section. When electrons are non-relativistic, $k_B T \ll m_e c^2$ or $T \ll 5.9 \times 10^9$ K, this is Thompson scattering and is independent of frequency ν. From radiation theory,

$$\sigma_e = \frac{8\pi}{3} \left(\frac{e^2}{m_e c^2} \right)^2 \propto \frac{1}{m_e^2}, \qquad (4.37)$$

where m_e and e are the mass and charge of the electron and c is the speed of light. The strong inverse dependence on mass means that scattering off other particles has a negligible effect when electrons are present. This cross-section $\sigma_e = 0.6652 \times 10^{-24}$ cm^2 $\ll \lambda_{max}^2$, where λ_{max} is the wavelength at the peak of the black-body distribution, so the incident direction is unimportant and electron scattering is isotropic. Now $\kappa_{es} = \sigma_e n_e / \rho$ so for a fully ionised mixture, in which

$$n_e = \frac{\rho N_A (1 + X)}{2 m_H}, \qquad (4.38)$$

where N_A is Avogadro's number and m_H is the atomic mass unit, the opacity owing to electron scattering

$$\kappa_{es} = 0.2\,(1 + X)\,\mathrm{cm^2g^{-1}}, \tag{4.39}$$

which is independent of temperature and density. Electron scattering is the dominant source of opacity at high temperatures, when all atoms are ionised, but its contribution falls off below 10^4 K because, when the temperature drops below this, the hydrogen recombines and so n_e falls significantly.

When electrons move relativistically, $k_B T > m_e c^2$, we must turn to the more general theory of relativistic Compton scattering which becomes frequency dependent and anisotropic. Degeneracy also has an effect but it is not important because conduction by the degenerate electrons comes to dominate the transport of energy over radiation.

4.2.1.2. *Bound–bound κ_{bb} and bound–free κ_{bf} absorption*

Electrons bound to positively charged nuclei provide an important source of opacity because they can be unbound or excited to a higher energy level by the absorption of a photon. The processes of exciting electrons from one energy level to a higher level, bound–bound absorption, and those that ionise the atom, bound–free absorption, are illustrated in Fig. 4.6. To calculate the cross-section for a particular transition we must evaluate the quantum mechanical overlap integrals between the initial and final state and the photon,

$$\sigma \propto \langle \psi_2 | \exp(i\boldsymbol{k} \cdot \boldsymbol{x}) | \psi_1 \rangle, \tag{4.40}$$

where ψ_1 and ψ_2 are the wave functions of the initial and final states and we must integrate over photons of all momenta characterised by $\boldsymbol{k}$ which also takes account of polarisation. Bound–bound absorption is associated with particular frequencies or spectral lines while bound–free absorption occurs over a continuum of frequencies above an edge when $h\nu = \chi_i - E_n$ (Fig. 4.7). The calculation is complex and lengthy because it must be repeated for each element, such as $H, He, C, \ldots$ in each ionisation stage, H, H^+, $He, He^+, He^{++}, \ldots$ and for each excitation state within each stage, such as H with

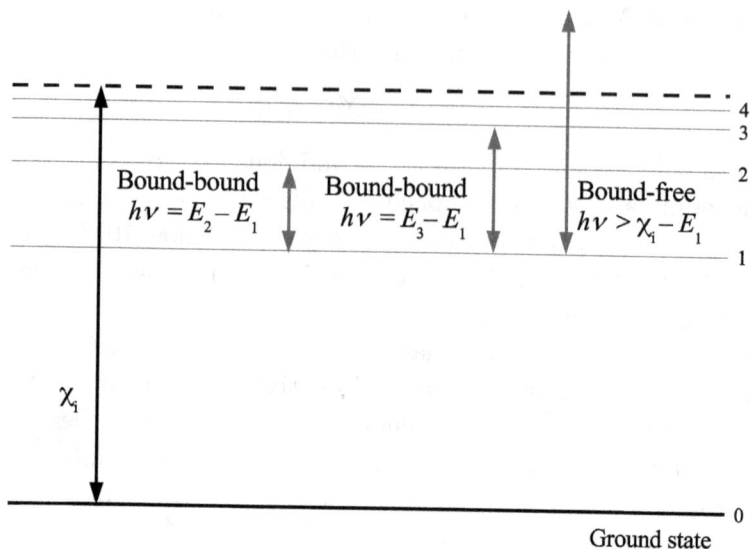

Fig. 4.6. The energy levels E_j of electrons in an atom with ionisation energy χ_i. Bound-bound transitions occur when the electrons move between energy levels both less than χ_i. Bound-free absorption occurs when an electron is freed from its nucleus completely.

$n = 1, 2, 3, \ldots$. The contributions must be added, weighted according to the number of species in each state. For photons of frequency ν

$$\kappa_\nu = \sum_{n,j,k} \sigma_{\nu,k,j,n} \left(\frac{N_{j,n}}{N_j}\right) \left(\frac{N_j}{\sum_j N_j}\right) \frac{1}{A_k m_\mathrm{H}} X_k, \qquad (4.41)$$

N_j is the total number of atoms in ionisation stage j, of which $N_{j,n}$ are in excitation state n, and A_k is the atomic mass of isotope k with mass fraction X_k. In LTE the relative excitation states obey a Boltzmann distribution so that

$$\frac{N_{j,n}}{N_{j,n'}} = \frac{g_{j,n}}{g_{j,n'}} e^{(E_{j,n} - E_{j,n'})/k_\mathrm{B}T}, \qquad (4.42)$$

where $E_{j,n}$ is the energy of the state with degeneracy $g_{j,n}$. Then

$$\frac{N_{j,n}}{N_j} = \frac{g_{j,n}}{u_j} e^{(E_{j,n} - E_{j,1})}, \qquad (4.43)$$

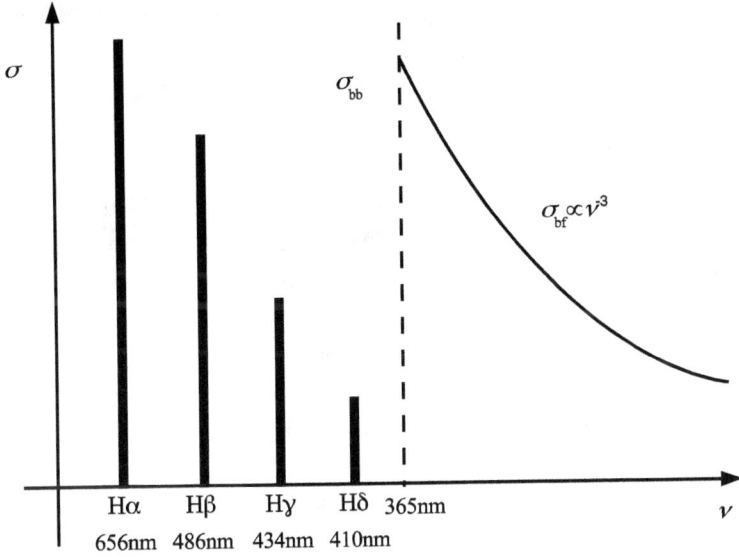

Fig. 4.7. A schematic representation of the cross-section of interaction for bound–bound, Hα, Hβ, ..., the Balmer lines and bound–free, above the Balmer break at 365 nm. In absorption all these transitions are from the $n = 2$ state of a hydrogen atom.

where the partition function

$$u_j = \sum_n g_{i,n} e^{(E_{j,n} - E_{j,1})}. \tag{4.44}$$

Whereas for the equation of state it was possible to approximate this by the degeneracy of the ground state a more detailed calculation is required for opacity. We then integrate over frequency.

These opacity sources are important at temperatures above 10^4 K when hydrogen is fully ionised up to temperatures at which electron scattering becomes important. In this range of temperature the major contributors are partially ionised metals for which, typically, $\kappa_{\rm bf} \approx 10\kappa_{\rm bb}$ and the overall opacity is proportional to both Z and the total fraction of free electrons. Fitting to the results of full calculations yields

$$\kappa_{\rm bf} \propto \frac{Z}{\mu_{\rm e}} \rho T^{-3.5}, \tag{4.45}$$

where μ_e is the mean molecular weight of electrons. For a fully ionised mixture

$$\frac{1}{\mu_e} = \frac{1+X}{2} \qquad (4.46)$$

because hydrogen contributes one electron for each nucleon while helium and metals contribute one electron for every two nucleons. As the temperature falls electrons first recombine into the tightly bound s-orbitals of the heaviest elements while their outer orbitals remain ionised. Because of their closed electron shells the noble gases are the first elements to fully recombine and, because of its high abundance, it is the loss of two electrons to helium that first makes a significant difference. For singly ionised helium

$$\frac{1}{\mu_e} \approx \frac{1}{2}\left(1 + X - \frac{Y}{2}\right), \qquad (4.47)$$

a little smaller because of the loss of electrons to the inner orbitals of metals. Once helium is fully recombined

$$\frac{1}{\mu_e} < \frac{1 + X - Y}{2} \qquad (4.48)$$

falling further as more electrons recombine with the metals and eventually hydrogen itself begins to recombine as the temperature falls further.

4.2.1.3.   *Free–free absorption κ_{ff}*

Electrons can also be scattered from one free state to another by the absorption of a photon in the presence of a nucleus, required to conserve momentum. This is the reverse process of Bremsstrahlung, or braking radiation, in which electrons are deflected by nuclei accompanied by the emission of radiation. The dependence on density and temperature was deduced by Kramers and takes the same form as bound–free absorption. There is less direct dependence on metallicity but the dependence on ionisation is stronger because the absorption cross-section for an ion of charge Z_i^+ and atomic weight

A_i, $\sigma_{\text{ff}} \propto Z_i^2/A_i$. Combining these dependencies we have

$$\kappa_{\text{ff}} \propto (1+X) \sum_i \left(\frac{X_i Z_i^2}{A_i}\right) \rho T^{-3.5}, \qquad (4.49)$$

where the sum is over all ionised or partially ionised species with mass fraction X_i. Typically $\kappa_{\text{ff}} > \kappa_{\text{bf}}$ when $Z < 0.01$.

4.2.1.4. H^- opacity

Below 10^4 K hydrogen atoms recombine and photons are no longer energetic enough to excite their electrons. However free electrons are available from the partially filled outer orbitals of metals and these can be captured by neutral hydrogen to form negative hydrogen ions H^-, in the mildly endothermic reaction

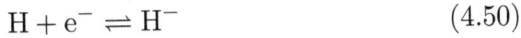

$$\text{H} + \text{e}^- \rightleftharpoons \text{H}^- \qquad (4.50)$$

with an ionisation energy of $-0.75\,\text{eV}$. These H^- ions make a major bound–free contribution to the opacity to optical photons. Together with a free–free contribution, this dominates stellar opacity down to about 3000 K and so is important in the atmospheres of G, F and K stars, including the Sun. Because the free electrons are supplied by the still partially ionised metals there is a very strong dependence on temperature as well as a dependence on metallicity. In the temperature range $3000 \leq T/\text{K} \leq 10000$ and density range $10^{-10} \leq \rho/\text{g cm}^{-3} \leq 10^{-5}$,

$$\kappa_{\text{H}^-} \propto Z\rho^{\frac{1}{2}}T^9. \qquad (4.51)$$

4.2.1.5. Molecules

When $T \leq 3000\,\text{K}$ molecules take over as the major contributors to the opacity. Infrared and optical photons excite vibrational and rotational modes and this gives rise to broad molecular bands in stellar spectra. Important molecules are H_2, TiO, H_2O, CN and CO. The calculation of opacity becomes very complex and has often been treated rather approximately in the past. Today good tables are available and stellar models of low-mass cool stars have much

improved in a large part because of improvements in the calculation
of molecular opacity.

4.2.2. *Combined opacity tables*

Opacity is almost always included in stellar evolution calculations by
interpolation in tables derived from detailed calculations that include
all these contributions for a wide range of compositions. Figure 4.8
shows typical stellar opacities for a composition $X = 0.7$ and $Z =
0.02$. The regions where the various sources of opacity dominate can
be identified. At the highest temperatures, at any density, opacity

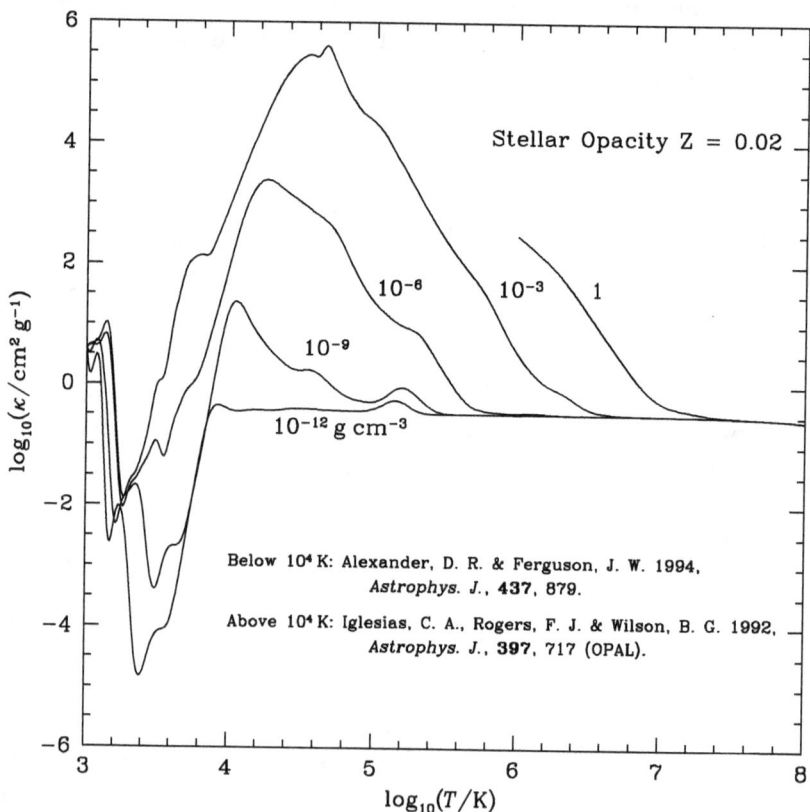

Fig. 4.8. Opacity as a function of temperature at various densities. The composition has $X = 0.7$ and $Z = 0.02$ with the metals distributed as in the solar envelope.

is dominated by electron scattering and appears constant but notice that it drops off at very high temperatures when electrons become relativistic. Moving to lower temperatures bound–free absorption and free–free absorption begin to dominate. There are significant bumps caused by atomic transitions and ionisation of iron, particularly just above 10^5 K. Around 10^4 K, higher temperatures at higher densities, hydrogen recombines and opacity is provided by H^- ions, falling off as T^9 until molecular opacity dominates below about 3000 K.

It is often useful to employ a power-law opacity of the form

$$\kappa \propto \kappa_0 \rho^\alpha T^\beta. \tag{4.52}$$

We can do this for similar stars, or parts of stars, because each opacity source, with its particular dependence on the state of the material, dominates over a particular region of the material state. When $T > T_{\text{crit}} = 2 \times 10^7$ K $(\rho/\text{g cm}^{-3})(Z/0.02)^{2/7}$, when electron scattering dominates,

$$\kappa = \kappa_{\text{es}} = \text{const} \tag{4.53}$$

and $\alpha = \beta = 0$. This is the case in the hot interiors of massive stars. For $T_{\text{H+}} \approx 10^4$ $K < T < T_{\text{crit}}$, while hydrogen is fully ionised and free–free or bound–free absorption dominates, the opacity takes Kramers' form

$$\kappa \propto \rho T^{-3.5}, \tag{4.54}$$

with $\alpha = 1$ and $\beta = -3.5$. Such an opacity law is typical in the interior of the Sun and similar stars. For 3000 K $< T < T_{\text{H+}}$, H^- opacity dominates with

$$\kappa \propto \rho^{\frac{1}{2}} T^9, \tag{4.55}$$

so $\alpha = 1/2$ and $\beta = 9$. For $T < 3000$ K molecules dominate the opacity with only a very weak density dependence and

$$\kappa \propto T^{-30}, \tag{4.56}$$

or $\alpha \approx 0$ and $\beta \approx -30$ though the particular conditions must be examined in more detail to obtain a useful law for application to the cool atmospheres of M dwarfs and giants.

4.3. Conduction

Conduction is most important in degenerate material when the mean free path of the electrons becomes large. The actual calculation is complex, requiring a knowledge of plasma and solid-state physics but the theory is now well understood and accurate tables of effective opacity are available. Within a star electron conduction provides a contribution to the energy transport flux F_{cond} that can be added to the radiative flux of energy carried by photons F_{rad} to give the total energy flux

$$F = F_{\mathrm{rad}} + F_{\mathrm{cond}}. \tag{4.57}$$

These energy transport mechanisms operate in parallel and give the conduction an effective opacity κ_{cond} which is inversely proportional to the conductivity of the stellar material. The combined opacity κ is given by

$$\frac{1}{\kappa} = \frac{1}{\kappa_{\mathrm{rad}}} + \frac{1}{\kappa_{\mathrm{cond}}}, \tag{4.58}$$

where κ_{rad} would be the opacity if heat transport were by radiation alone. Conduction is important in white dwarfs, neutron stars and the degenerate cores of giants. Once matter becomes degenerate we can often practically assume that

$$\kappa_{\mathrm{cond}} \to 0 \tag{4.59}$$

so that

$$\frac{\mathrm{d}T}{\mathrm{d}r} \to 0 \tag{4.60}$$

and the stellar material becomes almost isothermal even when energy sources and sinks are active.

4.4. Convection

Convection remains the most uncertain and debated aspect of the physics of stellar interiors. It is the transport of heat by bulk flows of the fluid itself. It occurs within a stellar interior when energy cannot be transported by radiation or conduction quickly enough. Whenever it becomes more efficient to transport energy by moving it within the hot material then convection sets in. We shall demonstrate that deep within a convection zone the flow is such as to uniformly distribute entropy and modelling convective regions as isentropic gives us a very good approximation. However at the edges of convection zones this approximation breaks down and a better theory is required. Because the flow is turbulent, with a very high Reynolds number, detailed numerical models require similarly very high resolution and it is still not possible to make a full three-dimensional model of every star. Instead we resort to a simpler phenomenological model of the flow that is calibrated to give the correct radius for the Sun. This is possible because the vast majority of the solar convection zone is isentropic and a single parameter in the model suffices to fix the adiabat on which this majority of the Sun's convective envelope lies. A very good model in excellent agreement with helioseismological measurements can still be obtained for all but a thin surface layer.

4.4.1.   *Stability*

The first, and most important, step is to determine a local criterion within the star that determines whether or not material is convectively unstable. Given an underlying stellar structure with known temperature $T(r)$, pressure $P(r)$ and density $\rho(r)$ as functions of radius r, consider a small packet of material displaced antiparallel to the direction of gravity $\boldsymbol{g}$, radially outwards in a star, by a small distance δz (Fig. 4.9). We shall also assume for now that the material within the region of interest has a uniform composition so that its state can be expressed with just two state variables. Outside the packet

$$T \to T + \delta z \frac{\mathrm{d}T}{\mathrm{d}r}, \quad P \to P + \delta z \frac{\mathrm{d}P}{\mathrm{d}r} \quad \text{and} \quad \rho \to \rho + \delta z \frac{\mathrm{d}\rho}{\mathrm{d}r}. \quad (4.61)$$

Inside the pressure P' equalises with that of the surroundings on the very short sound crossing time-scale of the packet so that

$$P' = P + \frac{\mathrm{d}P}{\mathrm{d}r} \delta z. \quad (4.62)$$

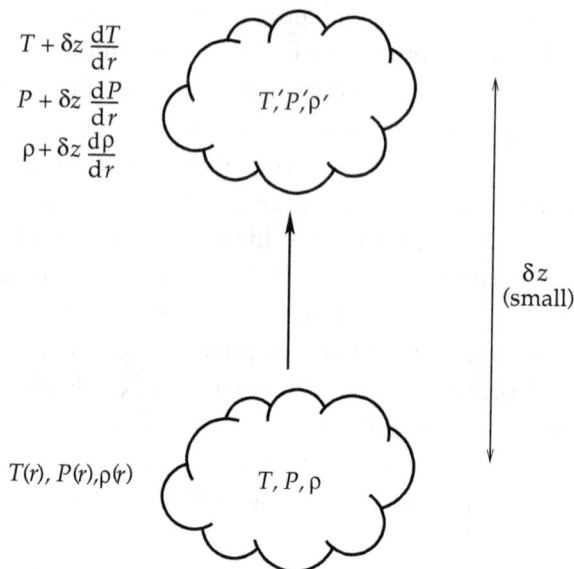

Fig. 4.9. Convective stability. A packet of material is displaced upwards by a small distance δz, against gravity, adiabatically and reaches pressure equilibrium with its surroundings so that $P' = P(z) + \delta z (dP/dr)$. If its density $\rho' < \rho(z) + \delta z (d\rho/dr)$, smaller than that of its surroundings, it continues to rise and so the material is convectively unstable.

The time-scale for heat transport by radiation or conduction is much longer so the interior of the packet behaves adiabatically. From the equation of state of the material inside the packet, we can calculate its new density ρ' and compare with the surroundings. If

$$\rho' < \rho + \frac{dP}{dr} \delta z, \qquad (4.63)$$

the packet is buoyant and continues to rise against gravity. For an adiabatic process, at constant entropy S, a change in pressure δP corresponds to a change in density $\delta \rho$ related by

$$\frac{\delta P}{P} = \Gamma_1 \frac{\delta \rho}{\rho}, \qquad (4.64)$$

where

$$\Gamma_1 = \left(\frac{\partial \log P}{\partial \log \rho} \right)_S \qquad (4.65)$$

can be determined from the equation of state (Sec. 3.4). Thus

$$\rho' - \rho = \delta\rho = \frac{1}{\Gamma_1} \frac{\rho}{P} \frac{dP}{dr} \delta z. \qquad (4.66)$$

We can define a similar derivative for the underlying structure of the star by

$$\Gamma = \frac{d\log P}{d\log \rho} \qquad (4.67)$$

so that

$$\frac{d\rho}{dr} = \frac{d\log \rho}{d\log P} \frac{\rho}{P} \frac{dP}{dr} = \frac{1}{\Gamma} \frac{\rho}{P} \frac{dP}{dr}. \qquad (4.68)$$

The displacement is unstable, because the packet is buoyant, when

$$\delta\rho < \delta z \frac{d\rho}{dr}. \qquad (4.69)$$

For our adiabatic packet in pressure equilibrium with its surroundings this can be rewritten as

$$\frac{1}{\Gamma_1} \frac{\rho}{P} \frac{dP}{dr} \delta z < \frac{1}{\Gamma} \frac{\rho}{P} \frac{dP}{dr} \delta z. \qquad (4.70)$$

For a star in hydrostatic equilibrium

$$\frac{dP}{dr} < 0 \qquad (4.71)$$

so this becomes

$$\frac{1}{\Gamma_1} > \frac{1}{\Gamma} \qquad (4.72)$$

or

$$\frac{\Gamma - \Gamma_1}{\Gamma\Gamma_1} > 0, \qquad (4.73)$$

for instability. A Fourier analysis of perturbations of different wavelengths demonstrates that the converse, of stability when this relation is violated, is also true. Noting that $\Gamma_1 > 0$ we find that relation (4.72) is satisfied either when

$$\Gamma < 0 \tag{4.74}$$

or when

$$\Gamma > \Gamma_1. \tag{4.75}$$

The first of these is the criterion for the Rayleigh–Taylor instability that occurs when a denser fluid is supported against gravity above a sparser fluid. The second is the Schwarzschild criterion for convection and it is this that is of most interest to us here.

To make a model we must determine whether a star that were to transport its energy by radiation or conduction would be convectively unstable anywhere. Equation (4.35) gives us the temperature gradient within the star were it to remain in radiative equilibrium so we recast the criterion Eq. (4.75) in terms of a temperature gradient. By expressing its equation of state as $\rho = \rho(P, T)$ we may write the density change in the packet as

$$\mathrm{d}\rho = \left(\frac{\partial \rho}{\partial P}\right)_T \mathrm{d}P + \left(\frac{\partial \rho}{\partial T}\right)_P \mathrm{d}T \tag{4.76}$$

so that

$$\frac{1}{\Gamma} = \frac{\mathrm{d}\log\rho}{\mathrm{d}\log P} = \frac{P}{\rho}\frac{\mathrm{d}\rho}{\mathrm{d}P} = \frac{P}{\rho}\left[\left(\frac{\partial \rho}{\partial P}\right)_T + \left(\frac{\partial \rho}{\partial T}\right)_P \frac{\mathrm{d}T}{\mathrm{d}P}\right] \tag{4.77}$$

in the background stellar material and

$$\frac{1}{\Gamma_1} = \left(\frac{\partial \log\rho}{\partial \log P}\right)_S = \frac{P}{\rho}\left(\frac{\partial \rho}{\partial P}\right)_S = \frac{P}{\rho}\left[\left(\frac{\partial \rho}{\partial P}\right)_T + \left(\frac{\partial \rho}{\partial T}\right)_P \left(\frac{\partial T}{\partial P}\right)_S\right] \tag{4.78}$$

in the packet. The first term in our instability relation (4.72) cancels to leave

$$\left(\frac{\partial \rho}{\partial T}\right)_P \frac{T}{P}\nabla_a > \left(\frac{\partial \rho}{\partial T}\right)_P \frac{T}{P}\nabla, \tag{4.79}$$

for which we have defined an adiabatic temperature gradient

$$\nabla_a = \left(\frac{\partial \log T}{\partial \log P}\right)_S = \frac{\Gamma_2 - 1}{\Gamma_2} \qquad (4.80)$$

and a structural temperature gradient

$$\nabla = \frac{d \log T}{d \log P} = \frac{\Gamma - 1}{\Gamma}. \qquad (4.81)$$

Now, generally for stellar material at constant pressure, and not completely cold, density must fall as temperature rises so

$$\left(\frac{\partial \rho}{\partial T}\right)_P < 0 \qquad (4.82)$$

and the star is unstable to convection whenever the temperature gradient in the star is superadiabatic. That is when

$$\nabla > \nabla_a. \qquad (4.83)$$

Now let us suppose we have a star in radiative equilibrium. Combining Eq. (2.10) of hydrostatic support with Eq. (4.35) for radiative heat transport we can define a radiative temperature gradient

$$\nabla_r = \left(\frac{d \log T}{d \log P}\right)_r = \frac{3\kappa}{16\pi acG} \frac{P}{T^4} \frac{L_r}{m}. \qquad (4.84)$$

Then if

$$\nabla = \nabla_r > \nabla_a \qquad (4.85)$$

the star cannot remain in radiative equilibrium and must instead become convective. Essentially if neither radiation nor conduction can transport the energy fast enough convection, by transporting heat with the material itself, takes over. The form of the radiative gradient (4.84) together with the condition (4.85) indicates that convection can be driven by

(i) increase in L_r, when there is more energy to transport,
(ii) increase in κ, when radiation cannot transport energy so readily
   or
(iii) decrease of ∇_a, when the material can carry energy more efficiently.

In massive main-sequence stars the temperature is high enough for hydrogen burning by the CNO cycle (see Sec. 6.2.1.2). Burning is concentrated at the very centre of the star, because of the high temperature sensitivity (see Sec. 6.3.4.1) of the nuclear reactions, so that L_r becomes large when m is still small and consequently ∇_r is large and the core is convective. As m increases ∇_r falls to reach ∇_a at the base of a radiative envelope. For an ideal gas $\nabla_a = (\gamma - 1)/\gamma$ where $\gamma = C_P/C_V$ is the ratio of specific heat capacities at constant pressure and at constant volume (Sec. 3.4). During ionisation heat must be supplied to free the electrons even if temperature is fixed. Both C_V and C_P are large and similar so $\gamma \to 1$ and so $\nabla_a \to 0$ and convection is much more likely in ionisation zones. This is the case in the envelopes of cool stars, including red giants and low-mass main-sequence stars, such as the Sun.

4.4.2.   *Convective energy transport*

Having established when and where a star's interior is convective we now turn our attention to the heat flux carried by convecting fluid. This is a much more difficult problem because it requires a model of an extremely turbulent fluid. We are also keen to keep our equations of stellar structure local in the sense that the derivatives of quantities such as temperature depend only on conditions at the local radius in the star. Typically to obtain F_{conv} we use a *mixing length theory* (MLT). The uncertainties of convection are encapsulated in such models which critically affect stellar evolution calculations. This remains the greatest uncertainty in the physics of stellar interiors. To begin, consider a packet, now of finite dimensions and mass δm, that has risen a finite distance δr within the star and is rising at speed v (Fig. 4.10). Suppose the packet has risen fast enough that we still consider the state of the material within to have altered adiabatically. The temperature excess of the packet over its new surroundings is then

$$\delta T = \left(\frac{\partial \log T}{\partial \log P} \right)_S \frac{T}{P} \frac{dP}{dr} \delta r - \frac{dT}{dr} \delta r. \qquad (4.86)$$

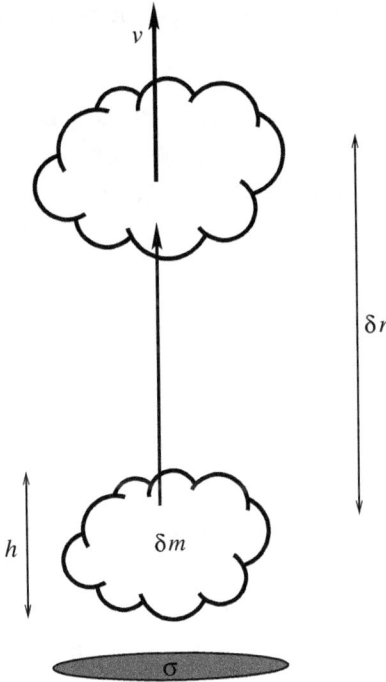

Fig. 4.10. The basis of a Mixing Length Theory. A packet of material of mass δm, of length h in the direction of $\boldsymbol{g}$ and cross-sectional area σ perpendicular to it, has risen a finite distance δr and continues to rise at speed v.

This we may write as

$$\delta T = (\nabla_a - \nabla)\frac{T}{P}\frac{\mathrm{d}P}{\mathrm{d}r}\,\delta r = \Delta\nabla T\,\delta r, \qquad (4.87)$$

where we define the superadiabatic gradient $\Delta\nabla T$. It represents the excess of the temperature gradient over what it would be if the stars were isentropic. Now suppose the packet were to dissipate its excess energy at this point while maintaining a constant pressure P in equilibrium with its surroundings. The heat, specifically enthalpy because $P\,\mathrm{d}V$ is not zero, delivered up to the surroundings would be

$$\delta H = c_p\,\Delta\nabla T\,\delta r\,\delta m, \qquad (4.88)$$

where c_p is the specific heat capacity per unit mass of material at constant pressure. Let the length of the packet in its direction of travel be h and its cross-section perpendicular to this be σ. We can then say that this particular packet of mass $\delta m = \rho h \sigma$ is carrying a heat flux

$$F = \frac{\delta H}{\sigma} \frac{v}{h} = \Delta \nabla T \, \delta r \, c_P \rho v. \qquad (4.89)$$

The same flux is carried by similar rising or falling packets because $v \delta r = (-v)(-\delta r)$. Suppose such packets can be used to represent rising and falling cells in a stellar convection zone. This flux F is the convective flux F_{conv} if δr is the mean distance cells move, v is their mean vertical speed and the cells are closely packed.

To estimate the speed v consider the buoyancy force acting on the rising packet. The density deficit in the packet compared with its surroundings is

$$\delta \rho = \frac{d\rho}{dr} \delta r - \frac{1}{\Gamma_1} \frac{\rho}{P} \frac{dP}{dr} \delta r = \frac{\rho}{T} \Delta \nabla T \, \delta r \qquad (4.90)$$

and the buoyancy force is $g \, \delta \rho$, where g is the local gravitational acceleration. Suppose this buoyancy force increased linearly with δr as the packet rose, and we can neglect any energy lost to drag of the surrounding material, then the kinetic energy of the packet is

$$\frac{1}{2}\rho v^2 = \frac{1}{2} \delta \rho \, g \, \delta r = \frac{1}{2} \frac{\rho g}{T} \Delta \nabla T \, (\delta r)^2 \qquad (4.91)$$

and we can express the convective flux as

$$F_{\text{conv}} = c_P \rho \left(\frac{g}{T}\right)^{\frac{1}{2}} (\Delta \nabla T)^{3/2} (\delta r)^2. \qquad (4.92)$$

However we do not actually know how far a typical convective cell rises or falls before it dissipates and returns its transported enthalpy to its surroundings. To find this we would need to model the details of the fluid flow. This is the subject of ongoing numerical investigations of convection in various fluids including realistic stellar materials and these promise to eventually provide us with an accurate model of convection that can be included in stellar models. For now as a simple, and generally successful, approximation we define a mixing length l to be the average distance travelled by a cell before it dissipates and write

$$\overline{\delta r} = \frac{1}{2}l. \tag{4.93}$$

It is important to remember that l is unknown. It is a free parameter that can be calibrated so that models fit particular stars. It is often adjusted in solar models to obtain the correct radius $R_\odot$. As we have already mentioned, and shall justify in the next section, MLT theory only affects the very outer parts of the solar convection zone and so, for the Sun, this can be considered as a calibration of the adiabat on which most of the envelope lies and it is this that determines the radius of the Sun.

Laboratory experiments on incompressible fluids reveal that l is usually as large as it can be in the experimental apparatus. Because stellar material is highly compressible the concept of a rigid wall does not apply and instead we suppose that a significant change in pressure from one end of a cell to the other defines the distance over which it can maintain its structure. Therefore the mixing length is often expressed as a multiple of the pressure scale height,

$$l = \alpha \frac{P}{|dP/dr|}, \tag{4.94}$$

with $\alpha \approx 1$. The free parameter α is calibrated so that a solar model fits the Sun and then is usually used recklessly for stars of any mass in any phase of evolution.

4.4.3. *The temperature gradient*

The energy flux carried by convection usually considerably exceeds that carried by radiation or conduction in convective regions and we can replace Eq. (4.35) by

$$\frac{dT}{dr} = \nabla_a \frac{T}{P} \frac{dP}{dr} + \Delta \nabla T. \tag{4.95}$$

To understand how convection typically operates let us consider the Sun's convective envelope. It has a depth of about $0.3 R_\odot$ over which the pressure increases by a factor of ten or so we estimate a mixing length $l \approx 0.03 R_\odot$. The gravitational acceleration

$$g \approx \frac{GM_\odot}{R_\odot^2} \approx 3 \times 10^4 \, \text{cm} \, \text{s}^{-2},$$
(4.96)

the density

$$\rho \approx 1 \, \text{g} \, \text{cm}^{-3},$$
(4.97)

the specific heat capacity

$$c_P = \frac{5}{2} \frac{\mathcal{R}}{\mu} = 2 \times 10^9 \, \text{erg} \, \text{g}^{-1} \, \text{K}^{-1}$$
(4.98)

and the energy flux

$$F \approx \frac{L_\odot}{4\pi R_\odot^2} \approx 6 \times 10^{10} \, \text{erg} \, \text{cm}^{-2} \, \text{s}^{-1}.$$
(4.99)

So we require a superadiabatic gradient

$$\Delta \nabla T \approx 10^{-10} \text{K} \, \text{cm}^{-1} \ll \frac{\text{d}T}{\text{d}r} \approx \frac{T_\text{c}}{R_\odot} \approx 3 \times 10^{-4} \, \text{K} \, \text{cm}^{-1}$$
(4.100)

and

$$\nabla = \nabla_\text{a}$$
(4.101)

to a very good approximation. In general this equality holds extremely well deep within any stellar convection zone. In particular for an ideal gas with $\gamma = 5/3$ a convection zone is extremely well modelled as a polytrope with $n = 3/2$.

However near the surface of the Sun the density is much lower nor can cells travel as far so l must become small. Consequently $\Delta \nabla T$ is not negligible. It is only in such regions that MLT becomes important and conditions at the surface determine the adiabat of the underlying deep convective zone. Calibration of l, or α, for the Sun could equally well be a calibration of the adiabatic constant K under the assumption that $P = K\rho^\gamma$ through the entire convection zone. Because calibrations of α, the ratio of the mixing length to pressure scale height, made for the Sun are usually used for all stars we must keep in mind the error we may make in calculating their radii. For example variation around $\alpha = 1$, from 2 to 0.5 can more than double the radius of a typical giant.

4.4.4. *Other points of interest for convection*

(1) The typical temperature excess of a rising convective cell over its surroundings is

$$\delta T = \Delta \nabla T \, \delta r \approx 0.1 \mathrm{K} \ll T \qquad (4.102)$$

so temperature remains uniform on spherical shells to a very good approximation even in significantly superadiabatic regions.

(2) The typical velocity of a convective cell

$$v \approx 10^3 \, \mathrm{cm \, s^{-1}} \ll c_{\mathrm{s}} = \sqrt{\gamma P / \rho} \approx 3 \times 10^7 \, \mathrm{cm \, s^{-1}}, \qquad (4.103)$$

the speed of sound. This justifies our assumption that cells remain in pressure equilibrium with their surroundings throughout their existence.

(3) The convective turnover time-scale

$$t_{\mathrm{conv},\odot} = \frac{l}{v} \approx 20 \, \mathrm{d} \ll \tau_{\mathrm{thermal}} \ll \tau_{\mathrm{nuclear}}. \qquad (4.104)$$

So convective cells turn over many times before the stellar structure changes significantly. It is for this reason that we avoid a full hydrodynamic treatment of convection in models of stars. To do so would severely increase the time taken to evolve a star.

(4) Convective zones are constantly mixed by the motion of the cells. So generally convective regions have uniform composition except in extreme conditions when $\tau_{\mathrm{nuclear}} \approx \tau_{\mathrm{conv}}$. This usually does not affect the major sites of energy production which determine stellar structure but it can be very important for nucleosynthesis of other isotopes. The hot bottom burning at the base of the convective envelopes of massive asymptotic giant branch stars is one such instance. Mixing length theory lends itself to a simple theory of diffusive convective mixing, akin to the kinetic theory

of gases, in which the diffusion coefficient

$$D \propto vl. \tag{4.105}$$

We return to this in Sec. 7.1.

(5) The Reynolds number

$$\mathcal{R}_e = \frac{lv}{\nu} \approx 2 \times 10^{10} \tag{4.106}$$

so the flow is turbulent and difficult to model in detail even on short time-scales.

4.4.5. *Refinements to MLT*

Much can be done to add to the physical nature of mixing length theories. However none of these refinements removes the need for the calibration of a free parameter. Most often added are radiative losses from the rising packet as it moves a finite distance in a finite time. Sometimes added are viscous drag, increasing as a packet accelerates, a full turbulent spectrum for the sizes and speeds of cells and their geometry. Indeed the solar granulation appears to be due to large slowly rising cells alongside small rapidly falling flows (Fig. 4.11).

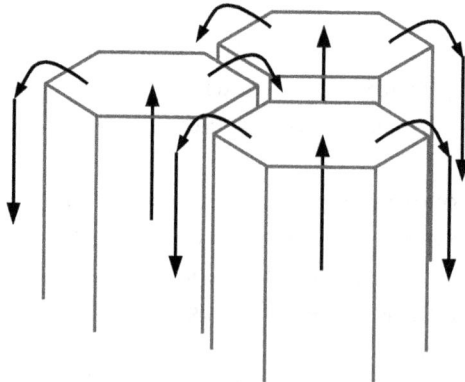

Fig. 4.11. A simple model of the geometrical structure of convective cells near the surface of the Sun where they give rise to the observed granulation. The material rises slowly in large cells and falls back more quickly in the narrow space between them.

High resolution models in three dimensions can deal with all these effects and are beginning to give us a much better idea of how to properly deal with the superadiabatic regions in stars.

4.4.6. *Convective overshooting*

At the boundary of a convective zone, $\dot{v} \to 0$ but $v \nrightarrow 0$. So a convective zone may extend into a neighbouring radiative zone. The effects on energy transport are usually minor but the consequent mixing of material from a radiative region into a convective region can be quite profound. For instance massive stars burning hydrogen in a convective core can last significantly longer because overshooting at the convective boundary entrains hydrogen-rich material and consequently refuels the burning core. Comparison of models with actual stars, particularly the coevolving members of binary systems indicates that a better fit is found when some convective overshooting is included. However, in practice, the deceleration of any overshooting material is very rapid and the size of any overshoot region ought to be much smaller than what is necessary to obtain the better fit. Nevertheless, for now, convective overshooting is used in models as a proxy for whatever extra mixing mechanism actually operates.

4.4.7. *Semi-convection*

Semi-convection is used to describe convective processes in the presence of a composition gradient. Consider burning in a growing convective core (Fig. 4.12). This is more typical during core helium burning but can be illustrated hypothetically for hydrogen. At the centre of the star $\nabla_r > \nabla_a$. As r increases ∇_r falls eventually dropping below ∇_a at the boundary to the convective core m_1. Conditions at this boundary are such that electron scattering dominates the opacity κ so that

$$\nabla_r \propto \kappa_{es} \propto (1 + X). \tag{4.107}$$

Just outside the core, where hydrogen has not been depleted, opacity jumps up so ∇_r also jumps up before continuing its gradual fall with r. As a consequence a second convective region, $m_1 < m < m_2$, appears

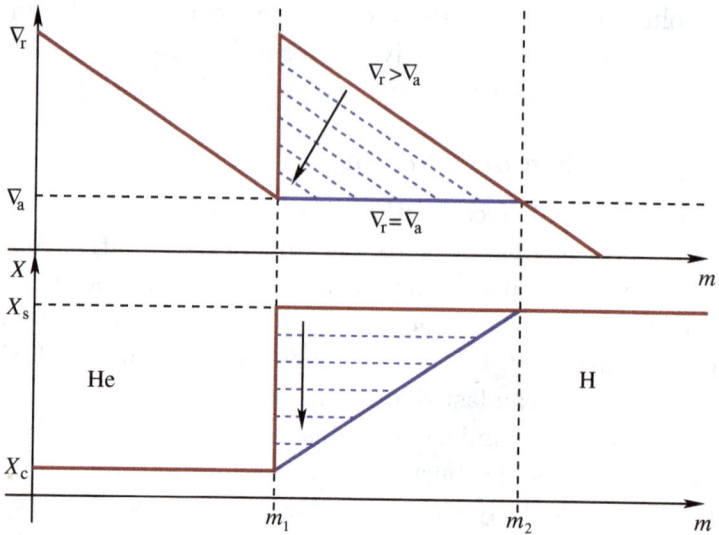

Fig. 4.12. Semi-convection outside a convective hydrogen-burning core. The hydrogen abundance X in the core has been depleted relative to the surface of the star. Material is hot and fully ionised with $\nabla_a = \text{const}$ and opacity dominated by electron scattering so that $\nabla_r \propto \kappa_{es} \propto 1 + X$. In the helium-rich core ∇_r falls monotonically with radius until at mass m_1 the Schwarzschild criterion is satisfied at the edge of the convective core. The red solid lines show the state when the Schwarzschild criterion is strictly applied. No helium is mixed from the core to $m > m_1$. Just outside the core the opacity is higher, because X is larger, and there is another convective region $m_1 < m < m_2$. However a small amount of mixing across the boundary at m_1 reduces ∇_r in this region and its outer boundary moves inwards. Small bursts of repeated mixing leave behind the equilibrium composition profile shown in blue that ensures $\nabla_r = \nabla_a$.

immediately outside the core. Now imagine that, by whatever mechanism, there is a small amount of mixing across the boundary between these two convective regions. The hydrogen abundance in the outer region would fall and its outer boundary would move inwards to $m < m_2$. Mix a bit more and X falls more and the outer boundary moves in further. No matter how small the amount of mixing, maintenance of the two separate zones is unstable until the composition gradient in the whole of the region $m_1 < m < m_2$ is such that

$$\nabla = \nabla_r = \nabla_a. \tag{4.108}$$

To properly model stars we need to know how quickly this equilibrium is reached. The mixing across the boundary between the two convective regions is a form of double diffusion. Convective packets must diffuse across the boundary and they must disperse and mix before buoyancy returns them. Numerical models indicate that this process is in fact fast compared with stellar evolution time-scales and so can easily be included by extending the diffusive mixing predicted by MLT into the semi-convective region with $D \neq 0$ at m_1.

4.4.8. *The Ledoux criterion*

In Sec. 4.4.1 we derived the Schwarzschild criterion for convective stability under the assumption that the composition in the material surrounding a rising or falling packet is uniform. If there is an underlying composition gradient in the star it can have an effect on stability. To investigate this let us add the mean molecular weight μ as a third state variable to account for changes in composition. We now have $\rho = \rho(P, T, \mu)$ and so

$$d\rho = \left(\frac{\partial \rho}{\partial P}\right)_{T,\mu} dP + \left(\frac{\partial \rho}{\partial T}\right)_{P,\mu} dT + \left(\frac{\partial \rho}{\partial \mu}\right)_{P,T} d\mu. \qquad (4.109)$$

Our infinitesimal rising packet shares neither its heat nor its composition with its surroundings when displaced infinitesimally but remains in pressure equilibrium. So its change in density is

$$\delta\rho = \left(\frac{\partial \rho}{\partial P}\right)_{S,\mu} \frac{dP}{dr} \delta z = \left[\left(\frac{\partial \rho}{\partial P}\right)_{T,\mu} + \left(\frac{\partial \rho}{\partial T}\right)_{P,\mu}\left(\frac{\partial T}{\partial P}\right)_{S,\mu}\right] \frac{dP}{dr} \delta z,$$
$$(4.110)$$

where $dP/dr < 0$ is the pressure gradient in the surrounding material where the change in density is

$$\frac{d\rho}{dr} \delta z = \frac{d\rho}{dP}\frac{dP}{dr} \delta z = \left[\left(\frac{\partial \rho}{\partial P}\right)_{T,\mu} + \left(\frac{\partial \rho}{\partial T}\right)_{P,\mu}\frac{dT}{dP} + \left(\frac{\partial \rho}{\partial \mu}\right)_{P,T}\frac{d\mu}{dP}\right]\frac{dP}{dr} \delta z.$$
$$(4.111)$$

Again because $dP/dr < 0$ this packet is now unstable only if

$$\left(\frac{\partial \rho}{\partial T}\right)_{P,\mu} \left(\frac{\partial T}{\partial P}\right)_{S,\mu} > \left(\frac{\partial \rho}{\partial T}\right)_{P,\mu} \frac{dT}{dP} + \left(\frac{\partial \rho}{\partial \mu}\right)_{P,T} \frac{d\mu}{dP}. \qquad (4.112)$$

Keeping the definitions in common use we write

$$\delta = \left(\frac{\partial \log \rho}{\partial \log T}\right)_{P,\mu} < 0, \qquad (4.113)$$

$$\varphi = \left(\frac{\partial \log \rho}{\partial \log \mu}\right)_{P,T} \qquad (4.114)$$

and

$$\nabla_\mu = \frac{d \log \mu}{d \log P} \qquad (4.115)$$

and obtain the Ledoux criterion for convective instability in the presence of a composition gradient in the form

$$\nabla_a < \nabla + \frac{\varphi}{\delta} \nabla_\mu. \qquad (4.116)$$

For an ideal gas $\varphi = 1$ and $\delta = -1$ and the criterion becomes

$$\nabla > \nabla_a + \nabla_\mu \qquad (4.117)$$

from which we can see that a composition gradient that has $\nabla_\mu > 0$, μ increasing towards the centre of the star, stabilises convection even in regions that would be unstable to the Schwarzschild criterion.

As an example, consider a region in which nuclear burning has left behind a composition profile in which the mass fraction of hydrogen X increases outwards so that μ falls as P falls $\nabla_\mu > 0$. Figure 4.13 illustrates how a composition gradient might stabilise the outer parts of a convective core. In this case mixing between m_2 and m_3 is unstable because this region would be convective in the absence of a composition gradient. This is again a double-diffusive mixing depending on mixing of composition between a displaced packet and its surroundings before buoyancy restores it. Again the time-scale is uncertain but numerical models are beginning to throw light on how quickly it can operate. This situation is further complicated by

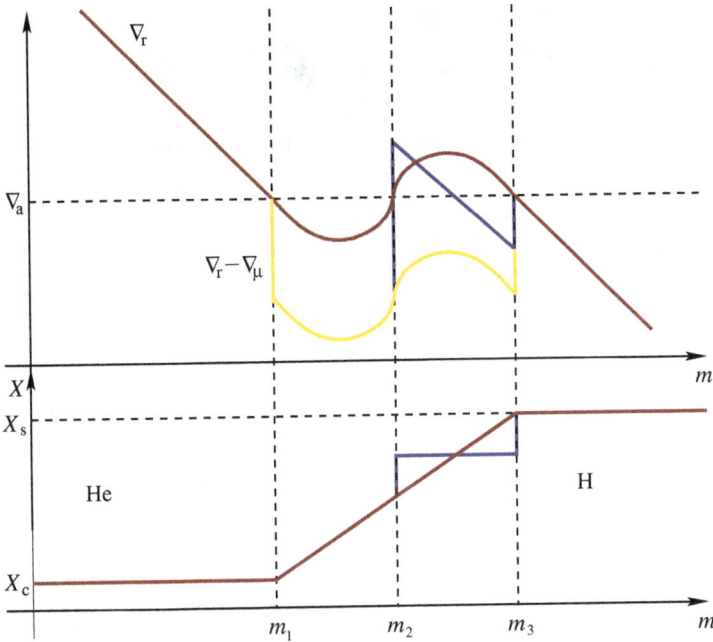

Fig. 4.13. Hydrogen is burning in a core convective out to m_1 whence there is a composition gradient out to m_3 beyond which material has been unaffected by earlier burning. In the lower panel the dark red line is hydrogen abundance X before any extra mixing and the dark red line in the upper panel shows the corresponding ∇_r. There is uniform composition for $m < m_1$ and $m > m_3$ and the straightforward Schwarzschild criterion applies. For $m_1 < m < m_2$ the hydrogen abundance X is increasing with m so μ decreases with m as does P and $\nabla_\mu > 0$ making $\nabla_r - \nabla_\mu < \nabla_a$. So this region is stable to convection under the Ledoux criterion, as illustrated by the dark yellow line. However for $m_2 < m < m_3$ we would find $\nabla_r > \nabla_a$ were there no composition gradient. Consequently any mixing that smooths out the composition gradient in this region is unstable. The dark blue line shows the mixed composition in the lower panel and the corresponding $\nabla_r = \nabla_r - \nabla_\mu$ in the upper panel.

the fact that the new mixed composition profile would then lead to jumps in ∇_r that would be unstable to any small amount of mixing in a similar way to the case we discussed first in Sec. 4.4.7. When the time-scale is short compared to that for structural changes in the stars again an application of the Schwarzschild condition with rapid diffusive mixing can be used to model the equilibrium state.

4.5.   Thermohaline Mixing

Two forms of double-diffusive mixing are commonly encountered in terrestrial oceans when layers of different salinity coexist. As in stars buoyancy is the driver of the mixing but the rates depend on the different rates of diffusion of heat and salt, hence the designation as thermohaline mixing. Generally warm water is less dense than cold water and so a layer of warm water under a layer of cold water would rise. Salty water is more dense than fresh water and so a layer of salty water on top of a layer of fresh water at the same temperature would sink. In both these cases the mixing is of a Rayleigh–Taylor nature and proceeds dynamically. However it is possible to have on top a layer of warm salty water which is less dense than cold fresh water underneath it and so is not Rayleigh–Taylor unstable. Now suppose a packet of the hot salty water is displaced downwards into the cold fresh water. Because heat diffuses faster than salt the packet can lose enough of its heat fast enough to become more dense than its surroundings and continue to sink. In practice this leads to descending salt fingers across the boundary. Because the process depends on diffusion of heat it runs on the thermal time-scale of the small fingers. Similarly rising fingers of cool fresh water can warm up fast enough to continue rising. Alternatively consider cool fresh water above warm salty water. In this case diffusion of heat cannot cause instability so a displaced packet can only mix on the time-scale for salt to diffuse. The diffusion of salt takes longer than the diffusion of heat so this is a slower process. Both these processes are referred to as thermohaline mixing. They have been analysed extensively in the laboratory and more recently numerically.

In stars thermohaline mixing is used to refer to the analogous situations where molecular weight differences take the place of salinity. For hydrostatic equilibrium pressure must decrease with increasing radius within a star. Usually this is accompanied by falling temperature and density. If density rises with radius a dynamical Rayleigh–Taylor instability rapidly mixes the material. Usually nuclear fusion tends to increase the mean molecular weight of stellar material and the most advanced nuclear burning takes place at the hot centre of a star with earlier burning phases in shells moving outwards. Thus μ usually decreases with radius in a star too. The term thermohaline mixing is applied to stellar interiors when μ increases with radius. It is particularly important in binary star evolution when material from the deep interior of an evolved star overflows on to the unprocessed surface of a companion. Suppose we have a stable interface across which both pressure and density decrease outwards with higher molecular weight material supported by lower molecular weight material. Suppose the material is also stable to convection in the sense that a packet of material displaced downwards adiabatically and without mixing has a higher density than its surroundings and so rises back up. Now allow such a packet to reach temperature as well as pressure equilibrium with its surroundings. For an ideal gas $P = \rho \mathcal{R} T / \mu$ so because μ is larger ρ becomes larger and the packet continues to sink. This behaviour is analogous to the hot salty water layer above a cold fresh water layer and modelling is rather similar. It probably operates on a thermal time-scale or faster so that we can expect mixing to remove the molecular weight gradient on a time-scale much shorter than that on which the star evolves. Heat transport by such processes is not likely to be as efficient as radiation so the only important effect is to homogenise the accreted material with the stellar envelope.

An interesting situation also occurs during hydrogen burning by the ppI-chain (Sec. 6.2.1.1). In the course of the reaction

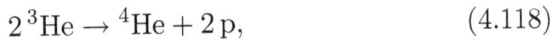

$$2\,^3\mathrm{He} \rightarrow {}^4\mathrm{He} + 2\,\mathrm{p}, \tag{4.118}$$

in which two $^3\mathrm{He}$ nuclei are converted to one alpha particle and two protons the molecular weight μ falls from 1 to 6/7. Because this reaction requires a higher temperature than that required to produce the $^3\mathrm{He}$, a $^3\mathrm{He}$-rich region is generated outside a $^4\mathrm{He}$-rich layer when hydrogen is burning radiatively. In low-mass red giants excess $^3\mathrm{He}$ left behind by core hydrogen burning is mixed into a deep convective envelope at first dredge up. The convective envelope then retreats, leaving behind a homogeneous region with excess $^3\mathrm{He}$, ahead of a growing hydrogen-burning shell. Reaction (4.118) operates at lower temperatures than the CNO cycle in the advancing burning shell so destruction of ^{3}He creates a mild molecular weight inversion, in the radiative region between the burning shell and the base of the convective envelope, sufficient to drive extra mixing down to regions where partial CNO burning is possible. The ^{3}He abundance in the envelope falls as does the ^{12}C/^{13}C isotopic ratio both of which would otherwise be too high to reconcile with observations.

4.6.   Questions

1. Define the critical luminosity $L_{\text{crit}}(r) = 4\pi c G m/\kappa$. Show that

$$\frac{dP_{\text{g}}}{dP_{\text{r}}} = \frac{(L_{\text{crit}} - L_{\text{rad}})}{L_{\text{rad}}},$$

where L_{rad} is the radiatively transferred luminosity.

What happens if $L_{\text{crit}} < L_{\text{rad}}$ (a) inside the star and (b) at the surface?

What is L_{crit} for electron scattering opacity?

2. Show that the Schwarzschild stability criterion for convection can be written in the form

$$n + 1 < N(r) + 1,$$

where $\gamma \equiv (n + 1)/n$ is the adiabatic index and $N(r)$ is the local polytropic index defined by $N + 1 = TdP/PdT$.

A box of pure hydrogen is compressed adiabatically. Sketch the behaviour of P with T as T varies between $3000\,\text{K}$ and $30\,000\,\text{K}$. Sketch the polytropic index n you would deduce from your graph.

What can you say about the convective stability of ionisation zones?

3. In the outer layers of a star we can take $L = \text{const}$, $m = \text{const}$ and the opacity can be assumed to be of the form

$$\kappa \propto P^{\alpha-1}/T^{\beta-4},$$

where α and β are constants. Write down, without detailed justification, a differential equation and a boundary condition that serve to determine temperature T as a function of pressure P through the outer layers. Investigate the asymptotic behaviour of $T(P)$ at great depth and show that the atmosphere must be convective (except possibly for a thin surface layer) if

$$\alpha > \frac{2}{5}\beta > 0$$

when the adiabatic gradient is $2/5$. Discuss the cases

$$\kappa = \text{const},$$

$$\kappa \propto \rho T^{-3.5}$$

and

$$\kappa \propto \rho^{1/2}T^{10}.$$

What is their relevance?

Under what opacity laws would the atmosphere tend to be isothermal at great depth?

4. For an ideal gas with radiation pressure show that the Ledoux criterion for convective instability in the presence of a composition gradient takes the form

$$\nabla > \nabla_a + \frac{\beta}{4 - 3\beta}\nabla_\mu,$$

where $\beta = P_g/P$ is the fraction of the total pressure P provided by the gas.

5. Suppose that all forms of double-diffusive mixing are fast compared with the time-scale on which a massive star evolves. Consider the example illustrated in Fig. 4.13. What is the final equilibrium composition profile?

Chapter 5

Stellar Atmospheres

From the surface of the star, the photosphere, photons escape to infinity so that their mean free path can no longer be considered short. Neither can the radiation field be expected to be in local thermodynamic equilibrium (LTE), nor is it isotropic because photons moving inwards are still absorbed or scattered after travelling only a short distance. However particles typically do have short enough mean free paths to remain in LTE to a somewhat greater height in the atmosphere. In this chapter we consider the radiation transport through the atmosphere more carefully and use this to derive more correct boundary conditions for the stellar interior.

5.1. Specific Intensity

We define the specific intensity I_ν of radiation in an analogous way to the black-body intensity B_ν (Eq. 3.25) so that

$$I_\nu \cos\theta \, d\Sigma \, d\Omega \, dt \, d\nu \qquad (5.1)$$

is the energy carried by photons with frequencies between ν and $\nu + d\nu$ crossing a surface of area $d\Sigma$ into a cone of solid angle $d\Omega$ at an angle θ to the vector area $d\Sigma$ in time dt (Fig. 5.1). We align $d\Sigma$ with a radial vector so that, for a spherically symmetric star, $I_\nu = I_\nu(r, \theta)$ is a function only of radius and angle to the vertical. The total radiative flux carried by photons with frequencies between

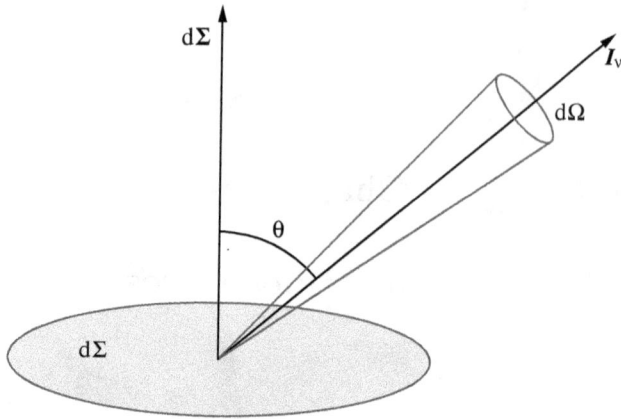

Fig. 5.1. The specific intensity I_ν is such that $I_\nu \cos\theta \, d\Sigma \, d\Omega \, dt \, d\nu$ is the energy carried by photons with frequencies between ν and $\nu + d\nu$ crossing a surface of area $d\Sigma$ into a cone of solid angle $d\Omega$ at an angle θ to the vector area $d\Sigma$ in time dt.

ν and $\nu + d\nu$ through $d\Sigma$ is then

$$F_\nu = \int_{\text{sphere}} I_\nu \cos\theta \, d\Omega = 2\pi \int_0^\pi I_\nu \cos\theta \sin\theta \, d\theta. \qquad (5.2)$$

5.2. Absorption

As the radiation traverses the atmosphere it can be scattered or absorbed. We briefly postpone scattering and first consider processes of true absorption, bound–bound, bound–free and free–free in which photons are removed from the radiation field. We begin by defining an absorption opacity κ_ν for photons with frequencies between ν and $\nu + d\nu$ such that the attenuation of the intensity over a distance δs in the material of density ρ is

$$\delta I_{\nu_{\text{absorption}}} = -\rho\kappa_\nu I_\nu \, \delta s. \qquad (5.3)$$

We integrate over a finite distance s to find

$$I_\nu(s) = I_\nu(0) \exp\left(-\int_0^s \rho\kappa_\nu \, ds\right) = I_\nu(0) \exp\left(-\tau_\nu\right), \qquad (5.4)$$

defining an optical depth

$$\tau_\nu = \int_0^s \rho \kappa_\nu \, ds. \tag{5.5}$$

5.3. Emission Coefficient

Radiation is also emitted by the converse processes of bound–bound, bound–free and free–free emission and we define an emission coefficient j_ν such that

$$j_\nu \, d\Omega \, dt \, d\nu \, \delta m \tag{5.6}$$

is the energy emitted by a mass of material δm into solid angle $d\Omega$ in time dt as photons with frequencies between ν and $\nu + d\nu$. This adds to the intensity I_ν so that over a distance δs

$$\delta I_{\nu_{\text{emission}}} = +\rho j_\nu \delta s. \tag{5.7}$$

5.3.1. *Detailed balance and stimulated emission*

By thinking about energy conservation between processes of absorption and emission in local thermodynamic equilibrium, Einstein was able to deduce the need for two processes of emission, spontaneous and stimulated. Spontaneous emission is isotropic and proceeds independently of the external radiation field. Stimulated emission proceeds at a rate and direction proportional to the intensity. Consider an atomic transition from a jth to an ith state in which a photon of frequency ν_{ji} is emitted with energy

$$h\nu_{ji} = E_j - E_i > 0, \tag{5.8}$$

where E_k is the energy of the electron in the kth state. Now define two Einstein coefficients $\mathcal{A}_{ji}$ and $\mathcal{B}_{ji}$ to describe the probability of emission by an atom such that for spontaneous emission

$$h\nu_{ji}\mathcal{A}_{ji} \, d\Omega \, dt \tag{5.9}$$

is the energy emitted isotropically into solid angle $d\Omega$ in time dt per atom in the jth state and for stimulated emission

$$h\nu_{ji}\mathcal{B}_{ji}I_{\nu_{ji}}\,d\Omega\,dt \qquad (5.10)$$

is the energy emitted in the direction of I_ν into solid angle $d\Omega$ per atom in time dt. Stimulated emission is then only isotropic when $I_{\nu_{ji}}$ is isotropic. The total emission of energy by this transition into solid angle $d\Omega$ in time dt per unit volume is

$$j_{\nu_{ji}}\,d\Omega\,dt = n_j(\mathcal{A}_{ji} + \mathcal{B}_{ji}I_{\nu_{ji}})h\nu_{ji}\,d\Omega\,dt, \qquad (5.11)$$

where n_j is the number of atoms in state j per unit volume.

Kirchoff's law of radiation states that in LTE processes of absorption and emission must balance. So

$$j_\nu = \kappa_\nu I_\nu \qquad (5.12)$$

and the radiation field is that of a black body

$$I_\nu = B_\nu = \frac{2h\nu^3}{c^3}\frac{1}{\exp(h\nu/k_{\mathrm B}T) - 1} \qquad (5.13)$$

which depends only on temperature T. Define a probability of absorption $\mathcal{B}_{ij}$ such the rate of absorption of photons of frequency ν_{ji} by transitions $i \rightarrow j$ per unit mass is

$$n_i\mathcal{B}_{ij}I_{\nu_{ji}} \qquad (5.14)$$

and this must equal the rate of emission of photons per unit mass

$$n_j\left(\mathcal{B}_{ji}I_{\nu_{ji}} + \mathcal{A}_{ji}\right). \qquad (5.15)$$

In LTE the ratio of n_j to n_i depends only on temperature T according to a Boltzmann distribution,

$$\frac{n_j}{n_i} = \frac{g_j}{g_i}\exp\left(-\frac{h\nu_{ji}}{k_{\mathrm B}T}\right), \qquad (5.16)$$

where g_k is the degeneracy of the kth state. Thus equating rates (5.14) and (5.15) we arrive at the detailed balance

$$I_{\nu_{ji}} = B_{\nu_{ji}} = \frac{2h\nu_{ji}^3}{c^2} \frac{1}{\exp\left(h\nu_{ji}/k_\mathrm{B}T\right) - 1}$$

$$= \frac{n_j \mathcal{A}_{ji}}{n_i \mathcal{B}_{ij} - n_j \mathcal{B}_{ji}} = \frac{\mathcal{A}_{ji}}{\mathcal{B}_{ij}\frac{g_i}{g_j}\exp\left(h\nu_{ji}/k_\mathrm{B}T\right) - \mathcal{B}_{ji}}. \qquad (5.17)$$

It was this detailed balance, with its particular temperature dependence, that led Einstein to the need for stimulated emission. From it we find the ratio of the probability of spontaneous to the probability of stimulated emission

$$\frac{\mathcal{A}_{ji}}{\mathcal{B}_{ji}} = \frac{2h\nu_{ji}^3}{c^2}, \qquad (5.18)$$

the ratio of the probability of absorption to the probability of stimulated emission

$$\frac{\mathcal{B}_{ij}}{\mathcal{B}_{ji}} = \frac{g_j}{g_i}, \qquad (5.19)$$

and the ratio of the rate of stimulated emission to the rate of absorption is

$$\frac{n_j \mathcal{B}_{ji}}{n_i \mathcal{B}_{ij}} = \exp\left(\frac{h\nu_{ji}}{k_\mathrm{B}T}\right). \qquad (5.20)$$

5.4. Equation of Transfer

Combining the factors that add to and subtract from the radiation as it traverses material over a distance δs we can write an equation of transfer for the intensity I_ν,

$$\frac{1}{\rho}\frac{dI_\nu}{ds} = j_\nu - \kappa_\nu I_\nu. \qquad (5.21)$$

Because the rate of stimulated emission depends on the radiation intensity in the same way as absorption, it is convenient to include stimulated emission as a negative opacity and write

$$\kappa'_\nu = \kappa_\nu \left[1 - \exp\left(-\frac{h\nu}{k_{\rm B}T} \right) \right] \tag{5.22}$$

and

$$j'_\nu = j_\nu - \exp\left(-\frac{h\nu}{k_{\rm B}T} \right) \kappa_\nu I_\nu. \tag{5.23}$$

In this way we separate out an isotropic spontaneous emission coefficient j'_ν and the equation of transfer can then be written as

$$\frac{1}{\rho}\frac{{\rm d}I_\nu}{{\rm d}s} = j'_\nu - \kappa'_\nu I_\nu. \tag{5.24}$$

5.5. Scattering

We now turn our attention to scattering. In general this is complicated in an anisotropic radiation field because the distribution of scattered photons can depend on the incoming photon and the scatterer. In practice in stars we are usually dealing only with electron scattering which, at the temperatures of normal stellar atmospheres, is isotropic and independent of the direction of the incoming photon. This ceases to be true at high temperatures when relativistic Compton scattering that is neither isotropic nor elastic must be considered.

To proceed generally we define a scattering coefficient σ_ν analogous to the absorption opacity κ_ν so that

$$\delta I_{\rm scattering-} = -\rho\sigma_\nu I_\nu\,\delta s. \tag{5.25}$$

Radiation is scattered into I_ν from all directions so that

$$\delta I_{\rm scattering+} = \rho\sigma_\nu \frac{1}{4\pi} \int_{\rm sphere} P(\theta,\theta') I_\nu(\theta')\,{\rm d}\Omega', \tag{5.26}$$

where $P(\theta,\theta')$ is the probability of scattering from ${\rm d}\Omega'$ at an angle θ' to ${\rm d}\Omega$ at an angle θ. So the equation of transfer generally becomes an awkward to solve integral equation. However, when scattering is isotropic, we can average over a sphere to obtain

$$\delta I_{\text{scattering}+} = \rho \sigma_\nu \overline{I}_\nu. \tag{5.27}$$

Once again it is convenient to define new opacity and emission coefficients

$$\kappa_\nu'' = \kappa_\nu' + \sigma_\nu \tag{5.28}$$

and, the isotropic,

$$j_\nu'' = j_\nu' + \sigma_\nu \overline{I}_\nu. \tag{5.29}$$

The equation of transfer becomes,

$$\frac{1}{\rho}\frac{dI_\nu}{ds} = j_\nu' + \sigma \overline{I}_\nu - \kappa_\nu' I_\nu - \sigma_\nu I_\nu = j_\nu'' - \kappa_\nu'' I_\nu. \tag{5.30}$$

The first term on the right-hand side includes spontaneous emission and scattering and the last term includes absorption, stimulated emission and scattering.

5.6. Surface Boundary Condition

We can solve the equation of transfer for the intensity I_ν through the atmosphere and thence obtain a boundary condition for the interior of a star in which complete LTE is an excellent approximation. A full solution for a real stellar atmosphere is complex and each model often requires as much computation as the entire evolution of the interior of a star. However with further assumptions we can obtain reasonable model atmospheres, particularly for hot stars.

First we assume that the atmosphere is grey in the sense that the opacities $\kappa_\nu = \kappa$, $\kappa_\nu' = \kappa'$, $\kappa_\nu'' = \kappa''$ and $\sigma_\nu = \varsigma$ are independent of ν. We can then easily work with a total intensity

$$I = \int_0^\infty I_\nu \, d\nu, \tag{5.31}$$

a total mean intensity,

$$J = \frac{1}{4\pi} \int_{\text{sphere}} I \, d\Omega = \int_0^\infty \overline{I}_\nu \, d\nu \tag{5.32}$$

and emissivities j, j' and j'' of the form

$$j = \int_0^\infty j_\nu \, d\nu. \tag{5.33}$$

For our grey atmosphere we can integrate the equation of transfer (5.30) over frequency to obtain

$$\frac{dI}{ds} = \rho j' + \rho \varsigma J - \rho \kappa' I - \rho \varsigma I. \tag{5.34}$$

5.6.1.   *Plane-parallel atmosphere*

Next suppose that the atmosphere (Fig. 5.2) contains negligible mass compared with the stellar interior and forms only a very thin, locally flat, layer on the surface of the star, a very good approximation for most stars. We can then neglect curvature and consider an atmosphere of flat layers perpendicular to a vertical direction $\boldsymbol{z}$, aligned with a radius of the star and antiparallel to a constant gravitational acceleration $\boldsymbol{g}$ with magnitude

$$g = \frac{GM}{R^2}, \tag{5.35}$$

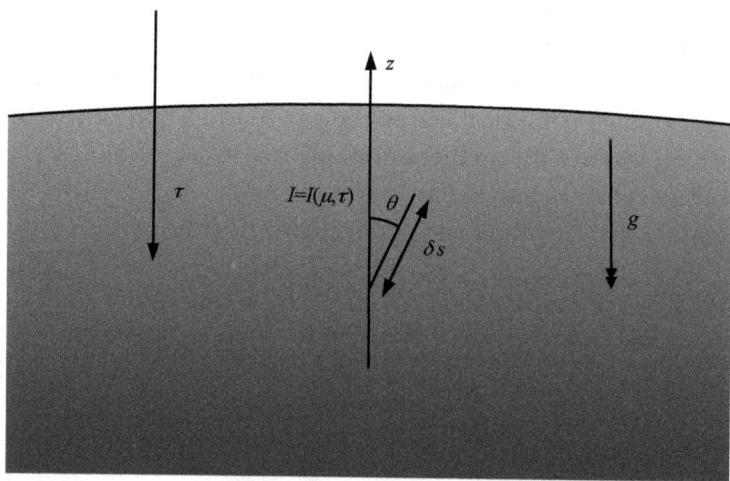

Fig. 5.2. A plane-parallel atmosphere forms a thin flat layer at the surface of a star. The vertical direction z is aligned with the local stellar radius.

where M is the total mass of the star and R the radius of its photosphere. For convenience we write

$$\mu = \cos\theta \tag{5.36}$$

so that $\mu\,ds = dz$ and write $I = I(\mu, \tau)$. Thence

$$J = \frac{1}{2}\int_{-1}^{1} I\,d\mu. \tag{5.37}$$

We define optical depth by $d\tau = -\rho\kappa''\,dz$ so that it accounts for absorption, scattering and stimulated emission. Optical depth is measured downwards into the star and we choose $\tau \to 0$ as $z \to \infty$, noting that τ actually becomes vanishingly small just outside the photosphere. Then for a general direction

$$ds = -\frac{1}{\mu\rho\kappa''}\,d\tau. \tag{5.38}$$

The equation of transfer can now be recast as

$$\mu\frac{dI}{d\tau} = \frac{\kappa'}{\kappa''}I - \frac{j'}{\kappa''} + \frac{\varsigma}{\kappa''}I - \varsigma\frac{J}{\kappa''}, \tag{5.39}$$

with $\kappa'' = \kappa' + \varsigma$. The flux at optical depth τ is

$$F = \int_{\text{sphere}} I(\theta, \tau)\cos\theta\,d\Omega = \int_{-1}^{1} 2\pi\mu I(\mu, \tau)\,d\mu. \tag{5.40}$$

Notice that this is a first moment of the intensity. Integrating the equation of transport (5.39) over a sphere we arrive at

$$\frac{dF}{d\tau} = \frac{\kappa'}{\kappa''}4\pi J - 4\pi\frac{j'}{\kappa''}, \tag{5.41}$$

in which the two scattering contributions have cancelled. In general energy is generated deep in the interior of stars so that, in our plane-parallel atmosphere $F(\tau) = F(0) = \text{const}$ and $dF/d\tau = 0$ so that

$$J = \frac{j'}{\kappa'}. \tag{5.42}$$

Now κ' and j' account for absorption and emission by material that is in LTE to which we can apply Kirchoff's law (5.12) and a black-body radiation field (5.13) and then integrate our grey atmosphere over frequency to obtain

$$j' = \frac{\sigma T^4}{\pi} \kappa',$$

(5.43)

where σ is the Stefan–Boltzmann constant. Hence

$$J = \frac{\sigma T^4}{\pi}.$$

(5.44)

Recall that the effective temperature T_{eff} of a star is the temperature of a black body that would emit the same total radiation flux as the star so

$$F = \sigma T_{\text{eff}}^4.$$

(5.45)

At this point we make Eddington's first approximation for anisotropy of the radiation intensity by splitting it into constant outward and inward components such that

$$I(\tau, \theta) = \begin{cases} I_1(\tau), & 0 \le \theta < \frac{\pi}{2} \quad \text{outward} \\ I_2(\tau), & \frac{\pi}{2} < \theta \le \pi \quad \text{inward} \end{cases}$$

(5.46)

so

$$J = \frac{1}{2}(I_1 + I_2)$$

(5.47)

and

$$F = \pi(I_1 - I_2).$$

(5.48)

Far outside the star there is no inward radiation so

$$I_2(0) = 0,$$

(5.49)

$$I_1(0) = \frac{\sigma T_{\text{eff}}^4}{\pi}$$

(5.50)

and

$$J(0) = \frac{\sigma T_{\text{eff}}^4}{2\pi}.$$

(5.51)

The radiation pressure $P_{\text{r}}(\tau)$ is the momentum flux through the surface at optical depth τ. Photons of momentum p have energy

$E = pc$ so the outward momentum carried by photons travelling at an angle θ to the vertical is $I(\theta, \tau)\cos\theta/c^2$ and the outward speed of such photons is $c\cos\theta$. So, integrating over a sphere, we have

$$P_r(\tau) = \int_{\text{sphere}} \frac{I(\theta, \tau)\cos\theta}{c^2} c\cos\theta \, d\Omega = \int_{-1}^{1} \frac{2\pi}{c} I\mu^2 \, d\mu, \qquad (5.52)$$

a second moment of intensity I. With Eddington's first approximation for $I(\tau)$

$$cP_r(\tau) = \frac{2}{3}\pi(I_1 + I_2) = \frac{4}{3}\pi J(\tau). \qquad (5.53)$$

What follows actually depends only on $cP_r(\tau) = \frac{4}{3}\pi J(\tau)$ and not the precise form of $I(\theta, \tau)$. This is Eddington's closure approximation and is true for a more general set of intensities $I(\theta, \tau)$. Now our equation of transfer (5.30) can also be integrated over frequency to give

$$\mu\frac{dI}{d\tau} = I - \frac{j''}{\kappa''}, \qquad (5.54)$$

where j'' is isotropic. Multiplying by $2\pi\mu$ and integrating over μ we arrive at

$$c\frac{dP_r(\tau)}{d\tau} = F = \text{const} \qquad (5.55)$$

which we integrate from $\tau = 0$ to obtain

$$P_r(\tau) = \frac{F\tau}{c} + P_r(0). \qquad (5.56)$$

Applying Eddington's closure we have

$$\frac{4}{3}\pi J(\tau) = F\tau + \frac{4}{3}\pi J(0) \qquad (5.57)$$

and so, from Eqs. (5.44) and (5.51), we can now deduce that

$$\frac{4}{3}\sigma T^4 = \sigma T_{\text{eff}}^4\left(\tau + \frac{2}{3}\right) \qquad (5.58)$$

or

$$T^4 = \frac{3}{4}\left(\tau + \frac{2}{3}\right)T_{\text{eff}}^4 \tag{5.59}$$

and so

$$T = T_{\text{eff}} \tag{5.60}$$

when

$$\tau = \frac{2}{3}. \tag{5.61}$$

At the outer edge of the atmosphere where $\tau = 0$

$$T = \frac{1}{2^{\frac{1}{4}}}T_{\text{eff}} \approx 0.841\,T_{\text{eff}} \tag{5.62}$$

so that we can often treat the whole atmosphere as isothermal at the effective temperature of the star and this is why the spectral lines, which are actually generated throughout the atmosphere, appear to originate at a single effective temperature. We take the definition of effective temperature as a first boundary condition on the stellar interior of the form

$$L_r = 4\pi r^2 \sigma T_{\text{eff}}^4 \tag{5.63}$$

when $\tau = \frac{2}{3}$, where we also have mass $m = M$ and determine the radius of the photosphere $R = r$.

5.7. Second Surface Boundary Condition

To arrive at a second surface boundary condition we assume that the atmosphere is in hydrostatic equilibrium with

$$\frac{\mathrm{d}P}{\mathrm{d}z} = -\rho g = -\rho\frac{GM}{R^2}. \tag{5.64}$$

Though good for most stars, this particular assumption breaks down when there are very strong winds in which the material is expanding through the photosphere at a significant rate. Combining this with our definition of optical depth, now dropping the double primes on $\kappa'' \to \kappa$,

$$\frac{\mathrm{d}\tau}{\mathrm{d}z} = -\kappa\rho, \tag{5.65}$$

we obtain

$$\frac{\mathrm{d}P}{\mathrm{d}\tau} = \frac{g}{\kappa}. \tag{5.66}$$

This is constant in our thin grey atmosphere of negligible mass, both still good approximations for hot stars, so integrating we find

$$P(\tau) = \frac{g}{\kappa}\tau + P(0). \tag{5.67}$$

We write pressure $P(\tau) = P_g(\tau) + P_r(\tau)$ as a combination of gas pressure P_g, which includes the contribution of electrons, and radiation pressure P_r. Outside the atmosphere there is no more material, the density $\rho \to 0$ and so the gas pressure $P_g \to 0$. There is however still radiation pressure and $P(0) = P_r(0)$. We subtract Eq. (5.56) to obtain

$$P_g(\tau) = \left(\frac{g}{\kappa} - \frac{F}{c}\right)\tau. \tag{5.68}$$

So at the photosphere, when $\tau = 2/3$, we can apply a second boundary condition to the stellar interior of the form

$$P_g = \frac{2}{3}\left(\frac{g}{\kappa} - \frac{F}{c}\right) = \frac{2}{3}\frac{g}{\kappa}\left(1 - \frac{L_r}{L_{\mathrm{Edd}}}\right), \tag{5.69}$$

where

$$L_{\mathrm{Edd}} = \frac{4\pi G c M}{\kappa} \tag{5.70}$$

is the Eddington luminosity at which radiation forces on atoms balance gravity. For many stars, including the Sun, the second term is negligible but for hot stars the luminosity can approach the Eddington luminosity but not exceed it if the star is to remain in hydrostatic equilibrium. As an example consider an atmosphere of ionised hydrogen atoms for which the main source of opacity is electron scattering. Gravity acts most strongly on the positively charged ions with a force on each of

$$\frac{GMm_{\mathrm{H}}}{R^2} \qquad (5.71)$$

while each electron feels a radiation force

$$\kappa m_{\mathrm{H}}\frac{F}{c} = \kappa m_{\mathrm{H}}\frac{L}{4\pi R^2 c}. \qquad (5.72)$$

Electrostatic forces between electrons and ions couple the equivalent of one electron to each ion and so the forces on each atom balance when $L = L_{\mathrm{Edd}}$. Recall that $\kappa = 0.2\,(1+X)\,\mathrm{cm}^2\mathrm{g}^{-1}$ (Eq. 4.39) so for hot stars $L_{\mathrm{Edd}} = 6.5 \times 10^4 (1 + X)(M/M_\odot)L_\odot$ independent of their radii.

5.8.   Breakdown of Assumptions

Though these boundary conditions are a significant improvement on simple zeroth-order conditions of $P = 0$ and $\rho = 0$ when $m = M$ and $r = R$ they are far from perfect for many stars. When an accurate model is required a full atmosphere integration, though time consuming, is often necessary. We enumerate some of the problems.

(i) The opacities κ_ν etc. are not independent of ν except for hot, but not too hot, stars for which Rayleigh electron scattering dominates. Indeed the appropriate mean for the opacity in Eq. (5.69) is not the Rosseland mean but

$$\kappa = \frac{\int_0^\infty \kappa_\nu B_\nu \,d\nu}{\int_0^\infty B_\nu \,d\nu} \qquad (5.73)$$

and a set of opacities calculated in this way, that differs from the tables used for the interior of the star, should be employed for the surface boundary condition. This is rarely done in practice for pure stellar interior calculations.

(ii) Nor is κ independent of T and ρ. Some progress can be made for power-law forms of opacity in which the atmosphere is cool enough to ignore radiation pressure but other problems with cool atmospheres make the improvements less worthwhile.

(iii) In giant stars, with low surface gravity g, the thickness of the atmosphere becomes significant, the plane-parallel assumption fails and curvature must be taken into account.

(iv) In very hot atmospheres the density of photons can be sufficiently high that radiation processes rather than collisions dominate the particles so the LTE assumption breaks down. In extreme cases the atmosphere can become stratified according to the ionisation stages of various elements. The opacity and temperature gradient are then interrelated in a complex manner.

(v) In stars with strong stellar winds the wind itself can be optically thick so that the assumption of hydrostatic equilibrium breaks down. When such winds are radiatively driven, as for the Wolf–Rayet stars, inevitably collisions between particles are again dominated by radiation processes and LTE fails again. Acceleration of an expanding atmosphere also requires careful treatment of the Doppler shift of one layer relative to another.

(vi) The solar granulation indicates that its photosphere lies below the top of its convection zone and so the assumption of a fully radiative atmosphere breaks down. For cooler stars convection can extend to very small optical depths $\tau \ll 2/3$.

Generally a full atmosphere model constructed for appropriate effective temperature T_{eff} and surface gravity g can be used to construct boundary conditions for a stellar interior model. The base of the atmosphere must be in LTE and hydrostatic equilibrium to match the assumptions used in the interior of the star. However for many stars, particularly those with convective envelopes, assumptions made for the mixing length and thence convective efficiency are often far more drastic than the use of a simplified stellar atmosphere. Much has to be done in these areas before we can claim that our models of such stars, other than the Sun, are accurate.

5.9. Line Formation

To understand the spectrum of a star and its relation to effective temperature we can return to our equation of transfer (5.30) which, with our definition of optical depth Eq. (5.38) at a specific frequency

τ_ν can be written as

$$\mu \frac{dI_\nu}{d\tau_\nu} = I_\nu - \frac{j_\nu''}{\kappa_\nu''}. \tag{5.74}$$

This has an integrating factor $\exp(-\tau_\nu/\mu)$ such that

$$\frac{d}{d\tau_\nu}\left[I_\nu \exp\left(-\frac{\tau_\nu}{\mu}\right)\right] = \frac{j_\nu''}{\mu\kappa_\nu''}\exp\left(-\frac{\tau_\nu}{\mu}\right). \tag{5.75}$$

Integrating we can relate the intensity at two optical depths τ_{in} and $\tau_{\text{out}} < \tau_{\text{in}}$ within the atmosphere by

$$I_\nu(\tau_{\text{out}}, \mu) = I_\nu(\tau_{\text{in}}, \mu)\exp\left(-\frac{\tau_{\text{in}} - \tau_{\text{out}}}{\mu}\right)$$
$$+ \int_{\tau_{\text{out}}}^{\tau_{\text{in}}} \frac{j_\nu''}{\mu\kappa_\nu''}\exp\left(-\frac{\tau_\nu - \tau_{\text{out}}}{\mu}\right) d\tau_\nu. \tag{5.76}$$

This illustrates how the intensity at some optical depth τ_{out} is related to the intensity deeper in the atmosphere at τ_{in} reduced by an exponential factor owing to absorption through the atmosphere and augmented with the positive contributions from emission processes. In the limits $\tau_{\text{out}} \to 0$ and $\tau_{\text{in}} \to \infty$ the first term vanishes and we can write the intensity emerging from the star's atmosphere as

$$I_\nu(0, \mu) = \int_0^\infty \frac{j_\nu''}{\mu\kappa_\nu''}\exp\left(-\frac{\tau_\nu}{\mu}\right) d\tau_\nu. \tag{5.77}$$

The exponential factor in the integral further means that the contribution of emission is most important when $\tau/\mu < 1$ but we cannot ignore the fact that different frequencies are affected by material at different optical depths and therefore different temperatures in the atmosphere. Near the frequency of a spectral line associated with a specific atomic transition the opacity is larger so that $\tau/\mu \approx 1$ higher in the atmosphere where the temperature T is smaller. Now $j_\nu''/\kappa_\nu'' \approx B_\nu$ because emission and absorption are dominated by atoms in LTE. But B_ν is smaller for smaller T and the contribution to the intensity is therefore smaller than at frequencies away from the line. The line appears dark.

Close to the edge of a star the intensity directed towards an observer emerges at large angle θ to the line of sight or small μ. Thence $\tau/\mu \approx 1$ again at lower temperature further out in the atmosphere where T and consequently j_ν''/κ'' are smaller. The edges or limbs of stars are darkened compared to most of the disc for which $\mu \approx 1$. In the case of very extended atmospheres it is possible for the line of sight to pass completely through the atmosphere so that the only emergent intensity is at frequencies close to spectral lines which then appear in emission.

When the atmosphere is expanding so that these lines have a blue-shifted component both emission and absorption can be seen combined as a P Cygni profile named after the prototypical example. Figure 5.3 illustrates the spectral line of a thin spherical shell expanding with velocity v. With the line of sight to the right, material at point C absorbs the continuum light from the star. This produces a narrow blue-shifted absorption line because this part of the absorbing shell is moving towards us at v. Everywhere else the shell is seen in emission. At point A the material is receding from

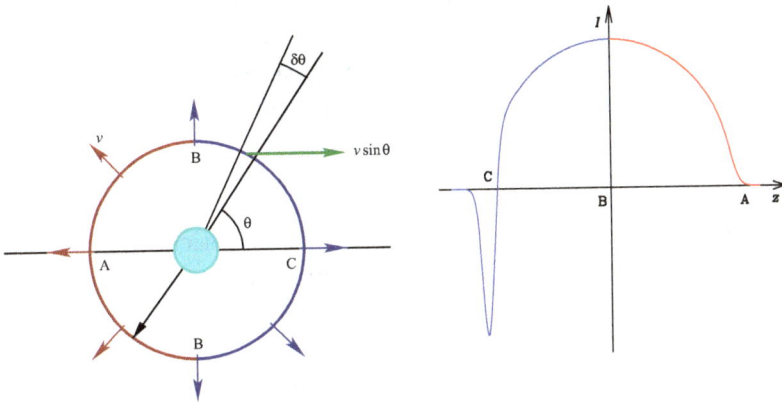

Fig. 5.3. The left panel is a schematic diagram of a thin spherical shell of material expanding away from a star at speed v. At point C any atomic transition absorbs the light of the star to produce a narrow blue-shifted absorption line relative to the stellar continuum. Everywhere else the line is seen in emission relative to the continuum, symmetrically blue and red-shifted relative to the star. The right panel is the intensity I of the combined spectral line as a function of redshift z relative to the central star, convoluted with a Gaussian velocity distribution centred on v in the shell.

us at v and emits with maximum redshift. At B the material in a ring expanding perpendicular to the line of sight absorbs without redshift. At a general angle θ to the line of sight the redshift, $z \propto -v \cos \theta$, the projected velocity away from us and the size of the emitting surface $d\Sigma \propto \sin \theta \, d\theta$. Combining these and convoluting with a narrow Gaussian velocity distribution, to mimic local line broadening, centred on v in the expanding shell produces the spectral line in the figure. For a star losing mass in a continuous wind many infinitesimal shells moving at varying velocities contribute so that the blue-shifted absorption, in particular, becomes noticeably broader. From the shape of such lines we can find the velocity of the expanding shell or wind, the density and proximity to the star. Different lines are excited at different temperatures as expanding material cools. Though the models rapidly become rather complex it is possible to deduce stellar mass loss rates of some stars.

5.10.   Questions

1. In a plane-parallel atmosphere of negligible mass and containing no sources of energy the optical depth τ is defined by $d\tau = -\kappa\rho \, dz$, where $\kappa(\rho, T)$ is the total opacity of material of density ρ at temperature T, z is the height in the atmosphere and $\tau \to 0$ at large z. The equation of radiative transfer can be written in the form

$$\cos\theta \frac{dI}{d\tau} = I - \frac{j}{\kappa}, \qquad (*)$$

where $I(\tau, \theta)$ is the intensity of radiation at optical depth τ at an angle θ to the z-axis and j is the effective emissivity given by

$$\frac{j}{\kappa} = \frac{\sigma T^4}{\pi},$$

where σ is the Stefan–Boltzmann constant. Integrate $(*)$ over a sphere and use the fact that the flux F in the z direction is independent of τ to deduce that the mean intensity

$$J(\tau) = \frac{1}{4\pi} \int_{\text{sphere}} I(\tau, \theta) \, d\Omega = \frac{j}{\kappa}.$$

Show that the form

$$I(\tau, \theta) = A(\tau) + C(\tau) \cos \theta$$

satisfies the Eddington closure approximation

$$cP_{\mathrm{r}} = \frac{4}{3} \pi J$$

between radiation pressure P_{r}, the speed of light c and the mean intensity. Deduce that

$$C(\tau) = \frac{3F}{4\pi}$$

and that $(*)$ is satisfied if

$$\frac{\mathrm{d}A}{\mathrm{d}\tau} = C(\tau).$$

Use the fact that there is no flux into the star at $\tau = 0$ to find $A(\tau)$ and use the definition $F = \sigma T_{\mathrm{e}}$ of effective temperature T_{e} to deduce that

$$T^4 = \frac{3}{4} T_{\mathrm{e}}^4 \left(\tau + \frac{2}{3} \right)$$

and that $T = T_0 = 2^{-1/4} T_{\mathrm{e}}$ when $\tau = 0$.

2. In the atmosphere of a red dwarf the opacity

$$\kappa = \kappa_0 P^{\alpha-1} T^{4-4\beta}$$

and radiation pressure is negligible. From hydrostatic equilibrium show that the pressure P varies with temperature T according to

$$P^\alpha = \frac{2\alpha g}{3\beta \kappa_0 T_0^4} (T^{4\beta} - T_0^{4\beta}),$$

where g is the surface gravity of the star.

Hence deduce that an approximate surface boundary condition for the stellar interior is

$$\frac{P\kappa}{g} = \frac{4\alpha}{3\beta} (1 - 2^{-\beta})$$

at a radius r where the stellar luminosity $L_r = 4\pi \sigma r^2 T^4$.

3. In a stellar atmosphere radiative transfer theory can be used to express temperature T in the form

$$T^4 = \frac{3}{4}T_e^4\left(\tau + \frac{2}{3}\right),$$

where T_e is the effective temperature and

$$\tau = \int_r^\infty \kappa\rho\,dr$$

is the optical depth. The atmosphere of the cool star contains negligible mass and is in hydrostatic equilibrium. The opacity within the atmosphere is given by $\kappa = \kappa_0 P^{1/2}T^8$, where κ_0 is constant. Show that

$$P^{\frac{3}{2}} = \frac{8GM}{R^2\kappa_0 T_e^8}\left[\frac{1}{2} - \frac{1}{3\tau+2}\right].$$

Find $\tau = \tau_c$ at which convection sets in.

If the star is fully convective for $\tau > \tau_c$ so that the equation of state is $P = A\rho^{5/3}$ with A determined at $\tau = \tau_c$ deduce that

$$L \propto R^{98/47}M^{28/47}.$$

[*You may use the fact that for an $n = 3/2$ polytrope $R \propto AM^{-1/3}$.*]

Chapter 6

Energy Generation

One further equation, for energy generation, is required to close the set that describe stellar structure. The last physical ingredient required to make models of stellar structure and evolution is the source of energy that ultimately causes them to shine. Up until the start of the twentieth century the source of the Sun's energy remained a puzzle. Early in the nineteenth century geologists had established that the Earth must be at least ten million if not a hundred million years old. It was easy to calculate that the energy liberated by chemical burning is insufficient. For example, consider the combustion of one mole of hydrogen to water. This requires half a mole of oxygen, so a total mass of $\Delta m = 10\,\mathrm{g}$ and liberates $\Delta Q = 286\,\mathrm{kJ}$. Combustion of the entire mass of the Sun at a rate to maintain its current luminosity would take only $(M_\odot/\Delta m)/(\Delta Q/L_\odot) \approx 4\,800\,\mathrm{yr}$. In 1941 Julius Mayer proposed, as an alternative source, the gravitational energy released by the impact of meteors on the surface of the Sun but ruled this out himself based on his estimate of the rate they actually fall into the Sun. Hermann von Helmholtz took this idea a step further suggesting that the Sun itself is gradually contracting and releasing energy. He estimated that this self gravity could keep the Sun alight for some twenty million years. Towards the end of the nineteenth century Lord Kelvin refined Helmholtz' theory by asserting that the Sun had contracted from a very large gas cloud. During this contraction its core was heated to several million degrees and this slowed the contraction to its current state similar to that proposed by Helmholtz. With energy from gravitational contraction they calculated that the Sun could last some one hundred million years in a similar state. However the theory was already under threat because geologists were by then claiming an age of two thousand million years for the Earth and so at least that for the Sun. Our current understanding was born in the early twentieth century when Albert Einstein's theory of relativity hinted that matter could be converted to energy. In 1920, Arthur Stanley Eddington proposed that the fusion of hydrogen to helium was just the energy source that the Sun needs. However it was not until quantum mechanics and statistical physics were combined that we could develop a plausible theory of how fusion can proceed. We shall spend much of this chapter discussing nuclear fusion and its wide-ranging consequences but begin more generally by combining all energy sources and sinks into a single equation.

We first consider energy generation in a static star illustrated in Fig. 6.1. Within a stellar interior we define ε to be the energy generation rate per unit mass. In our static star ε is a function of the state of the material in the stellar interior. If we include thermal

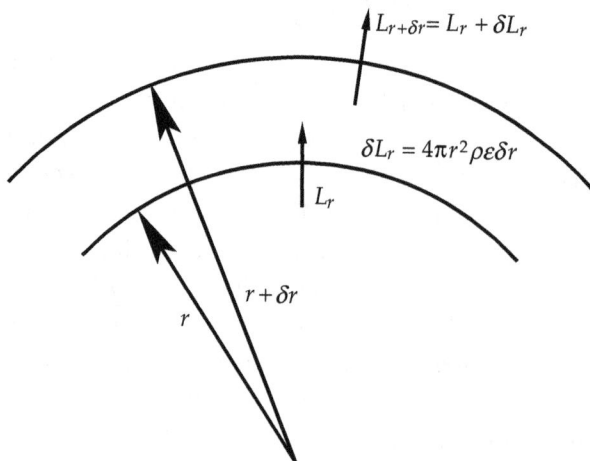

Fig. 6.1. Energy generation within stellar interiors. In a static star energy is generated at a rate ε per unit mass within the star. The luminosity passing through a spherical shell of thickness δr is augmented by $4\pi\rho r^2 \varepsilon \delta r$.

energy release in ε it hides a partial derivative with respect to time and we shall return to this matter shortly. For now let us write

$$\varepsilon = \varepsilon(\rho, T, X_i). \tag{6.1}$$

The fourth equation of stellar structure is then

$$\frac{dL_r}{dr} = 4\pi r^2 \rho\varepsilon. \tag{6.2}$$

There are three contributions to energy generation in stellar interiors:

(1) *Gravitational contraction.* This contribution is mostly small except for short periods of time, such as before nuclear reactions begin or when the structure changes rapidly as the major nuclear burning progresses from one reaction to another or from one part of the star to another. For example after core hydrogen burning is complete the core contracts supplying energy so $\varepsilon > 0$. The reverse process also occurs. Energy can be stored in the structure of an expanding star so that there is an effective $\varepsilon < 0$.

(2) *Nuclear fusion reactions.* These are the typical source of a star's energy, both supporting it against gravitational collapse and making it luminous to the rest of the Universe. The energy is released in three components, as high energy gamma rays released when the nuclear state changes, in kinetic energy of the recoiling particles and as neutrinos created in weak reactions. The first two thermalise locally because of the short mean free path of both photons and particles. Neutrinos on the other hand have a mean free path substantially longer than the radius of the star and so escape with their energy. For exothermic reactions $\varepsilon > 0$.

(3) *Neutrino losses.* At high temperatures and densities neutrinos can form spontaneously and, just as those created in nuclear reactions, escape with their energy. This is a negative contribution to the energy generation rate, $\varepsilon < 0$, that can lead to significant cooling in the hot cores of evolved stars.

6.1. Gravitational Contraction

In Chapter 2 we used the virial theorem to derive Eq. (2.52) to demonstrate that a bound star shrinks when its only source of energy is gravity. Indeed, for an evolving star, Eq. (6.2) is more correctly written as the partial differential equation

$$\frac{\partial L_r}{\partial r} = 4\pi r^2 \rho \left(\epsilon - T \frac{\partial s}{\partial t} \right), \tag{6.3}$$

where ϵ only includes the rate of energy generation by nuclear reactions less the neutrino energy loss rate and s is the entropy of the stellar material per unit mass. To keep the form of Eq. (6.2) for stellar structure, we can write

$$\epsilon_{\text{grav}} = -T \frac{\partial s}{\partial t} \tag{6.4}$$

and include it in ε. On time-scales much less than the Kelvin–Helmholtz τ_{KH} we may treat ϵ_{grav} as independent of t.

To calculate ϵ_{grav} we consider an element of stellar material of mass

$$\delta m = \rho \delta V, \tag{6.5}$$

density ρ and volume δV. Conservation of energy in the form of the first law of thermodynamics tells us that heat added to the element

$$\text{d}Q = T\text{d}S = \text{d}U + P\text{d}V, \tag{6.6}$$

where $S = s\rho\delta V$ is the entropy of the element and $U = u\rho\delta V$ is its internal energy at temperature T and pressure P. For an element of fixed mass δm,

$$\text{d}V = \text{d}\left(\frac{1}{\rho}\right)\delta m = -\frac{1}{\rho^2}\text{d}\rho\,\delta m \tag{6.7}$$

and we may write the first law of thermodynamics in its specific form

$$\text{d}q = T\text{d}s = \text{d}u - \frac{P}{\rho^2}\text{d}\rho, \tag{6.8}$$

where $dq = dQ/\rho$ is the heat exchanged per unit mass. The energy gained by the element is lost to the surroundings so

$$\epsilon_{\text{grav}} = -\left(\frac{dq}{dt}\right)_m = -T\left(\frac{ds}{dt}\right)_m, \qquad (6.9)$$

and the rate of heat loss per unit mass

$$\epsilon_{\text{grav}} = -\frac{\partial u}{\partial t} + \frac{P}{\rho^2}\frac{\partial \rho}{\partial t}, \qquad (6.10)$$

where now, and subsequently in this section, we take all time derivatives to be implicitly at constant mass m. Next consider material of constant composition so that

$$u = u(P, \rho) \qquad (6.11)$$

only and, by the chain rule, the exact differential

$$du = \left(\frac{\partial u}{\partial P}\right)_\rho dP + \left(\frac{\partial u}{\partial \rho}\right)_P d\rho. \qquad (6.12)$$

Taking the partial derivative with respect to time, setting $d \equiv (\partial/\partial t)$, we arrive at

$$\epsilon_{\text{grav}} = -\left(\frac{\partial u}{\partial P}\right)_\rho \frac{\partial P}{\partial t} - \left(\frac{\partial u}{\partial \rho}\right)_P \frac{\partial \rho}{\partial t} + \frac{P}{\rho^2}\frac{\partial \rho}{\partial t}. \qquad (6.13)$$

Returning to the first law of thermodynamics (Eq. 6.8) we take the partial derivative with respect to pressure at constant specific entropy and set $d \equiv (\partial/\partial P)_s$ to find

$$\left(\frac{\partial u}{\partial P}\right)_s - \frac{P}{\rho^2}\left(\frac{\partial \rho}{\partial P}\right)_s = T\left(\frac{\partial s}{\partial P}\right)_s = 0 \qquad (6.14)$$

from which

$$\left(\frac{\partial u}{\partial P}\right)_s = \frac{1}{\Gamma_1 \rho}, \qquad (6.15)$$

where, by definition,

$$\Gamma_1 = \left(\frac{\partial \log P}{\partial \log \rho}\right)_s = \frac{P}{\rho}\left(\frac{\partial P}{\partial \rho}\right)_s. \qquad (6.16)$$

We may also set $d \equiv (\partial/\partial P)_s$ in Eq. (6.12) to obtain

$$\left(\frac{\partial u}{\partial P}\right)_s = \left(\frac{\partial u}{\partial P}\right)_\rho + \left(\frac{\partial u}{\partial \rho}\right)_P \frac{\rho}{P\Gamma_1} \tag{6.17}$$

and with Eq. (6.15) we arrive at

$$\left(\frac{\partial u}{\partial \rho}\right)_P = \frac{P}{\rho^2} - \frac{\Gamma_1 P}{\rho}\left(\frac{\partial u}{\partial P}\right)_\rho \tag{6.18}$$

so Eq. (6.13) may be written as

$$\epsilon_{\text{grav}} = \left(\frac{\partial u}{\partial P}\right)_\rho \left(\frac{\Gamma_1 P}{\rho}\frac{\partial \rho}{\partial t} - \frac{\partial P}{\partial t}\right), \tag{6.19}$$

a simple looking form which illustrates the dependence on changes in the local pressure and density of the stellar material with time. A more enlightening form for ϵ_{grav} follows when we write

$$P\left(\frac{\partial u}{\partial P}\right)_\rho = P\left(\frac{\partial u}{\partial T}\right)_\rho \Big/ \left(\frac{\partial P}{\partial T}\right)_\rho = T\left(\frac{\partial u}{\partial T}\right)_\rho \Big/ \left(\frac{\partial \log P}{\partial \log T}\right)_\rho = \frac{c_V T}{\chi_T}, \tag{6.20}$$

where $c_V = (\partial u/\partial T)_\rho$ is the specific heat capacity at constant volume and $\chi_T = (\partial \log P/\partial \log T)_\rho$, and rearrange the term in the second set of parentheses to get

$$\epsilon_{\text{grav}} = -\frac{c_V T}{\chi_T}\left[\frac{\partial}{\partial t}\left(\log_e \frac{P}{\rho^{\Gamma_1}}\right) + \frac{\partial \Gamma_1}{\partial t}\log_e \rho\right], \tag{6.21}$$

which can be verified by expanding the first term derivative. For an ideal gas, with $P = \mathcal{R}\rho T/\mu$, $\chi_T = 1$, $\Gamma_1 = \gamma$ and $\partial\Gamma_1/\partial t = 0$ so the energy generation rate simplifies to

$$\epsilon_{\text{grav}} = -c_V T\frac{\partial}{\partial t}\left(\log_e \frac{P}{\rho^\gamma}\right) \tag{6.22}$$

and this makes it clear that adiabatic changes make no contribution.

In the above analysis we ignored changes in compositions X_i. This is usually satisfactory because ϵ_{grav} is only important when ϵ_{nuc} is not. However if convective boundaries move through a composition gradient or ionisation zones move in mass then entropy changes owing to changing compositions must be included. To do so we add this contribution to the first law of thermodynamics, Eq. (6.6), which becomes

$$\text{đ}Q = T\text{d}S = \text{d}U + P\text{d}V - \sum_i \mu_i \text{d}N_i, \qquad (6.23)$$

where μ_i is the chemical potential and $\text{d}N_i$ the change in the number of particles of species i.

6.2. Nuclear Energy Generation

The thermal time-scale of the Sun is 20 Myr, while the age of the solar system, inferred from radiometric dating is 4.54 Gyr. The typical energy source of the Sun, and indeed of nearly every star, is nuclear fusion, the creation of new heavier nuclei from lighter ones. The primary reaction is hydrogen burning that combines four protons and two electrons to make a single helium nucleus or α-particle. This releases energy because one ^{4}He nucleus weighs less than four ^{1}H nuclei plus two electrons both because the binding energy of the nucleons in ^{4}He is larger and two neutrons have replaced two pairs of protons and electrons. Though a free neutron is slightly more massive than a free proton plus free electron this is overwhelmed by the binding energy because the nucleons are in a more stable configuration in the ^{4}He nucleus. A simple estimate indicates that the Sun contains sufficient hydrogen fuel to burn at its current luminosity for one hundred thousand million years. In practice the Sun will not be able to consume more than about a tenth of this fuel but even then the lifetime problem is easily solved.

To understand nuclear energy production fully we need to know which nuclear reactions are possible, how much energy they release and at what rate they do so. First let us define the binding energy of a nucleus of a single isotope of a particular element

$$E_{\text{B}} = [Zm_{\text{p}} + (A - Z)m_{\text{n}} - m_{\text{nuc}}]c^2, \qquad (6.24)$$

where Z and A are the integer atomic number and mass of the isotope, m_{p} and m_{n} are the masses of free protons and neutrons,

m_{nuc} is the mass of the bound nucleus and c is the speed of light. For ^{1}H which consists only of free protons $E_B = 0$. We then let the binding energy per nucleon be

$$f = \frac{E_B}{A}. \tag{6.25}$$

In Fig. 6.2 we plot f as a function of A for the most stable isotope of each atomic weight. The nuclei of ^{56}Fe lie at the relatively flat peak along with the iron group elements such as nickel and cobalt, often found as awkward impurities in iron ores. Isotopes with $A < 56$ can be combined in exothermic fusion reactions to create heavier isotopes. If fusion reactions were to create elements heavier than the iron group they would be endothermic. There is no more nuclear energy to be

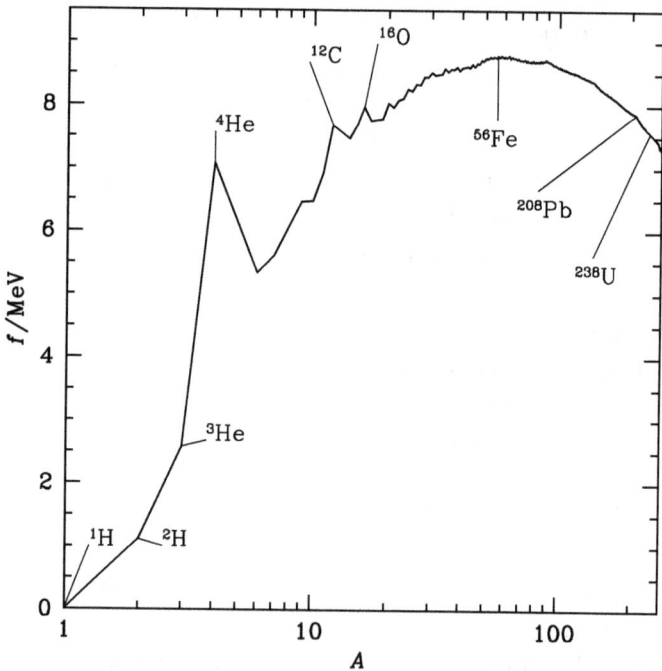

Fig. 6.2. The binding energy per nucleon f for the most stable isotope of each atomic weight A. Iron-56 occupies the site of the most stable nucleus. Energy can be obtained by fusion of nuclei of lower atomic weight or fission of nuclei of higher atomic weight. Notable peaks appear for helium-4, carbon-12 and oxygen-16. Lead-208 is the heaviest non-radioactive nucleus.

extracted from the stellar material. Heavier elements are created as byproducts of later stages of evolution and the death of stars. Above the iron group energy can be extracted by fission of heavy nuclei into lighter components. All known isotopes heavier than ^{208}Pb are unstable to some form of fission.

For fusion to take place charged nuclei must first overcome electrostatic repulsion, the strength of which is proportional to the product of their atomic numbers. Even for two protons the energy required to cross this barrier is enormous, much larger than the typical kinetic energy of protons at the centre of the Sun. It is only with the combination of quantum mechanical tunnelling and the high energy tail of the proton velocity distribution that the barrier can be overcome. Consequently fusion reactions are extremely sensitive to temperature, with the minimum temperature for a particular reaction depending on the product of the atomic numbers. Thus the lightest nuclei tend to react first and be completely consumed at a given depth in the star before the next reaction can begin. We shall return to this when we discuss reaction rates in Sec. 6.3. First we examine the important nuclear reactions and the energy they can liberate.

6.2.1. *Hydrogen burning*

The most important fusion reaction in a star's life is hydrogen burning. Essentially four protons and two electrons combine to form a single helium-4 nucleus or α-particle,

$$4\,^1\mathrm{H} + 2\mathrm{e} \rightarrow\,^4\mathrm{He}. \tag{6.26}$$

To conserve lepton number at least two neutrinos must also be liberated. The energy released can be calculated as the actual mass difference between the product and reactants multiplied by the square of the speed of light. The atomic mass unit m_u is defined as one-twelfth of the mass of a neutral carbon atom,

$$1\,m_\mathrm{u} = \frac{1}{12}\left(\text{mass of one }^{12}\mathrm{C}\text{ atom}\right) = 1.66054 \times 10^{-24}\,\mathrm{g}. \tag{6.27}$$

So the masses of the relevant particles are

$$m_p = 1.0072765 \, m_u, \tag{6.28}$$

$$m_e = 0.0005486 \, m_u, \tag{6.29}$$

$$m_{He} = 4.0015062 \, m_u. \tag{6.30}$$

In the production of a single ^{4}He nucleus the change in mass is $\Delta m = 0.02870 \, m_u$ which corresponds to an energy

$$\Delta E = \Delta mc^2 = 4.283 \times 10^{-5} \, \text{erg} \tag{6.31}$$

$$= 26.73 \, \text{MeV} = 6.68 \, \text{MeV per nucleon.} \tag{6.32}$$

This energy is released as photons, kinetic energy of particles and neutrinos. The first two forms of energy are thermalised locally because both the photons and particles have short mean free paths. They therefore contribute positively to ϵ. However neutrinos contribute very little to heating the interior because their interaction cross-section is tiny, $\sigma_{\nu_e} \approx 10^{-44} \text{cm}^2$. The mean free path of neutrinos in a stellar interior is

$$\lambda_{\nu_e} \approx \frac{1}{n\sigma_{\nu_e}} \approx 10^9 R_\odot \approx 23 \, \text{pc}. \tag{6.33}$$

Hence neutrinos escape from the stellar interior with all their energy throughout most of a star's life. There are exceptions during the collapse of the dense iron cores that give rise to supernovae and we shall revisit this in Sec. 8.10. So the fraction of the available fusion energy that actually contributes to ϵ depends on the nature of the reactions because different paths for the fusion of hydrogen to helium lose a variety of energies in escaping neutrinos.

6.2.1.1.   *The proton–proton chain*

The proton–proton, the pp-chains, make up the dominant process by which hydrogen is fused to helium at low temperatures. Table 6.1 lists the reactions. First two protons must fuse to form a deuterium nucleus, ^{2}H. This is the slowest and so rate determining step. The emitted positron immediately annihilates with an existing electron while the deuterium fuses rapidly with another proton to form ^{3}He.

Table 6.1. The reactions of the proton–proton chains. The ppI-chain dominates at low temperatures, with ppII branching off at its last step and ppIII at the second step of ppII at sufficiently high temperatures. The very first step of ppI is always the rate determining step. Kinetic energy of the recoiling nuclei and the photons γ is thermalised locally, while energy in the electron neutrinos ν_e is lost from the star. Neutrinos carry off progressively more energy in ppIII than ppII than ppI so the local energy contribution of each chain falls off.

	Reaction	Thermal energy (MeV)	Neutrino energy (MeV)	Notes
	$^1\mathrm{H} + {}^1\mathrm{H} \rightarrow {}^2\mathrm{H} + e^+ + \nu_e$	0.16	0.26	slow
	$e^+ + e^- \rightarrow \gamma$	1.02	–	
	$^1\mathrm{H} + {}^2\mathrm{H} \rightarrow {}^3\mathrm{He} + \gamma$	5.49	–	very rapid
	Energy per ^{3}He nucleus	**6.67**	–	
ppI	$^3\mathrm{He} + {}^3\mathrm{He} \rightarrow {}^4\mathrm{He} + 2\,{}^1\mathrm{H}$	12.86	–	
	Total Energy per ^{4}He nucleus	**26.21**		$= Q_{\mathrm{ppI}}$
ppII	$^3\mathrm{He} + {}^4\mathrm{He} \rightarrow {}^7\mathrm{Be} + \gamma$	1.59	–	dominates for
	$^7\mathrm{Be} + e^- \rightarrow {}^7\mathrm{Li} + \nu_e$	0.06	0.80	$T > 1.4 \times 10^7\,\mathrm{K}$
	$^7\mathrm{Li} + {}^1\mathrm{H} \rightarrow 2\,{}^4\mathrm{He}$	17.35	–	
	Total Energy per ^{4}He nucleus	**25.67**		$= Q_{\mathrm{ppII}}$
ppIII	$^7\mathrm{Be} + {}^1\mathrm{H} \rightarrow {}^8\mathrm{B} + \gamma$	0.13	–	dominates for
	$^8\mathrm{B} \rightarrow {}^8\mathrm{Be}^* + e^+ + \nu_e$	10.78	7.22	$T > 2.3 \times 10^7\,\mathrm{K}$
	$^8\mathrm{Be}^* \rightarrow 2\,{}^4\mathrm{He}$	0.09	–	
	Total Energy per ^{4}He nucleus	**19.26**		$= Q_{\mathrm{ppIII}}$

The energy carried off by the neutrinos is distributed between 0 and 0.42 MeV and averaging over all reactions we obtain 0.26 MeV per neutrino. At low temperatures burning proceeds predominantly by the ppI-chain in which two ^{3}He nuclei combine to form a single ^{4}He nucleus and return two protons. No further neutrinos are lost so this route is the most efficient with only 2% of the available energy lost from the star in neutrinos. As temperature rises so the ppII- and ppIII-chains become gradually more important. The weak reaction of ^{7}Be with an electron liberates a total of 0.861 MeV. For about 87% of reactions all this energy is lost to the neutrino but in the remaining 13% the neutrino carries off 0.383 MeV leaving the ^{7}Li in an excited state ^{7}Li* and the rest of the energy is released in γ-rays as the nuclei decay to stable states. So the average energy lost in neutrinos per reaction is 0.80 MeV. Together neutrinos carry off 4% of the available energy in the ppII-chain. In the ppIII-chain the neutrinos from the decay of ^{8}B can carry off up to 14.1 MeV with the net result that, on average, neutrinos carry off 28% of the available energy. There is no stable state of ^{8}Be and the excited state ^{8}Be* decays essentially instantaneously during hydrogen burning. It is interesting to note that the ppII-chain can be a source of ^{7}Li but only in circumstances when the ^{7}Be is removed, by some mixing process, to regions too cool for ^{7}Li to capture another proton, before it captures an electron. Otherwise most lithium is produced only in the Big Bang and subsequently destroyed by stars.

In the Sun, the fractions of energy derived from ppI:ppII:ppIII are 85%:15%:0.02%. Though not very important for the structure of the Sun, the ppIII-chain was responsible for the identification of the solar neutrino problem in the 1980s. The ^{8}B neutrinos, being the most energetic expected in any significant number from the Sun, are the easiest to detect in as much as neutrinos can be detected at all. The critical experiment required 10^7 kg of tetrachloroethylene C_2Cl_4 deep in the Homestake gold mine in South Dakota. Though small the detected flux was only about one-third of that expected and by 1989 this had become a highly significant result with a discrepancy of more than three standard deviations in the combined measurement and prediction errors. Though this could have been remedied with small changes in the temperature stratification in the core of the Sun without upsetting its total luminosity, which is predominately from the ppI-chain, helioseismological measurements of the sound speed ruled out this possibility by confirming the standard structure of the Sun. The paradox was finally resolved in the 1990s when further large experiments determined that neutrino mass states do not coincide with their flavour states. So the neutrinos oscillate between flavours on their journey from the centre of the Sun to the

Earth. Many of the electron neutrinos emitted by the Sun are not detected because they have oscillated to muon and tau neutrinos when they reach the detector. This instance of stellar physics driving a major advance in particle physics lends considerable credence to our modelling of the Sun and by extension some confidence in applying similar physics to other stars and stellar evolution in general. The even more minor proton–electron–proton or pep reaction,

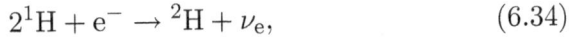

$$2^1\text{H} + \text{e}^- \rightarrow {}^2\text{H} + \nu_\text{e}, \tag{6.34}$$

which accounts for about 1/400th of the deuterium production, liberates relatively high energy neutrinos at 1.44 MeV and these too are detected at the expected rate.

The proton–proton chains are active in big bang nucleosynthesis in the first few minutes of the Universe but then free neutrons also play an important role. The major deuterium producing reaction is the combination of a neutron and proton which, with no Coulomb repulsion to overcome, proceeds much more rapidly than the proton plus proton route. However the half-life of a free neutron is only 615 s so they play no role on the nuclear time-scales for hydrogen burning in stars.

6.2.1.2. *The carbon, nitrogen, oxygen cycle*

Above 2×10^7 K hydrogen preferentially burns by catalytic reactions involving carbon, nitrogen and oxygen. This CNO cycle (Table 6.2) dominates in stars more massive than about $1.5\,M_\odot$. Most burning, particularly at lower temperatures, is by the CN cycle alone. The slowest reaction is the capture of a proton by ^{14}N and so much of any ^{12}C present before fusion begins is converted to ^{14}N. The catalytic elements reach an equilibrium fairly rapidly compared to the time-scale for hydrogen burning with about 98% of the carbon and nitrogen ending up as ^{14}N. The branch to the NO cycle occurs only about once in every 2,500 proton captures by ^{15}N during core hydrogen fusion. Higher temperatures in the burning shells of giants activate more of the NO cycle so that oxygen too is converted to nitrogen. As the temperature increases cycles involving neon,

Table 6.2. The CNO cycle. Four protons combine to form a helium ^{4}He nucleus in a catalytic cycle with a branch at ^{15}N. Each cycle is a series of proton captures and β decays. The slowest step is the capture of a proton by ^{14}N. Both Q_{CN} and Q_{NO} are the local thermalised energy released in the production of a single ^{4}He nucleus by all reactions in the cycle combined with annihilation of the positrons with electrons.

CN cycle	^{1}H + ^{12}C	$\rightarrow$	^{13}N + γ	
	^{13}N	$\rightarrow$	^{13}C + e$^+$ + ν_e	
	^{1}H + ^{13}C	$\rightarrow$	^{14}N + γ	
	^{1}H + ^{14}N	$\rightarrow$	^{15}O + γ	rate controlling step
	^{15}O	$\rightarrow$	^{15}N + e$^+$ + ν_e	
	^{1}H + ^{15}N	$\rightarrow$	^{12}C + ^{4}He + γ	$Q_{\mathrm{CN}} = 25.02\,\mathrm{MeV}$
NO cycle	^{1}H + ^{15}N	$\rightarrow$	^{16}O + γ	
	^{1}H + ^{16}O	$\rightarrow$	^{17}F + γ	
	^{17}F	$\rightarrow$	^{17}O + e$^+$ + ν_e	
	^{1}H + ^{17}O	$\rightarrow$	^{14}N + ^{4}He	$Q_{\mathrm{NO}} = 24.08\,\mathrm{MeV}$

sodium, magnesium and aluminium are also activated. Indeed at high enough temperatures protons may be captured by almost any nucleus. In stars the thermostatic nature of nuclear burning generally prevents capture by heavier nuclei because, by the time any part of a star is hot enough, all the free protons have been consumed. In uncontrolled hydrogen burning, such as that experienced in nova eruptions (Sec. 9.9.2) or whenever protons are ingested into a significantly hotter part of a star, these cycles lead to distinctive nucleosynthesis.

In all such catalytic cycles neutrino losses mean that the energy efficiency of hydrogen burning is lower than that of the ppI- and ppII-chains. In particular

$$Q_{\mathrm{NO}} < Q_{\mathrm{CN}} < Q_{\mathrm{ppII}} < Q_{\mathrm{ppI}}. \tag{6.35}$$

Thus hydrogen burning generally becomes less efficient as the temperature at which it burns rises. Consequently lower-mass stars are more conservative at extracting energy from hydrogen burning.

6.2.2. *Helium burning: The triple-α reaction*

With metals making up no more than a few per cent, and often much less, of the stellar material, hydrogen burning leaves behind almost pure helium-4 cores in stars. Once the temperature has risen sufficiently this ^{4}He can provide the next source of energy. It happens that there is no stable nuclear configuration with $Z = 4$ and $A = 8$. The reaction

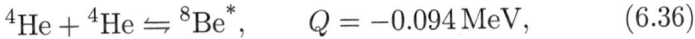

$$^4\text{He} + {}^4\text{He} \rightleftharpoons {}^8\text{Be}^*, \qquad Q = -0.094\,\text{MeV}, \qquad (6.36)$$

familiar from the last step of the ppIII-chain, is endothermic and so easily reversed at temperatures just hot enough to get it going. The unstable nucleus ^{8}Be* has a lifetime of only 3×10^{-16} s before it fissions back to two helium nuclei. In order for fusion to proceed, a rapid collision with a third nucleus is needed. In the 1950s Salpeter calculated that temperatures $T > 10^8$K are needed for the chance of a collision with a third α-particle before the fission. This can form an unstable nucleus of carbon-12,

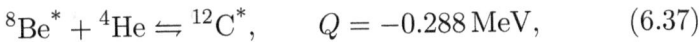

$$^8\text{Be}^* + {}^4\text{He} \rightleftharpoons {}^{12}\text{C}^*, \qquad Q = -0.288\,\text{MeV}, \qquad (6.37)$$

again an endothermic reaction. Energy is finally recovered when the ^{12}C* nucleus emits two photons, required to conserve nuclear spin angular momentum, to reach the ground state of ^{12}C,

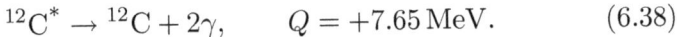

$$^{12}\text{C}^* \rightarrow {}^{12}\text{C} + 2\gamma, \qquad Q = +7.65\,\text{MeV}. \qquad (6.38)$$

No weak reactions are involved so all the released energy is thermalised locally. For the total reaction, commonly referred to as the triple-α reaction, this energy is $Q_{3\alpha} = 7.27\,\text{MeV}$ or $0.606\,\text{MeV}$ per nucleon. This is only about a tenth of the energy per proton available from hydrogen burning. Typically stars burn helium at a higher luminosity than hydrogen and so we can immediately deduce that a star is generally supported by core helium fusion for a somewhat shorter time than it spends on the main sequence burning hydrogen in its core.

Once enough carbon has been created by the triple-α reaction further alpha captures by ^{12}C can begin. The reaction

$$^{12}\text{C} + {}^4\text{He} \rightarrow {}^{16}\text{O} + \gamma, \qquad Q = +7.16 \,\text{MeV}, \qquad (6.39)$$

being only a two-particle reaction, proceeds relatively rapidly and

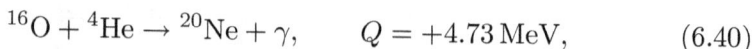

$$^{16}\text{O} + {}^4\text{He} \rightarrow {}^{20}\text{Ne} + \gamma, \qquad Q = +4.73 \,\text{MeV}, \qquad (6.40)$$

runs at a slow but non-negligible rate at temperatures suitable for the triple-α reaction. By the end of central helium burning the composition of a stellar core is typically only 20% carbon with about 80% oxygen along with heavier isotopes. The oxygen-to-carbon ratio increases with the mass of the star as the core temperature rises. Later in a star's evolution, whenever free α-particles are available, captures by heavier nuclei can take place but require increasingly higher temperatures to overcome the growing Coulomb barriers. Isotopes with equal numbers of protons and neutrons, ^{24}Mg, ^{28}Si, ^{32}S, ... are created in this way and are known as α-process isotopes.

Nitrogen that has accumulated during CNO hydrogen burning can also capture an α-particle at the temperatures required for the triple-α process,

$$^{14}\text{N} + {}^4\text{He} \rightarrow {}^{18}\text{F} + \gamma, \qquad (6.41)$$

but in this case the ^{18}F is unstable to a rapid β-decay,

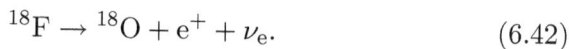

$$^{18}\text{F} \rightarrow {}^{18}\text{O} + e^+ + \nu_e. \qquad (6.42)$$

The ^{18}O can go on to capture another α-particle,

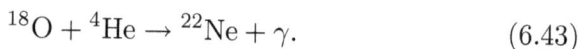

$$^{18}\text{O} + {}^4\text{He} \rightarrow {}^{22}\text{Ne} + \gamma. \qquad (6.43)$$

Importantly both ^{18}O and ^{22}Ne are neutron-rich, their nuclei contain more neutrons than protons. The overall neutron-to-proton ratio in a stellar core has consequences for the powering of supernovae (Sec. 8.10) and ^{22}Ne created from a star's original CNO elements is a significant contributor of neutrons. It is also the major source of neutrons for the s-process (Sec. 6.5) in massive asymptotic giant branch stars.

6.2.3. *Advanced burning stages*

Only another 0.8 MeV per nucleon is available from fusion reactions beyond ^{16}O. Each significant burning stage liberates less energy per nucleon and requires a higher temperature so they last for progressively shorter fractions of a star's lifetime. Once helium fuel is exhausted the next available fuel is carbon. At temperatures above about 5×10^8 K two ^{12}C nuclei combine to form an excited ^{24}Mg* nucleus. This very rarely decays to a stable state of ^{24}Mg. The first two reactions given in Table 6.3 are in fact the most likely. The ^{23}Na produced in the second would lead to a very significant quantity of this neutron-rich isotope were it not likely to recapture most of the free protons and end up as ^{20}Ne and an α-particle itself. The free α-particles are preferentially captured by carbon and oxygen. So the major product of carbon burning is in fact ^{20}Ne.

As the temperature rises photons in the high-energy tail of the radiation field can begin to penetrate nuclei and a new type of reaction, photodisintegration, becomes possible. The first major nucleus to succumb is ^{20}Ne. At about 10^9 K a photon can eject an α-particle from a ^{20}Ne nucleus. That α-particle can then be captured by another neon nucleus to form ^{24}Mg. Though the first

Table 6.3. A selection of significant reactions involved in advanced burning stages.

$T \approx 5 \times 10^8$ K	^{12}C + ^{12}C	$\rightarrow$	^{20}Ne + ^{4}He	$Q = +4.62$ MeV
		$\rightarrow$	^{23}Na + ^{1}H	$Q = +2.24$ MeV
		$\rightarrow$	^{23}Mg + n	$Q = -2.60$ MeV
$T \approx 10^9$ K	^{20}Ne + γ	$\rightarrow$	^{16}O + ^{4}He	$Q = -4.73$ MeV
	^{20}Ne + ^{4}He	$\rightarrow$	^{24}Mg + γ	$Q = +9.31$ MeV
$T \approx 2 \times 10^9$ K	^{16}O + ^{16}O	$\rightarrow$	^{28}Si + ^{4}He	$Q = +9.59$ MeV
$T \approx 3 \times 10^9$ K	^{28}Si + γ	$\rightarrow$	^{24}Mg + ^{4}He	$Q = -9.98$ MeV
	^{28}Si + ^{4}He	$\rightarrow$	^{32}S + γ	$Q = -6.94$ MeV
	...	$\rightarrow$	^{56}Ni	$Q = +19.15$ MeV

step is endothermic the combined result is overall exothermic. At about 2×10^9 K ^{16}O can react with itself. There are a number of possible outcomes but the major product is ^{28}Si. At this point there is only 0.34 Mev left to be gleaned from each nucleon. Further burning proceeds by a series of photodisintegrations and α captures. The photodisintegration of ^{28}Si is a limiting step and once it begins, when the temperature reaches about 3×10^9 K, burning proceeds to predominantly ^{56}Ni within a day or so. The actual isotopic composition of the iron-group elements, Ni, Co and Fe, as well as V, Cr and Mn, formed at the end of nuclear fusion is set by nuclear statistical equilibrium (Sec. 6.4.1) which depends on the ratio of neutrons to protons present. Because a large number of photons are absorbed during this phase there is also a significant addition to the stellar opacity to be included at high temperatures.

The formation of a core of iron-group elements marks the end of a star's nuclear burning life. No more energy is available from nuclear fusion to support the core which generally has a mass in excess of the Chandrasekhar limit and so collapses to a neutron star or black hole. The enormous release of gravitational energy causes an explosion visible as a supernova. We discuss these final events in stars' lives in Sec. 8.9.

6.3.   Nuclear Reaction Rates

To find the energy generation rate we must also calculate the rate at which each reaction proceeds. Let us first consider the fusion of two nuclei A and B, with rest masses m_A and m_B and atomic numbers Z_A and Z_B, to form a single stable nucleus C (Fig. 6.3). In their centre of momentum frame let the two nuclei approach one another with relative speed v. In LTE this relative velocity is drawn from a Maxwell–Boltzmann distribution that depends on temperature such that the fraction of pairs approaching each other with energy between E and $E + \mathrm{d}E$ is

$$f(E)\,\mathrm{d}E = \frac{2}{\sqrt{\pi}} \frac{E^{\frac{1}{2}}}{(kT)^{\frac{3}{2}}} \exp\left(-\frac{E}{kT}\right) \mathrm{d}E, \qquad (6.44)$$

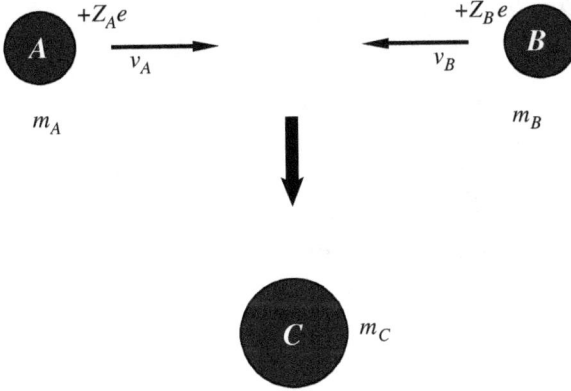

Fig. 6.3. Two nuclei A, of mass m_A and atomic number Z_A, and B, of mass m_B and atomic number Z_B fuse to form C, of mass m_C, by the reaction $A+B \to C$. In the centre of momentum frame A moves at $v_A = mv/m_A$ and B at $v_B = mv/m_B$, where v is the relative velocity and $m = m_A m_B/(m_A + m_B)$ is the reduced mass.

where $E = \frac{1}{2}mv^2$ and $m = m_A m_B/(m_A + m_B)$ is the reduced mass. Let the effective reaction cross-section for energies between E and $E + \mathrm{d}E$ be $\sigma(E)$ so that the reaction rate per unit volume for these energies is

$$R_{AB}(E)\,\mathrm{d}E = \frac{n_A n_B}{1 + \delta_{AB}}\sigma(E)v(E)f(E)\,\mathrm{d}E, \qquad (6.45)$$

where n_A is the number density of particles A and

$$\delta_{AB} = \begin{cases} 1 & \text{when } A \equiv B, \\ 0 & \text{otherwise} \end{cases} \qquad (6.46)$$

accounts for reactions between identical particles, such as in the first step of the pp-chains. When $A \equiv B$ the number of pairs is $N(N-1)/2 \approx N^2/2$ for a large number of particles N. The integrated rate per unit volume over all energies is then

$$R_{AB} = \frac{n_A n_B}{1 + \delta_{AB}}\langle \sigma v \rangle, \qquad (6.47)$$

with

$$\langle \sigma v \rangle = \int_0^\infty \sigma(E)v(E)f(E)\,\mathrm{d}E. \qquad (6.48)$$

6.3.1.   *The Coulomb barrier*

For a successful nuclear reaction A and B must get close enough for the short-range strong nuclear force to overcome the long-range Coulomb repulsion. This energy barrier is illustrated in Fig. 6.4. The nuclear radius of a nucleus with atomic mass A is $r_0 \approx 1.44\,A^{1/3} \times 10^{-13}$ cm. Classically, for the reactants to reach this separation, the reaction energy would need to exceed the Coulomb energy

$$E_C = \frac{Z_A Z_B e^2}{4\pi\epsilon_0 r_0} \approx Z_A Z_B \,\text{MeV}, \qquad (6.49)$$

where e is the electronic charge and ϵ_0 is the permittivity of free space. Nuclear spin is conserved during reactions and so often the

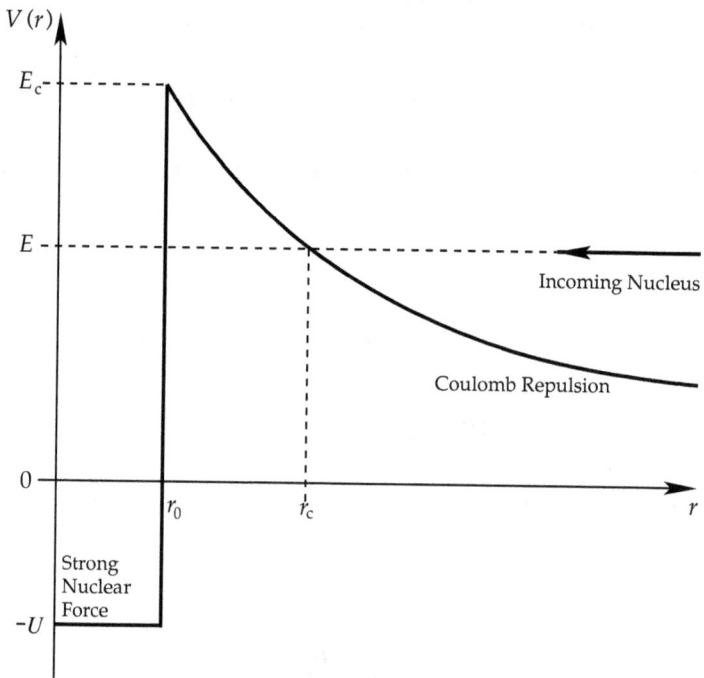

Fig. 6.4.   The potential energy of two nuclei of relative energy E separated by r in their centre of momentum frame. At small $r < r_0$, the effective nuclear radius, the short-range strong nuclear force dominates and $V \approx -U$. At large r the Coulomb repulsion dominates with $V(r) = Z_A Z_B e^2/(4\pi\epsilon_0 r)$. The classical Coulomb barrier can only be overcome when $E > E_c$. The radius at which $E = V(r > r_0)$ is r_c.

reacting particles must have non-zero angular momentum relative to one another. This adds an effective centrifugal potential to be overcome in addition to the Coulomb potential. When possible, interaction with zero angular momentum is always fastest and we neglect this complication in our simple description here. At 10^7 K, typical of the centre of the Sun, $E \approx kT \approx 10^3$ eV, from which we might conclude, as did Eddington, that the temperature at the centre of the Sun is too small by a factor of 1000 or so for the proton–proton chain. According to the Maxwell–Boltzmann distribution the fraction of interactions above this energy is only

$$\int_{E_C}^{\infty} f(E)\, dE \approx \exp(-1000) = 10^{-434}. \qquad (6.50)$$

So unless σ is large enough to compensate there is no hope of nuclear fusion in classical physics. This led Eddington to declare, "...go find a hotter place." for fusion to occur. With the advent of quantum mechanics Gamow, in 1928, was able to propose that quantum tunnelling is the solution. To calculate reaction rates we require the probability that a particle can penetrate the barrier rather than scattering off it. This probability of tunnelling is known as the penetration factor, usefully written in the form

$$P(E) = e^{-\omega(E)}. \qquad (6.51)$$

In the quantum mechanical description the incoming particle behaves as an evanescent wave inside the barrier. It has a small but finite possibility of penetrating to the nucleus. Though a full calculation would require knowledge of the uncertain behaviour of the short-range nuclear force the most important dependence can be obtained without it. In 1928 Gamow applied the new atomic theory to the nuclear physics of α-decay by treating the strong force with a spherically symmetric, deep square potential well of nuclear radius trapped within the repulsive Coulomb potential. The rate of decay then depends on the rate at which α-particles can tunnel through the barrier. A simple treatment of this gives excellent agreement with experimental data and represents a major success of quantum wave mechanics. The process of fusion of two charged particles is just the opposite of such a decay and we follow Gamow's approach in reverse to estimate our penetration factor $w(E)$.

6.3.2. *Barrier energy*

We consider the case of a nucleus A of atomic number Z_A being penetrated by a nucleus B of atomic number Z_B. The approximate potential experienced by the incoming nucleus is then

$$V(r) = \begin{cases} \dfrac{Z_A Z_B e^2}{4\pi\epsilon_0 r}, & r \geq r_0, \\[2mm] -U, & 0 \leq r < r_0, \end{cases} \tag{6.52}$$

where U represents the deep potential of the strong nuclear force of effective range $r_0 \ll r_c$. We write the time-independent Schrödinger equation as

$$-\frac{\hbar^2}{2m}\nabla^2\psi(\boldsymbol{r}) + V(r)\psi(\boldsymbol{r}) = E\psi(\boldsymbol{r}), \tag{6.53}$$

where $m = m_A m_B/(m_A + m_B)$ is the reduced mass and $\psi(\boldsymbol{r})$ is the wave function of the interacting nuclei. For spherically symmetric $\psi(\boldsymbol{r}) = \psi(r)$ this becomes

$$-\frac{\hbar^2}{2mr^2}\frac{\mathrm{d}}{\mathrm{d}r}\left(r^2\frac{\mathrm{d}\psi(r)}{\mathrm{d}r}\right) + V(r)\psi(r) = E\psi(r) \tag{6.54}$$

and we substitute

$$\psi(r) = \frac{\chi(r)}{r} \tag{6.55}$$

so that

$$-\frac{\hbar^2}{2m}\frac{\mathrm{d}^2\chi}{\mathrm{d}r^2} + V(r)\chi = E\chi. \tag{6.56}$$

Next consider $r \gg r_c$, far outside the classical turning point, where the potential $V(r) < E$ varies only slowly with r and write

$$k^2(r) = 2m\frac{E - V(r)}{\hbar^2}, \tag{6.57}$$

choosing $k(r) > 0$, to arrive at

$$\frac{d^2\chi}{dr^2} + k^2(r)\chi(r) = 0. \tag{6.58}$$

By analogy with the simple harmonic oscillator we presume that the wave function takes the form

$$\chi(r) = F(r)e^{i\phi(r)}, \tag{6.59}$$

where both $F(r)$ and $\phi(r)$ are slowly varying real functions of r. We differentiate twice to find

$$\frac{d^2\chi}{dr^2} = \left(F'' + 2iF'\phi' + iF\phi'' - F\phi'^2\right)e^{i\phi(r)}. \tag{6.60}$$

Comparing real and imaginary parts of Eq. (6.58) we obtain

$$F'' - F\phi'^2 + k^2 F = 0 \tag{6.61}$$

and

$$2F'\phi' + F\phi'' = \left(F^2\phi'\right)' = 0. \tag{6.62}$$

From the latter we obtain

$$F^2(r)\phi'(r) = \mathcal{C}^2 = \text{const.} \tag{6.63}$$

Now if $F(r)$ varies sufficiently slowly that

$$\frac{F''}{F} \ll \left(\phi'\right)^2 - k^2 \tag{6.64}$$

Eq. (6.61) yields approximate solutions

$$\phi'(r) = \pm k(r) \tag{6.65}$$

and

$$\phi(r) = \pm \int k(r)\,dr \tag{6.66}$$

so that

$$\chi(r) \approx \frac{\mathcal{C}}{\sqrt{k(r)}} \exp\left\{\pm i \int k(r)\,dr\right\}. \tag{6.67}$$

This is in fact the first term of the Wentzel–Kramers–Brillouin (WKB) asymptotic expansion for the solution that incorporates a rigorous justification of the condition (6.64). It turns out to be a very good approximation when $r > r_c$. For $r \gg r_c$ our solution becomes the incoming s-wave

$$\psi(r) \propto \frac{e^{-ikr}}{r} \qquad (6.68)$$

so we choose

$$\chi(r) \sim \frac{\mathcal{C}}{\sqrt{k(r)}} \exp\left\{ -i \int_{r_c}^{r} k(x) \, dx \right\}. \qquad (6.69)$$

A similar analysis applied to the tunnelling region when $r < r_c$ with

$$\kappa^2(r) = 2m \frac{V(r) - E}{\hbar^2} \qquad (6.70)$$

and $\kappa(r) > 0$ gives an asymptotic solution

$$\chi(r) \sim \frac{\mathcal{D}}{\sqrt{\kappa(r)}} \exp\left\{ \int_{r_0}^{r} \kappa(x) \, dx \right\}. \qquad (6.71)$$

Now the wave function $\psi(r)$ is continuous at $r = r_c$ but both (6.69) and (6.71) diverge there because $k(r_c) = \kappa(r_c) = 0$ and they cannot be matched directly. Fortunately the solutions are both otherwise analytic and can be continued in the complex plane. So we integrate along a contour that avoids the singularity at $r = r_c$, where $V = Z_A Z_B e^2/(4\pi\epsilon_0 r_c) = E$, by the adding a small imaginary part iy to r and matching at $r = r_c + iy$. There the potential

$$V(r_c + iy) = \frac{Z_A Z_B e^2}{4\pi\epsilon_0 (r_c + iy)} = \frac{Z_A Z_B e^2}{4\pi\epsilon_0 r_c} \left(1 - \frac{iy}{r_c} \right) + \mathcal{O}(y^2) \quad (6.72)$$

so that

$$k^2(r_c + iy) \approx -\kappa^2(r_c + iy) \approx \frac{2m}{\hbar^2} \left(\frac{iy Z_A Z_B e^2}{4\pi\epsilon_0 r_c^2} \right) \qquad (6.73)$$

and matching (6.69) and (6.71) at $r = r_c + iy$, where $k = e^{i\pi}\kappa + \mathcal{O}(y^2)$, and letting $y \to 0$ gives us

$$\mathcal{D} = \mathcal{C} e^{i\pi/4} \exp\left\{ \int_{r_0}^{r_c} \kappa(x) \, dx \right\}. \qquad (6.74)$$

At some point close to $r = r_0$ the potential $V \approx 0$ so $|\kappa(r)| = |k(r)|$, and the particle has tunnelled through the barrier. The probability current through a spherical surface is proportional to $r^2|\psi|^2$ so the penetration factor

$$P(E) = e^{-\omega(E)} = \frac{|\mathcal{D}|^2}{|\mathcal{C}|^2} = \exp\left\{-2\int_{r_0}^{r_c} \kappa(x)\,\mathrm{d}x\right\} \tag{6.75}$$

so that

$$\omega(E) = 2\int_{r_0}^{r_c} \kappa(x)\,\mathrm{d}x = 2\frac{\sqrt{2mE}}{\hbar}\int_{r_0}^{r_c} \left(\frac{r_c}{r} - 1\right)^{\frac{1}{2}}\,\mathrm{d}r. \tag{6.76}$$

This can be integrated with the substitution $r_c/r = \sec^2\theta$ to obtain

$$\omega(E) = 2\frac{\sqrt{2mE}}{\hbar}r_c\left[\theta - \frac{\sin 2\theta}{2}\right]_0^{\sec^{-1}(r_c/r_0)}. \tag{6.77}$$

Because $r_0 \ll r_c = Z_AZ_Be^2/(4\pi\epsilon_0 E)$ we can take the upper limit to be $\pi/2$ and then, defining the barrier energy E_B such that

$$\omega(E) = 2\sqrt{\frac{E_B}{E}}, \tag{6.78}$$

we find

$$E_B = \frac{e^4 Z_A^2 Z_B^2\, m}{16\epsilon_0^2\hbar^2\,2} = \frac{1}{2}\left(\pi\alpha_{\mathrm{fs}}Z_AZ_B\right)^2 mc^2 \approx 0.25Z_A^2 Z_B^2 A\mathrm{MeV}, \tag{6.79}$$

where $A = A_AA_B/(A_A + A_B)$ is the reduced nuclear mass of the reacting particles, α_{fs} is the fine structure constant and c is the speed of light. When the angular momentum of the interaction is non-zero the calculation is complicated by the centrifugal potential but proceeds along the same lines.

6.3.3. *Cross-section factor*

Once inside the nucleus the incoming particle is faced with a very deep potential well created by the strong nuclear force. In quantum mechanics such a deep well reflects as efficiently as a high wall so unless there exists an excited nuclear state of similar energy to

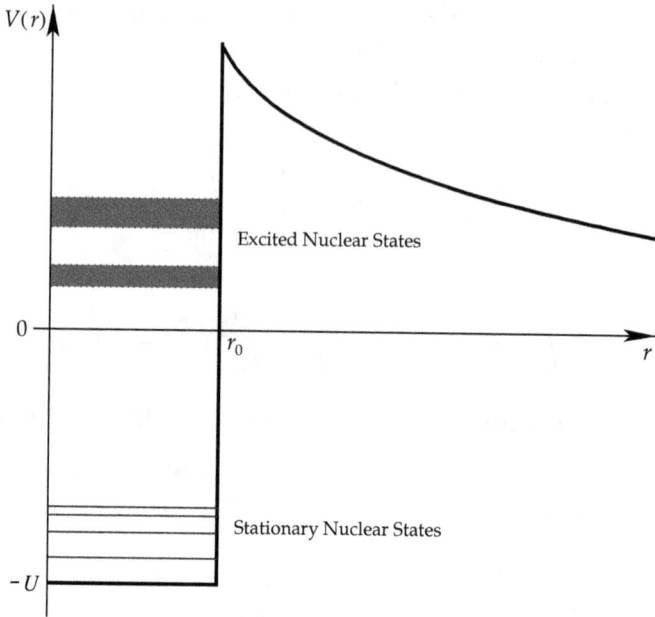

Fig. 6.5. Excited and bound states of the compound nucleus. Stable nuclei exist in narrow stationary states with $-U < E < 0$. There also exist excited states with $0 \leq E < E_c$ which have a finite width and are subject to radioactive decay by fission, typically the loss of an α-particle, or by emission of γ-rays to reach a stationary state. The deep potential well of the strong nuclear force repels an incoming nucleus and so reactions tend to proceed by first entering an excited state. When the incoming energy matches that of an excited state resonance greatly increases the cross-section for fusion.

the incoming particle it cannot remain inside. An unstable nucleus C^* must be formed initially and this can then decay to the stable nucleus C. Excited nuclear states (Fig. 6.5) have a short lifetimes τ, and hence large natural widths $\Gamma_{\mathrm{res}} = \hbar/\tau$, because they are unbound with energy $E_{\mathrm{res}} > 0$. Let us express the wave function of a resonant state as $\psi_{\mathrm{res}}(\boldsymbol{r}, t)$ and assume this can be separated into spatial and time-dependent parts. Suppose the state is created at $t = 0$ and decays exponentially on the time-scale τ such that the probability of finding the state at time t is $\exp(-t/\tau)$. We may then write

$$\psi_{\mathrm{res}}(\boldsymbol{r}, t) = \begin{cases} \chi(\boldsymbol{r}) \exp\left(-t/2\tau\right) \exp\left(-iE_{\mathrm{res}}t/\hbar\right), & t \geq 0, \\ 0, & t < 0, \end{cases} \qquad (6.80)$$

where

$$\int_{\text{all space}} \chi(\boldsymbol{r}) \, dV = 1. \tag{6.81}$$

We may also express the wave function as a Fourier transform with respect to the interaction energy E and write

$$\psi(\boldsymbol{r}, t) = \chi(\boldsymbol{r}) \int_{-\infty}^{\infty} \Psi(E) \exp\left(-\frac{iEt}{\hbar}\right) \, dE, \tag{6.82}$$

with

$$\Psi(E) \propto \int_{0}^{\infty} \exp\left(\frac{i(E - E_{\text{res}})t}{\hbar} - \frac{t}{2\tau}\right) \, dt \tag{6.83}$$

$$\propto \frac{\hbar}{i(E - E_{\text{res}}) - \hbar/(2\tau)}. \tag{6.84}$$

The probability of forming C^* depends on the overlap of the wave function of the reacting particles and that of the excited state. This adds a factor proportional to the reaction cross-section

$$|\Psi(E)|^2 \propto \frac{1}{(E - E_{\text{res}})^2 + \left(\frac{\Gamma_{\text{res}}}{2}\right)^2}. \tag{6.85}$$

As the interaction energy approaches that of a resonance so the cross-section rises. Now angular momentum must be conserved so the impact parameter b of the incoming particle must be such that the quantised angular momentum $L = bp = b\hbar/\lambda_{\text{dB}}$, where

$$\lambda_{\text{dB}} = \frac{\hbar}{\sqrt{2mE}} \tag{6.86}$$

is the de Broglie wavelength, lies approximately in the range $l\hbar \le L < (l+1)\hbar$, where $l \in \{0, 1, 2, 3, \ldots\}$ is the angular momentum state of the interaction. Thence $l < b/\lambda_{\text{dB}} < l + 1$ and the cross-section approaches $(2l + 1)\pi\lambda_{\text{dB}}^2 \propto 1/E$. We therefore remove this inverse dependence on E too and write the overall reaction cross-section

in the form

$$\sigma(E) = \frac{S(E)}{E} P(E), \tag{6.87}$$

defining a cross-section factor $S(E)$ which takes the form

$$S(E) = S_0(E) + \sum_i \frac{S_i}{(E - E_i)^2 + \left(\frac{\Gamma_{\text{res},i}}{2}\right)^2} \tag{6.88}$$

where $S_0(E)$ is due to distant resonances and is approximately constant while the sum is calculated over resonant states with energies close to the interaction energy E. Typical variation of $S(E)$ with E is depicted in Fig. 6.6.

In the next section we shall see that reaction rates are significant over only a small range of interaction energy E. Often, as indeed for the proton–proton chain, there are no local resonances. When local resonances are important usually only one or at most two

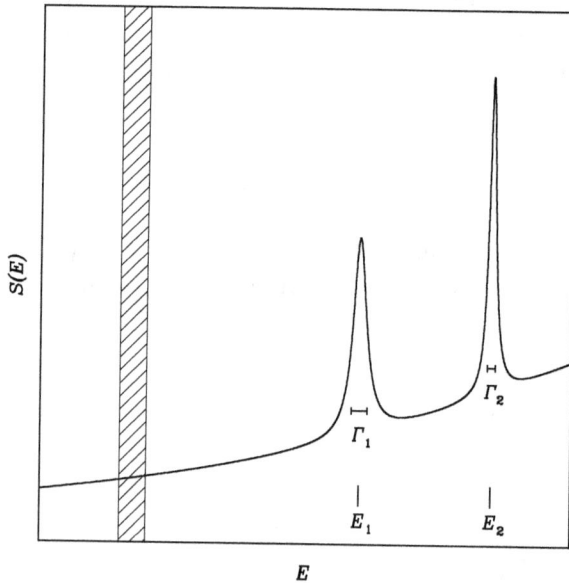

Fig. 6.6. The cross-section factor $S(E)$ as a function of E. Two resonances of energies E_1 and E_2 and widths Γ_1 and Γ_2 are illustrated. When the Gamow window (Sec. 6.3.4), shaded, is far from a resonance $S(E) \approx \text{const}$.

matter for a given reaction at a given temperature. In particular the triple-α process proceeds as fast as it does only because the unstable ^{8}Be* has an excited state close to the energy of two α-particles and because the excited ^{12}C* nucleus also resonates at the typical energy of the ^{8}Be* plus the third α-particle. Working from the abundance of carbon relative to oxygen in the solar system Fred Hoyle deduced the existence of this particular excited state of ^{12}C in 1953. It was then looked for and found by Fowler's team at the Kellogg Radiation Laboratory in California in 1957.

Generally $S(E)$ cannot be reliably calculated theoretically. Instead it must be measured in the laboratory but this is only easily done at high energies ($E > 0.1\,\mathrm{MeV}$) much greater than the $E \approx 1\,\mathrm{keV}$ at which hydrogen burning takes place in the Sun. Fortunately $S(E)$ can be accurately extrapolated as long as there are no unknown resonances in between where it is measured and the target energy. We can be fairly certain that the rate for the proton–proton reaction is well established because of the success of stellar models. However rates of some other reactions, for instance, that of the α-particle capture by ^{12}C are less certain.

6.3.4. *Gamow energy*

We can now combine the various factors to determine reaction rates. First consider the case where no nearby resonances are important so that $S(E)$ is a rather slowly varying function of E. Now $P(E) \propto \exp(-2\sqrt{E_B/E})$ and $f(E) \propto E^{1/2}\exp[-E/(kT)]$ and, at non-relativistic energies, $v(E) \propto E^{1/2}$ while $\sigma(E) \propto 1/E$. This leaves the dependence on E only in the exponential terms,

$$\sigma(E)v(E)P(E)f(E) \propto \exp\left(-\frac{E}{kT} - 2\sqrt{\frac{E_B}{E}}\right). \qquad (6.89)$$

This is illustrated for the proton–proton chain at $2 \times 10^7\,\mathrm{K}$, when $E_B = 73\,kT$, in Fig. 6.7. The reaction rate is dominated by a small range of energies in the Gamow window where the high energy tail of the Maxwell–Boltzmann distribution overlaps with the rising penetration factor for quantum mechanical tunnelling. The

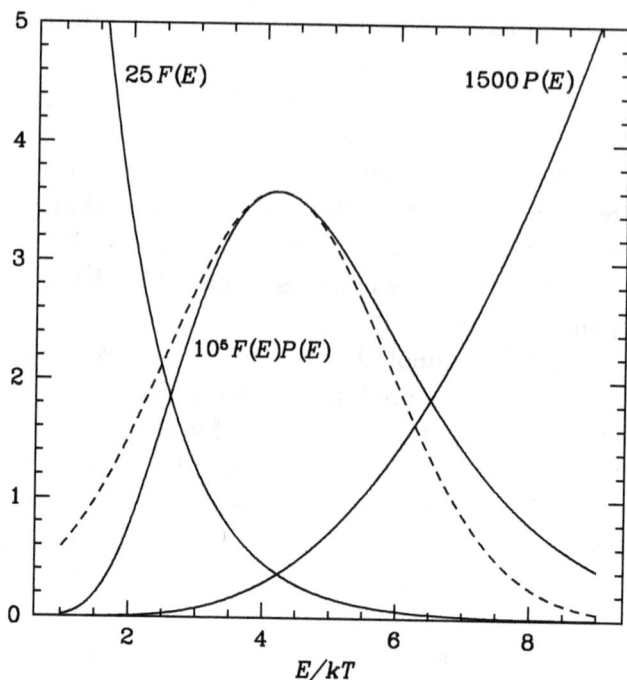

Fig. 6.7. The variation of the reaction rate with E for the pp reaction for which $E_B = 73\,kT$. Solid lines show the behaviour of the exponential factors $F(E) = \exp(-E/kT)$ that describes the high energy tail of the Maxwell–Boltzmann distribution, $P(E) = \exp(-2\sqrt{E_B/E})$ which describes the increasing penetration with E and their product which is proportional to the reaction rate for energies between E and $E + dE$. The integrated reaction rate is dominated by energies in the narrow region around $E_G = 4.2\,kT$. The dotted line is the Gaussian approximation used to integrate $\langle \sigma v \rangle$ by Laplace's method (Eq. (6.96)).

maximum occurs when

$$\frac{\mathrm{d}}{\mathrm{d}E}\left(\frac{E}{kT} + 2\sqrt{\frac{E_B}{E}} \right) = 0, \qquad (6.90)$$

at

$$E = E_G = E_B^{\frac{1}{3}}(kT)^{\frac{2}{3}}, \qquad (6.91)$$

the Gamow energy. For the pp-chain at $2 \times 10^7\mathrm{K}$, $E_G = 4.2\,kT$. We can calculate a good analytical approximation to the reaction rate by means of Laplace's method, a special real case of the method of steepest descents in the complex plane. We find the first two terms of the Taylor expansion in E and complete the square to obtain

$$\frac{E}{kT} + 2\sqrt{\frac{E_B}{E}} = 3\frac{E_G}{kT} + \left(\frac{E - E_G}{2\Delta}\right)^2 + \mathcal{O}(E^3), \qquad (6.92)$$

where

$$\Delta = \frac{1}{\sqrt{3}} E_B^{\frac{1}{6}} (kT)^{\frac{5}{6}} = 1.2\, kT \qquad (6.93)$$

and use this to approximate the rate by a Gaussian function which fits well within the Gamow window in the range

$$E_G - \Delta < E < E_G + \Delta. \qquad (6.94)$$

Again with no resonance within or very near to this window, as for the pp-chain, we have $S(E) = S_0(E) \approx$ const. Thence

$$\langle \sigma v \rangle = \int_0^\infty \frac{S_0(E)}{E} v(E) P(E) f(E) dE \qquad (6.95)$$

$$\sim \frac{S_0(E_G) v(E_G) 2 E_G^{\frac{1}{2}}}{E_G \sqrt{\pi} (kT)^{\frac{3}{2}}} \exp\left(-\frac{3 E_G}{kT}\right) \int_{-\infty}^{\infty} \exp\left(\frac{E_G - E}{2\Delta}\right)^2 dE \qquad (6.96)$$

$$= \frac{2\sqrt{2}}{\sqrt{3m}} \frac{E_B^{\frac{1}{6}}}{(kT)^{\frac{2}{3}}} S_0(E_G) \exp\left[-3\left(\frac{E_B}{kT}\right)^{\frac{1}{3}}\right] \qquad (6.97)$$

because the integral in Eq. (6.96) equates to $2\Delta\sqrt{\pi}$. The temperature and density dependence of the reaction rate then reduces to

$$R_{AB} \propto \frac{n_A n_B}{1 + \delta_{AB}} \frac{1}{(kT)^{\frac{2}{3}}} \exp\left[-3\left(\frac{E_B}{kT}\right)^{\frac{1}{3}}\right]. \qquad (6.98)$$

6.3.4.1. *The hydrogen burning energy generation rate*

The energy generation rate per unit mass for the pp-chain is just the product of the reaction rate and the energy released per reaction so

$$\epsilon_{pp} = R_{AB} Q_{AB}/\rho. \qquad (6.99)$$

It is often useful to be able to write a reaction rate in the power-law form

$$R_{AB} \propto n_A n_B T^{\nu}, \tag{6.100}$$

for a given T. The exponent

$$\nu = \frac{\partial \log R_{AB}}{\partial \log T} = -\frac{2}{3} + \frac{\tau}{3}, \tag{6.101}$$

where

$$\tau = 3\left(\frac{E_B}{kT}\right)^{\frac{1}{3}}. \tag{6.102}$$

For the pp-chain at 2×10^7 K, $\tau = 12.5$ so $\nu = 3.3$, while at 10^7 K, $\tau = 16$ so $\nu = 4.6$. Both $n_A = n_B \propto \rho$ in this case so the energy generation rate by the pp-chain at 2×10^7 K takes the form

$$\epsilon_{pp} = \epsilon_{0,pp} \rho T^{3.3}. \tag{6.103}$$

In the CNO cycle the proton capture by nitrogen $^{14}N(p,\gamma)^{15}O$ determines the rate. For this reaction the barrier energy $E_B \propto Z_A^2 Z_B^2 A$ is 46 times larger than for the pp-chain at any given temperature. So τ is 3.6 times larger and, at 2×10^7K, $\nu = 14$ and

$$\epsilon_{CNO} = \epsilon_{0,CNO} \rho T^{14}. \tag{6.104}$$

It is this stronger dependence on temperature that ensures that the CNO cycle eventually dominates hydrogen burning as the temperature rises. At solar metallicity the changeover is at about 2×10^7K. It is also this strong temperature dependence that is responsible for the convective cores of more massive stars (Sec. 4.4.1). The rate for the CNO cycle is proportional to the abundance of CNO nuclei and so at lower metallicities the pp-chain continues to dominate to correspondingly higher temperatures.

6.3.5. *Resonant reactions*

As a resonance approaches the Gamow window the increased cross-section at the resonant energy means that the dominant contribution occurs around the resonance rather than across the Gamow window.

The natural width of a resonance depends on the decay paths of the resonant state. The Coulomb barrier penetration makes charged particle decays unlikely and the electromagnetic emission of photons is usually even less likely so that $\Gamma_{\text{res}} \ll \Delta$ in many cases. Hence the cross-section factor $S(E)$ varies much more quickly than $f(E)$ and $v(E)$ and we can approximate the reaction rate by

$$R_{AB} \approx \frac{n_A n_B}{1 + \delta_{AB}} f(E_{\text{res}}) v(E_{\text{res}}) \int_0^\infty \sigma(E)\, \mathrm{d}E. \qquad (6.105)$$

The only temperature dependence enters through $f(E_{\text{res}})$ so

$$R_{AB} \propto n_A n_B \frac{1}{(kT)^{3/2}} \exp\left(-\frac{E_{\text{res}}}{kT}\right) \propto n_A n_B T^\eta, \qquad (6.106)$$

where

$$\eta = \frac{E_{\text{res}}}{kT} - \frac{3}{2}. \qquad (6.107)$$

6.3.5.1. *The helium burning rate*

For the triple-α reaction we expect,

$$R_{3\alpha} \propto (n_\alpha)^3, \qquad (6.108)$$

because three particles must combine together, and to have a very strong temperature dependence because of the resonances that are close to E_G in both reactions. A temperature of $T > 10^8$ K is required to get a significant burning rate. Both the formation of $^8\text{Be}^*$ and $^{12}\text{C}^*$ are dominated by their resonances at $92\,\text{keV}$ and $278\,\text{keV}$ respectively so the reaction rate

$$R_{3\alpha} \propto n_\alpha^2 \frac{1}{(kT)^{3/2}} \exp\left(-\frac{92\,\text{keV}}{kT}\right)$$
$$\times n_\alpha \frac{1}{(kT)^{3/2}} \exp\left(-\frac{278\,\text{keV}}{kT}\right) \propto n_\alpha^3 T^\eta, \qquad (6.109)$$

where

$$\eta = \frac{370\,\text{keV}}{kT} - 3. \qquad (6.110)$$

At $T = 10^8$ K, $\eta \approx 40$ so

$$\epsilon_{3\alpha} \propto \rho^2 T^{40}. \tag{6.111}$$

This temperature dependence is the strongest encountered for any physical process so far investigated. It is due to the two resonances entering the reaction window at just the temperature required to get the reaction going.

6.3.6.   *Thermostatic control*

Usually the strong dependence on temperature of nuclear fusion reactions leads to thermostatic control within a star. If the temperature rises the reaction rate rises. Increased energy generation causes the star to expand which in turn causes it to cool. This negative feedback means that burning in a star tends to reach an equilibrium. Material reaches the temperature required for a particular reaction to ignite and maintains a temperature close to that until the appropriate fuel is exhausted. Only then can the material heat up further until the next reaction begins. So only one major phase of burning takes place in the core of a star at any one time and once central fuel is exhausted that burning process moves to a shell outside the hotter core. Successive reactions continue to take place in an onion-like structure as each new reaction ignites at the centre. Stability by cooling on expansion requires an equation of state in which temperature falls when pressure falls. While this is true for an ideal gas it is not for cold degenerate matter, in which the temperature can rise without changing the pressure and reaction rates can run away, in some cases leading to a thermonuclear flash or explosion.

6.3.7.   *Electron screening*

At high densities nuclei do not experience full electrostatic repulsion until they are very close because the dense electron gas shields the nuclear charge. This reduces the required tunnelling distance and so increases the rate of nuclear reactions. Not surprisingly it is most important in degenerate matter where material eventually reaches densities at which neighbouring nuclei are sufficiently close for cold fusion. Such pycnonuclear reactions are responsible for igniting explosive carbon burning in accreting white dwarfs which become type Ia supernovae (Sec. 9.9.4). Question 1 further explores weak screening in a star like the Sun.

6.4. Reaction Equilibria

In any reaction cycle or chain, intermediate nuclei reach equilibrium abundances after sufficient time. For example the abundances of ^{2}H and ^{3}He in the ppI-chain are in equilibrium at the centre of the Sun. Consider the three reactions in the ppI-chain listed in Table 6.4 where R_{ij} is the reaction rate between species i and j with number densities n_i and n_j and $\lambda_{ij} = \langle \sigma v \rangle$ for the reactions of i with j. We label each species by setting i equal to its atomic weight. The equations for the rate of change of each species are then

$$\frac{dn_1}{dt} = -2R_{11} - R_{12} + 2R_{33}, \tag{6.112}$$

$$\frac{dn_2}{dt} = R_{11} - R_{12}, \tag{6.113}$$

$$\frac{dn_3}{dt} = R_{12} - 2R_{33} \quad \text{and} \tag{6.114}$$

$$\frac{dn_4}{dt} = R_{33}. \tag{6.115}$$

When both ^{2}H and ^{3}He have reached equilibrium

$$\frac{dn_2}{dt} = 0 \quad \text{and} \tag{6.116}$$

$$\frac{dn_3}{dt} = 0, \tag{6.117}$$

Table 6.4.   Reaction rates of the ppI-chain at the centre of the Sun.

	Reaction			Rate	
slow	^{1}H + ^{1}H	$\rightarrow$	^{2}H + e$^+$ + ν	R_{11}	$= \frac{1}{2}\lambda_{11}n_1^2$
very fast	^{1}H + ^{2}H	$\rightarrow$	^{3}He + γ	R_{12}	$= \lambda_{12}n_1 n_2$
fast	^{3}He + ^{3}He	$\rightarrow$	^{4}He + 2^{1}H	R_{33}	$= \frac{1}{2}\lambda_{33}n_3^2$

so that

$$R_{11} = R_{12} = 2R_{33} \tag{6.118}$$

from which we deduce that

$$\lambda_{11}\frac{n_1^2}{2} = \lambda_{12}n_1 n_2 = \lambda_{33}n_3^2 \tag{6.119}$$

and thence that

$$n_2 = \frac{\lambda_{11}}{2\lambda_{12}}n_1 \tag{6.120}$$

and

$$n_3 = \left(\frac{\lambda_{11}}{\lambda_{33}}\right)^{\frac{1}{2}} n_1. \tag{6.121}$$

As long as there are enough protons the deuterium and helium-3 abundances remain in equilibrium with a ratio dependent mostly on the temperature in the star through its effect on the reaction rates. Question 2 extends this analysis to the ppII reactions.

6.4.1. *Nuclear statistical equilibrium*

Once silicon burning begins temperatures are such that reaction rates for photodisintegrations and captures of protons, neutrons and α particles are similar and in particular much faster than any weak reactions. The abundances of various elements move to an energetic equilibrium that can be described in a similar way to ionisation equilibrium by the Saha equation (Sec. 3.6). Consider a reversible reaction in which a nucleus emits or absorbs a neutron,

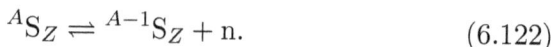

$$^A S_Z \rightleftharpoons {}^{A-1}S_Z + \text{n}. \tag{6.122}$$

In thermodynamic equilibrium the ratio of the number densities of nuclides $^{A-1}S_Z$ to $^A S_Z$ is given by

$$\frac{n(A-1,Z)n_{\text{n}}}{n(A,Z)} = \frac{2g(A-1,Z)}{g(A,Z)}\frac{(2\pi\mu kT)^{\frac{3}{2}}}{h^3}\exp\left(-\frac{Q_{\text{n}}}{kT}\right), \tag{6.123}$$

where $g(A,Z) = 2J(A,Z) + 1$ is the statistical weight of a nuclide $^A S_Z$ with a spin of $J(A,Z)$, n_{n} is the number density of free neutrons, of spin $1/2$ and so statistical weight 2, μ is the reduced mass of the nucleus $^{A-1}S_Z$ and the free neutron and

$$Q_n = [m(A, Z) - m(A - 1, Z) + m_n)]c^2 \qquad (6.124)$$

is the binding energy of the last neutron in the nucleus of $^A S_Z$ of mass $m(A, Z)$. We may write

$$\mu \approx \frac{A-1}{A} m_u, \qquad (6.125)$$

where m_u is the atomic mass unit, and define $\Theta(T)$ such that

$$\frac{(2\pi\mu kT)^{\frac{3}{2}}}{h^3} = \left(\frac{A-1}{A}\right)^{\frac{3}{2}} \Theta(T) = \left(\frac{A-1}{A}\right)^{\frac{3}{2}} \frac{(2\pi m_u kT)^{\frac{3}{2}}}{h^3} \qquad (6.126)$$

Similarly for the emission or absorption of a proton,

$$^A S_Z \rightleftharpoons {}^{A-1}S_{Z-1} + p, \qquad (6.127)$$

we have

$$\frac{n(A-1, Z-1)n_p}{n(A, Z)} = \frac{2g(A-1, Z-1)}{g(A, Z)} \left(\frac{A-1}{A}\right)^{\frac{3}{2}} \Theta(T) \exp\left(-\frac{Q_p}{kT}\right), \qquad (6.128)$$

where

$$Q_p = [m(A, Z) - m(A - 1, Z - 1) + m_p)]c^2. \qquad (6.129)$$

Combining the whole set of such equilibria for all less massive nuclides we find generally that

$$n(A, Z) = g(A, Z) A^{\frac{3}{2}} \frac{n_p^Z n_n^{A-Z}}{2^A} \Theta^{1-A} \exp\left[\frac{Q(A, Z)}{kT}\right], \qquad (6.130)$$

where

$$Q(A, Z) = [Zm_p + (A - Z)m_n - m(A, Z)]. \qquad (6.131)$$

So the abundance of any nuclide in nuclear statistical equilibrium (NSE) is determined entirely by the temperature and the number densities of free protons and neutrons. The density of the stellar material is

$$\rho = n_p m_p + n_n m_n + \sum_{A,Z} n(A, Z) m(A, Z), \qquad (6.132)$$

where the sum is taken over all nuclides so that the mass fraction of $^A S_Z$ is

$$X_{^A S_Z} = \frac{n(A, Z) m(A, Z)}{\rho}. \qquad (6.133)$$

When weak reactions are slow, the total number of neutrons and protons, both free and bound in heavier nuclei, are independently conserved so the ratio of protons to neutrons

$$\frac{\langle Z \rangle}{\langle N \rangle} = \frac{\sum_{A,Z} Z n(A, Z) + n_p}{\sum_{A,Z} (A - Z) n(A, Z) + n_n} \qquad (6.134)$$

is fixed and the abundance of any nuclide is determined by ρ, T and $\langle Z \rangle / \langle N \rangle$.

When $\langle Z \rangle / \langle N \rangle \approx 1$ the most abundant nuclei are ^{56}Ni and ^{54}Fe, both with zero spin, and by Eq. (6.128)

$$\frac{n(^{54}\text{Fe})}{n(^{56}\text{Ni})} n_p^2 = 4 \left(\frac{54}{56} \right)^{\frac{3}{2}} \Theta^2 \exp \left(\frac{Q(54, 26) - Q(56, 28)}{kT} \right), \quad (6.135)$$

where $Q(54, 26) - Q(56, 28) = -2.35 \, \text{MeV}$ can be measured in the laboratory. If we further assume that no other nuclides contribute we have

$$\frac{\langle Z \rangle}{\langle N \rangle} \approx \frac{26 \, n(^{54}\text{Fe}) + 28 \, n(^{56}\text{Ni}) + n_p}{54 \, n(^{54}\text{Fe}) + 56 \, n(^{56}\text{Ni})}. \qquad (6.136)$$

Together these last two equations may be solved to find n_p and the abundances of ^{54}Fe and ^{56}Ni (see Question 3). The result is important because it is the radioactive decay ^{56}Ni$(\beta, \gamma\nu)^{56}$Co$(\beta, \gamma\nu)^{56}$Fe that powers supernovae (Sec. 8.10) for months after the core collapse while ^{54}Fe is stable and so contributes no further energy. The peak luminosity of supernovae, including type Ia, therefore depends on the mass of ^{56}Ni produced and that itself depends on $\langle Z \rangle / \langle N \rangle$ at the time when NSE is established. The description of the various nuclear burning phases given in Sec. 6.2 reveals that $\langle Z \rangle / \langle N \rangle \approx 1$

is preserved because of the predominance of α-captures over weak reactions. The major contributors to excess neutrons are ^{22}Ne produced during helium burning of the ^{14}N created in earlier CNO burning and ^{23}Na produced during carbon burning though the latter is often destroyed by subsequent proton capture. Following silicon burning many of the heavy nuclides produced are, like ^{56}Ni, unstable to β-decays that tend to create more neutron-rich isotopes. When the time-scales for these become short compared with that to reach NSE, $\langle Z \rangle / \langle N \rangle$ falls. Weak reactions are not reversible in stars because the neutrinos escape too easily so there is no equivalent of NSE. Instead their effect can be included as a slow drift in $\langle Z \rangle / \langle N \rangle$.

6.5. The Origin of the Elements

During the first few minutes after the big bang protons and neutrons interacted through a network of reactions similar to those of the pp-chains but with the addition of free neutrons. Once all neutrons had decayed and the reaction rates had become negligible matter in the Universe consisted about 75% ^{1}H and 25% ^{4}He with trace amounts of ^{2}H, ^{3}He, ^{7}Li, ^{7}Be and ^{8}B. Nucleosynthesis could not proceed further because the Universe expanded and cooled too fast for the triple-α reaction to operate. It is stars that are responsible for the production of heavier elements and we have already discussed how fusion reactions can create elements up to the iron group (V, Cr, Mn, Fe, Co and Ni) while generating energy.

Heavier elements are generally created by neutron capture processes. Neutrons have no electric charge and so no Coulomb barrier to overcome when penetrating a nucleus. They have very high reaction cross-sections with nuclei that increase with the typical area of the nucleus, composed of A closely packed nucleons,

$$\sigma_{\rm n} \propto A^{\frac{2}{3}}. \tag{6.137}$$

So neutrons are more easily captured by iron and heavier elements than small light nuclei such as protons and α-particles. In this way heavy, neutron-rich isotopes can be built up. Consider a general nucleus ^A_ZS, with atomic number Z and atomic weight A that

captures a neutron,

$$^{A}_{Z}\text{S} + n \rightarrow {}^{A+1}_{Z}\text{S} + \gamma. \tag{6.138}$$

With typically a slightly larger cross-section the new nucleus can capture another neutron,

$$^{A+1}_{Z}\text{S} + n \rightarrow {}^{A+2}_{Z}\text{S} + \gamma \tag{6.139}$$

and similar neutron captures can continue until the new nucleus is so neutron-rich that it is unstable to a weak β-decay in which a neutron emits an electron and a neutrino to become a proton,

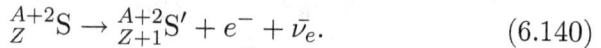

$$^{A+2}_{Z}\text{S} \rightarrow {}^{A+2}_{Z+1}\text{S}' + e^{-} + \bar{\nu}_{e}. \tag{6.140}$$

Now S' is a different element with one more proton than S. Whether or not the β-decay occurs before S can capture another neutron depends on both the lifetime of the unstable nuclei and the neutron flux. We distinguish the slow neutron capture or s-process, in which β-decays occur preferentially to further neutron capture, from the rapid neutron capture or r-process in which neutron captures occur preferentially. Each process generates a distinct set of isotopes in characteristic ratios.

Common s-process isotopes, such as yttrium, barium and lead, have magic numbers of neutrons, $A - Z \in \{2, 8, 20, 28, 50, 82, 126\}$, because, in the nuclear shell model, the binding energy of magic nuclei is increased when nucleons fill shells analogous to electron orbitals in atoms. Most of the relevant β-decays are rapid compared to the neutron capture rates but the s-process does create the unstable isotope of technetium-99 with a rather long half-life of about 2.1×10^{5} yr. Though relatively long this is still short when compared to the time-scales of stellar evolution. Technetium has no stable isotope at all so its presence in a star's atmosphere indicates that it must have been created recently and so the s-process must be currently active. Indeed detection of technetium in stars is direct evidence that nuclear reactions are ongoing in stars. The heaviest nucleus created in the s-process is $^{209}_{83}\text{Bi}$ because fission of heavier nuclei occurs faster than further neutron captures. Though it has been shown to decay by α-particle emission, with a half-life of

1.9×10^{19} yr, ^{209}Bi is effectively stable over the current age of the Universe. The lower the metallicity of a star the more massive are the s-process products because there are fewer heavy nuclei to absorb a similar number of neutrons. Abundance estimates from stellar spectra show that the s-process takes place in asymptotic giant branch (AGB) stars during thermal pulses (Sec. 8.3). The neutron source is either

$$^{13}\text{C} + {}^{4}\text{He} \rightarrow {}^{16}\text{O} + n \qquad (6.141)$$

or

$$^{22}\text{Ne} + {}^{4}\text{He} \rightarrow {}^{25}\text{Mg} + n. \qquad (6.142)$$

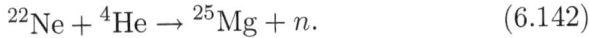

Recall that ^{22}Ne is formed during helium burning when α-particles are captured by the ^{14}N left over from CNO-burning of hydrogen. As a neutron source it requires high temperature helium burning and so is responsible for neutrons only in massive stars. The CNO cycle also includes ^{13}C but its equilibrium abundance is too small to make a viable neutron source. It can only be produced in sufficient quantities when the number of available protons is similar to the number of available ^{12}C nuclei. This seems to occur in the intershell region of low-mass AGB stars but the mechanism by which the appropriately small number of protons mixes into the intershell region is far from fully understood.

As the neutron flux is increased so more neutron captures are possible before the β-decays. At the extreme, in the r-process, neutrons are captured until nuclei reach the neutron drip line beyond which no more can be accommodated. Once the neutron flux falls off these heavy nuclei rapidly β-decay back to the valley of stability. The r-process is responsible for elements heavier than ^{208}Pb as well as a characteristic set of lighter isotopes. Europium, molybdenum and uranium are all typical products. The site of the r-process has not been categorically identified, though both supernovae and merging neutron stars are strong contenders. Indeed analysis of stellar abundances indicates that there are at least two sites, one present early in the evolution of our Galaxy and one activated somewhat later. Figure 6.8 shows part of the table of nuclides with

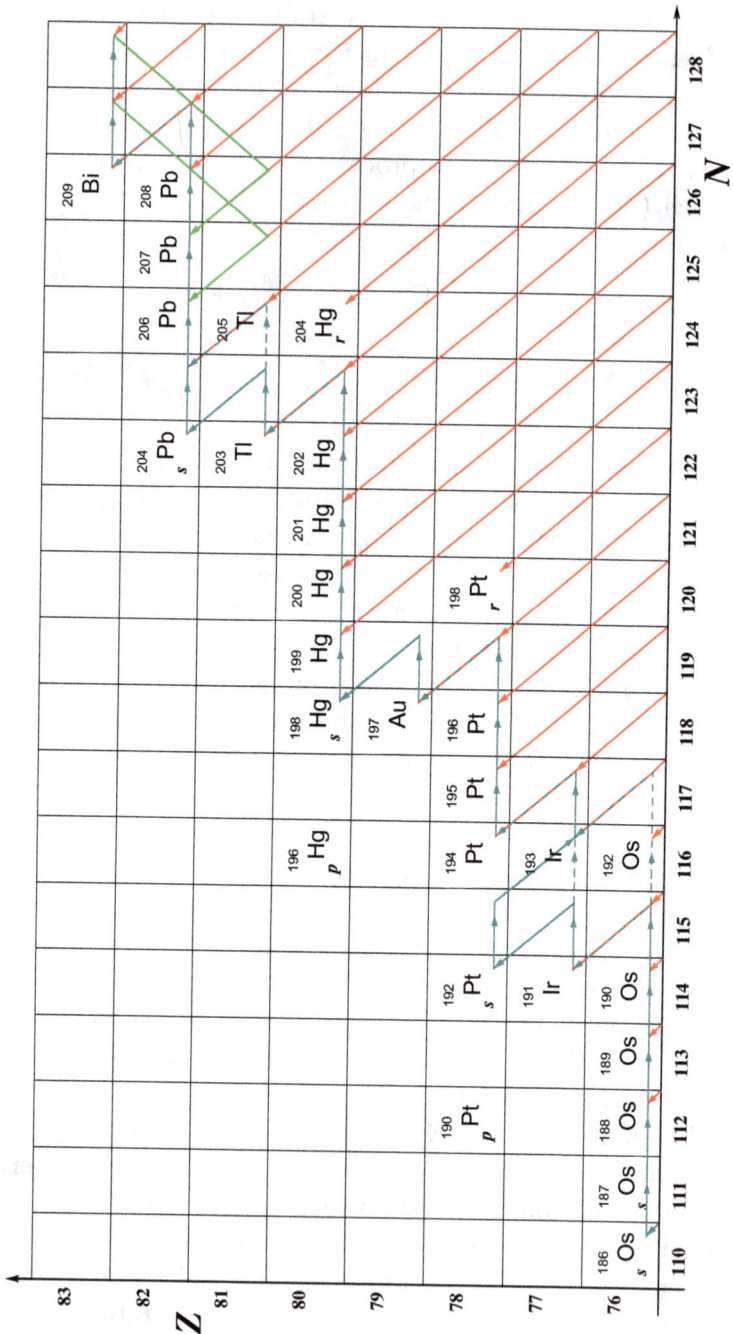

Fig. 6.8. A chart of nuclides in the neutron number–proton number plane for nuclides at the heavy end of the s-process production. All stable isotopes and those with half lives longer than the age of the Sun are shown. The s-process path (blue) proceeds horizontally through the stable isotopes of a single element as neutrons are captured. When a β-decay is fast enough a new element can be formed though, if the half-life of the nuclide is long enough, the s-process can branch (dashed routes), depending on the temperature and neutron flux, as at ^{191}Os, ^{192}Ir and ^{204}Tl. The s-process cannot proceed beyond ^{209}Bi because the next two isotopes in the sequence are unstable to rapid α-particle emission (green), ^{210}Bi decays to ^{206}Tl and ^{211}Bi to ^{207}Tl both of which proceed directly back to the lead isotopes of the same atomic weight by β-decay. In low-metallicity s-process environments, when neutrons outnumber heavy nuclei, lead can accumulate. In the r-process capture of neutrons proceeds without β-decays creating very neutron-rich isotopes up to the neutron-drip line beyond which no more neutrons can be added. Such nuclei are highly unstable to rapid β-decay which begins as soon as the neutron flux falls off. They approach the stable isotopes along paths (red) of constant atomic mass. Most of the stable elements shown can be reached by both the s- and r-processes. Except at very low metallicities the r-process is the dominant producer of these nuclides. Indeed nearly all the iridium and gold found in the solar system is thought to have an r-process origin. Some isotopes (marked r) such as ^{198}Pt and ^{204}Hg can only be formed by the r-process because β-decays are too fast for them to lie on the s-process route. Other isotopes (marked s) can only be formed by the s-process because they are protected from the r-process by a stable nucleus. For instance ^{192}Pt is protected by ^{192}Os while ^{204}Pb is protected by ^{204}Hg. There are also proton-rich isotopes (marked p), such as ^{190}Pt and ^{196}Hg, that cannot be reached by either the s- or the r-process. These p-nuclides are usually rare and their origin remains uncertain.

the s-process path marked. It can be seen that some isotopes can be formed by either process while others can only be formed by one. There are also some proton-rich isotopes that are protected by stable nuclei from both the s- and r-processes. These tend to be rare, making up only a tiny fraction of the terrestrial isotopic mix of a given element, and their site of production remains an interesting topic of research.

Stellar spectroscopy is not sufficiently sensitive to distinguish between the isotopes of such heavy elements but when they are trapped in tiny dust grains and brought to the Earth in meteorites from outside the solar system it is possible to measure relative abundances with high precision mass spectrometers. Each dust grain is likely to have formed in the atmosphere of a single star at a single time in its evolution. Thus the study of such grains is providing us with another powerful tool to understand the conditions in evolved stars.

6.6. Neutrino Losses

We have seen how neutrinos carry off part of the nuclear binding energy released in fusion reactions and escape with it from a star. At high temperatures and densities various other processes can spontaneously create neutrinos which again escape from the star making a negative energy generation contribution $\epsilon_\nu(\rho, T, X_i) < 0$. At high temperatures, $T > 10^9$ K, particle pairs form in equilibrium with radiation in the reaction,

$$\gamma + \gamma \leftrightharpoons e^+ + e^-. \tag{6.143}$$

But rarely, in about 1 in 10^{19} cases, the electron and positron can annihilate to form a neutrino and antineutrino pair,

$$e^+ + e^- \rightarrow \nu_e + \bar{\nu}_e \tag{6.144}$$

and both neutrinos escape from the stellar interior removing energy and therefore cooling the core. At the very high densities found in degenerate matter the material behaves as if it is a dielectric.

Travelling photons excite coherent charge oscillations and so acquire a dispersive energy–momentum relation of the form

$$\omega^2 = k^2 c^2 + \omega_0^2, \tag{6.145}$$

where ω is the angular frequency, related to energy by $E = \hbar\omega$, k is the wave number related to momentum by $p = \hbar k$ and ω_0 is the plasma frequency. Photons, in such cases called plasmons, can then behave as if they have mass and can decay directly to neutrino–antineutrino pairs

$$\gamma_{\text{plasmon}} \rightarrow \nu + \bar{\nu}, \tag{6.146}$$

a process that is forbidden by conservation of energy and momentum for non-dispersive photons. In a region in the temperature density plane not quite hot enough for pair production and not quite dense enough for plasmon decay the process of Compton scattering of photons by electrons can occasionally result in the emission of a neutrino-antineutrino instead of a photon,

$$\gamma + e^- \rightarrow e^- + \nu + \bar{\nu}. \tag{6.147}$$

Indeed, at high enough energies, any process involving electrons can be accompanied by neutrino losses. Bremsstrahlung scattering of electrons by nuclei becomes both important and dominant at the extremely high densities and relatively low temperatures of cold massive white dwarfs while recombination and ionisation of electrons in the most tightly bound atomic states of heavy elements contribute at lower temperatures and densities. The last electron bound to an iron nucleus ionises at about 2.6×10^6 K. By its nature this type of neutrino loss depends strongly on composition, becoming more important as nuclear burning proceeds.

Another mechanism that has been less well studied is known as the Urca process after a casino in Rio de Janeiro where Gamow and Schoenberg likened the inevitable energy loss to betting on a roulette table. It proceeds by two weak reactions that cycle between nuclides typically when $T > 10^8$K and $\rho > 10^7$g cm^{-3}. An example is the cycling from ^{20}Ne to ^{20}F and back,

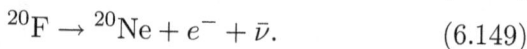

$$^{20}\text{Ne} + e^- \rightarrow {}^{20}\text{F} + \nu \tag{6.148}$$

$$^{20}\text{F} \rightarrow {}^{20}\text{Ne} + e^- + \bar{\nu}. \tag{6.149}$$

When the temperature is such that both reactions proceed at similar rates ^{20}Ne and ^{20}F remain in equilibrium but energy is constantly lost to neutrinos. The process is more complicated in convective regions where conditions for each reaction to proceed occur at different radii within the star. The correct neutrino loss rate in Urca processes requires a detailed knowledge of the composition of the stellar material as well as the reactions involved. Suitable neutrino loss rates have been calculated for hot dense material in NSE but rates at lower temperatures and densities are not yet routinely included in models of stellar evolution.

Because neutrino losses are larger in dense hot material they are most important in the cores of evolved stars. During and beyond shell helium burning the energy loss rates in a stellar core are such that it cools below the temperature of the burning shell leading to a region where $dT/dr < 0$, an inversion of the usual monotonic decrease of temperature from the core to the surface of the star.

6.7.   Questions

1. According to the Debye–Hückel theory the potential at a distance r from a nucleus of atomic number Z_i is

$$\phi_i = \frac{Z_i e}{4\pi\epsilon_0 r} \exp(-r/R_\mathrm{D}),$$

where the R_D is the Debye radius. In the case of weak screening

$$R_\mathrm{D} \gg r_\mathrm{c} \gg r_0,$$

where r_c is the Coulomb radius and r_0 is the nuclear radius. Show that the electrostatic potential energy between two nuclei with atomic numbers Z_i and Z_j, separated by r, can be approximated by

$$V(r) = \frac{Z_i Z_j e^2}{4\pi\epsilon_0 r} - V_0,$$

where V_0 is a constant.

The reaction rate between two nuclei is given by

$$R_{ij} \propto \int_0^\infty \sigma(E) v(E) P(E) f(E) \, \mathrm{d}E,$$

where $\sigma(E) = S(E)/E$ is the cross-section factor, with $S(E)$ a slowly varying function of E, $v(E) \propto E^{1/2}$ is the relative speed of the nuclei, $P(E)$ is the penetration factor and $f(E) \propto \exp(-E/kT)$ is the Maxwell–Boltzmann distribution for interaction energy E. The integrand is sharply peaked at the Gamow energy $E_\mathrm{G} > k_\mathrm{B} T \gg V_0$ in the weak screening case. Explain why, with such weak screening, the penetration factor is increased to

$$P(E) \approx P'(E + V_0),$$

where $P'(E)$ is the penetration factor in the absence of any screening. Deduce that weak screening increases the reaction rate by a factor

$$f_\mathrm{es} \approx e^{V_0 / k_\mathrm{B} T}.$$

Consider the central regions of a star composed entirely of protons and electrons. The potential felt by a proton at a distance r from another proton may be written as

$$\phi(r) = \frac{e}{4\pi\epsilon_0} \left(\frac{1}{r} + f(r) \right),$$

where the factor $f(r)$ accounts for the screening by the ambient charge distribution. Poisson's equation relates the perturbation to the potential to the mean ambient charge density $\bar{\rho}(r)$ by

$$\frac{e}{4\pi\epsilon_0} \nabla^2 f = -\frac{\bar{\rho}}{\epsilon_0}.$$

The mean ambient charge density is perturbed from neutrality because electrons find it more energetically favourable to be closer to the central proton than do other protons. In LTE this effect can be described by a Boltzmann distribution such that

$$\bar{\rho}(r) = n_e e \left\{ - \exp \left(\frac{+e\phi}{k_B T} \right) + \exp \left(\frac{-e\phi}{k_B T} \right) \right\},$$

where n_e is the number density of electrons when unperturbed. Use the fact that $e\phi \gg k_B T$ to deduce that

$$\nabla^2 f \approx \frac{2n_e e^2}{\epsilon_0 k_B T} \left(\frac{1}{r} + F \right).$$

State the boundary conditions $f(r)$ must satisfy as $r \to \infty$ and $r \to 0$ and deduce that

$$f(r) = \frac{1}{r} \left\{ \exp \left(-\frac{r}{R_D} \right) - 1 \right\},$$

with

$$R_D^2 = \frac{\epsilon_0 k_B}{2n_e e^2}.$$

Calculate R_D at the centre of such a pure hydrogen star at temperature $T = 1.56 \times 10^7$ K and mass density $148\,\mathrm{g\,cm^{-3}}$, similar to the Sun, and thence show that electron screening increases the rate of the proton–proton reaction by a little more than 5%.

2. In solar-like stars nuclear burning is dominated by the ppI- and ppII-chains,

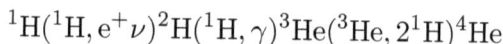

$$^1\mathrm{H}(^1\mathrm{H}, e^+\nu)^2\mathrm{H}(^1\mathrm{H}, \gamma)^3\mathrm{He}(^3\mathrm{He}, 2^1\mathrm{H})^4\mathrm{He}$$

and

$$^3\mathrm{He}(^4\mathrm{He}, \gamma)^7\mathrm{Be}(e^-, \nu)^{7\mathrm{Li}}(^1\mathrm{H}, {}^4\mathrm{He})^4\mathrm{He}.$$

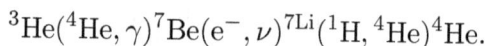

The reaction rate between species i and j is $\lambda_{ij} n_i n_j / (1 + \delta_{ij})$, where n_i is the number density of species i, δ_{ij} is the Kronecker delta and $\lambda_{ij} \propto \eta^2 e^{-\eta}$, where $\eta = 42.48(A Z_i^2 Z_j^2 T_6^{-1})^{1/3}$, $A = A_i A_j / (A_i + A_j)$ is the reduced atomic mass of the two reacting nuclei, Z_i is the atomic number of species i and the T_6 is related to temperature T by $T_6 = T/10^6$ K. The beta decay of $^7\mathrm{Be}$ is fast compared to all other reactions so that ^{7}Li is the predominant species of atomic mass 7 and all major species can be identified by $i \approx A_i$. Show that the temperature dependence of the rate r_{11} at the centre of the Sun, where $T_6 \approx 15$, of the reaction ^{1}H(^{1}H,$e^+\nu$)^{2}H can be written as $r_{11} \propto T^\alpha$, where $\alpha = \frac{1}{3}(\eta - 2) \approx 4$. Also show that β and γ are approximately 16 (with $\gamma > \beta$) in the expressions $r_{33} \propto T^\beta$ and $r_{34} \propto T^\gamma$.

Show that the rate of change of protons obeys

$$\frac{dn_1}{dt} = -\lambda_{11}n_1^2 - \lambda_{21}n_2n_1 + \lambda_{33}n_3^2 - \lambda_{17}n_1n_7,$$

and obtain the equivalent equations for n_2, n_3 and n_4.

At the centre of the Sun the characteristic time-scale of r_{11} is about 10^{10} yr while that of r_{12} is about 1 s. The characteristic time-scale for n_3 to reach equilibrium is $\tau \approx 6 \times 10^5$ yr. By making an appropriate approximation, which you should explain, show that

$$\frac{dn_1}{dt} \approx -\frac{3}{2}\lambda_{11}n_1^2 + \lambda_{33}n_3^2 - \lambda_{17}n_1n_7$$

and

$$\frac{dn_3}{dt} \approx \frac{1}{2}\lambda_{11}n_1^2 - \lambda_{33}n_3^2 - \lambda_{34}n_3n_4$$

near the centre of the Sun.

Show further that in equilibrium $n_3 \approx n_{3e}$ where

$$n_{3e} = -\frac{\lambda_{34}n_4}{2\lambda_{33}} + \sqrt{\left(\frac{\lambda_{34}n_4}{2\lambda_{33}}\right)^2 + \frac{\lambda_{11}n_1^2}{2\lambda_{33}}}.$$

Consider a small perturbation of the form $n_3 = n_{3e} + x$ about this equilibrium and linearise the evolution equation for n_3 to obtain $dx/dt = -x/\tau$, where $\tau = (2\lambda_{33}n_{3e} + \lambda_{34}n_4)^{-1}$. Estimate the temperature at which τ is comparable to the age of the Sun. Sketch the abundances X_1 and X_3 of ^{1}H and ^{3}He as a function of radius in the Sun today.

3. At very high temperatures nuclear burning leads to an equilibrium distribution of isotopes around the iron group. In this nuclear

statistical equilibrium, the abundance by number of an isotope ${}^A_Z S$ is

$$N(A, Z) \propto n_{\mathrm{p}}^Z n_{\mathrm{n}}^{A-Z} T^{3(1-A)/2} \exp\left[\frac{Q(A, Z)}{kT}\right],$$

where $Q(A, Z)$ is the binding energy of the isotope, n_{p} is the number density of free protons and n_{n} that of free neutrons. Derive an expression for the ratio of the abundances of ^{56}Ni and ^{54}Fe that transmute according to the reversible reaction

$$ {}^{56}_{28}\mathrm{Ni} \rightleftharpoons {}^{54}_{26}\mathrm{Fe} + 2\mathrm{p}. $$

As the reactions proceed the ratio of the mean number of protons to the mean number of neutrons $\langle Z \rangle / \langle N \rangle$ is preserved. When $\langle Z \rangle / \langle N \rangle = 1$ and all material is converted to either ^{56}Ni, ^{54}Fe or protons what is n_{p}? Estimate the temperature at which $N({}^{56}\mathrm{Ni})/[N({}^{54}\mathrm{Fe})]^3$ is a maximum.

4. How many *r*-only nuclei are there of a given atomic mass number?

Chapter 7

Stellar Models

We have assembled the necessary physics to make detailed models of stars. To do so we solve the corresponding set of nonlinear partial differential equations gathered together in this chapter. We also discuss numerical techniques in common use. In Chapter 8 we describe typical results of such calculations in detail while in the final sections of this chapter we look at analytical approximations that can give us significant insight into both the structure and evolution of stars without the need of detailed solutions.

For stellar evolution it is more useful to use the mass coordinate m, rather than the radius r, because, when there is no mixing, each mass shell carries its material with it as the star expands or contracts. So we write each of the variables as functions of m and time t. The radius $r(m,t)$ is then the radius of the sphere that encloses mass m. Partial time derivatives are Lagrangian derivatives of quantities moving with each mass shell. They can be related to Eulerian derivatives by

$$\left(\frac{\partial}{\partial t}\right)_m \equiv \frac{D}{Dt} = \left(\frac{\partial}{\partial t}\right)_r + u_r \left(\frac{\partial}{\partial r}\right)_t, \qquad (7.1)$$

where u_r is the radial speed of the shell enclosing mass m. The assumption of hydrostatic equilibrium requires that u_r necessarily be small and certainly not changing at an important rate. So the mass equation becomes

$$\left(\frac{\partial r}{\partial m}\right)_t = \frac{1}{4\pi r^2 \rho} \tag{7.2}$$

and can be used to convert derivatives with respect to radius to derivatives with respect to mass at constant time. Note that this Eq. (7.2) is singular as $r \rightarrow 0$ and so care must be taken with the central boundary conditions. Hereinafter when we write partial derivatives with respect to m they are implicitly at constant t and similarly partial derivatives with respect to t are at constant m. The equation of hydrostatic equilibrium can be written as

$$\frac{\partial P}{\partial m} = -\frac{Gm}{4\pi r^4}, \tag{7.3}$$

the equation for the temperature gradient becomes

$$\frac{\partial T}{\partial m} = \begin{cases} -\dfrac{3\kappa\rho L_r}{64\pi^2 acr^4 T^3} & \text{radiative,} \\[2ex] \nabla_a \dfrac{T}{P}\dfrac{\partial P}{\partial m} + \dfrac{\Delta\nabla T}{4\pi r^2 \rho} & \text{convective} \end{cases} \tag{7.4}$$

and the luminosity gradient

$$\frac{\partial L_r}{\partial m} = \epsilon - T\frac{\partial s}{\partial t}. \tag{7.5}$$

These are supplemented with the equation of state

$$P = P(\rho, T, \{X_i\}), \quad s = s(\rho, T, \{X_i\}), \quad \ldots, \tag{7.6}$$

the radiative opacity

$$\kappa = \kappa(\rho, T, \{X_i\}) \quad \text{radiative} \tag{7.7}$$

or the superadiabatic gradient

$$\Delta\nabla T = \Delta\nabla T(\rho, T, \{X_i\}, L_r) \quad \text{convective,} \tag{7.8}$$

and the energy generation rate

$$\epsilon = \epsilon(\rho, T, \{X_i\}), \tag{7.9}$$

all of which can be evaluated locally. Central boundary conditions, at $m = 0$, are

$$r = 0, \tag{7.10}$$

$$L_r = 0 \quad \text{and} \tag{7.11}$$

$$\frac{\partial P}{\partial m} = 0. \tag{7.12}$$

The equations for $\partial r/\partial m$ and $\partial P/\partial m$ are singular at the centre so, in a numerical solution, Taylor expansions are used to apply the boundary conditions at a small but finite mass (see Question 1). At the surface of the star

$$m = M \tag{7.13}$$

and various levels of sophistication can be applied to the atmospheric boundary condition. In the simplest case

$$P = T = 0 \tag{7.14}$$

but more often the Eddington closure form derived in Chapter 5,

$$L_r = 4\pi r^2 \sigma T^4 \quad \text{and} \tag{7.15}$$

$$P_{\rm g} = \frac{2}{3}\frac{g}{\kappa}\left(1 - \frac{L}{L_{\rm Edd}}\right), \tag{7.16}$$

is applied. More precisely a full model atmosphere that relates the variables at the surface to effective temperature and surface gravity may be used. When mass loss or gain is included, Eq. (7.13) must be replaced by

$$\frac{dm}{dt} = \dot{M}, \tag{7.17}$$

where $-\dot{M}$ is the mass loss rate from the surface of the star, or an even more sophisticated non-static atmosphere boundary condition could be used.

7.1.   Time Dependence and Stellar Evolution

Without the time dependence we have a set of ordinary differential equations that may be used to model the structure of a static star. With time, evolution is driven both by the gravitational energy generation term in Eq. (6.2), which dominates before nuclear fusion can support the star, and by composition changes once fusion has begun. Once the composition is non-uniform we must include the possibility of mixing over mass. To formulate the equations governing composition changes, we consider a volume V co-moving with mass m, and enclosed by a surface S, in which the number density of particles of species i is n_i. The total number of such particles in the volume varies both because nuclear reactions destroy and create particles and because various mixing processes move particles in and out of V. We may write

$$\frac{\mathrm{d}}{\mathrm{d}t} \int_V n_i \, \mathrm{d}V = \int_V (-n_i R_i + s_i) \, \mathrm{d}V - \int_S n_i \boldsymbol{v}_i \cdot \mathrm{d}\boldsymbol{S}, \qquad (7.18)$$

where $n_i R_i$ is the rate per unit volume at which nuclear reactions destroy species i summed over all such reactions, s_i is the rate per unit volume at which nuclear reactions produce species i summed over all such reactions and $\boldsymbol{v}_i$ is the typical diffusion velocity of particles of species i. Both R_i and s_i are functions of state and composition. Applying the divergence theorem to the last term and shrinking V to infinitesimal size we deduce that locally

$$\frac{\partial n_i}{\partial t} = -n_i R_i + s_i - \boldsymbol{\nabla} \cdot (n_i \boldsymbol{v}_i). \qquad (7.19)$$

Electrons diffuse with nuclei so as to maintain charge neutrality so the mass fraction X_i of species i is related to n_i by

$$n_i = \frac{\rho X_i}{A_i m_{\mathrm{H}}}, \qquad (7.20)$$

where A_i is the atomic mass of species i and m_H is the mass of the hydrogen atom. In general particles of species i diffuse down a composition gradient at a velocity that increases monotonically with the magnitude of that gradient so we define a diffusion coefficient D_i

such that

$$\boldsymbol{v}_i = -\frac{D_i}{X_i}\boldsymbol{\nabla}X_i, \tag{7.21}$$

and rewrite Eq. (7.19) as

$$\frac{\partial \rho X_i}{\partial t} = -\rho X_i R_i + \rho S_i + \boldsymbol{\nabla}\cdot(\rho D_i \boldsymbol{\nabla}X_i), \tag{7.22}$$

where we have defined $S_i = A_i m_{\rm H} s_i/\rho$. The change in total mass of particles owing to nuclear reactions is negligible and diffusion acts in such a way that the local density is unchanged so that

$$\frac{\partial \rho}{\partial t} = 0 \tag{7.23}$$

and we may divide by ρ to arrive at the equations for the evolution of mass fractions in a spherically symmetric star of the form

$$\frac{\partial X_i}{\partial t} = -X_i R_i + S_i + \frac{1}{\rho}\boldsymbol{\nabla}\cdot(\rho D_i \boldsymbol{\nabla}X_i) \tag{7.24}$$

$$= -X_i R_i + S_i + \frac{1}{\rho r^2}\frac{\partial}{\partial r}\left(\rho r^2 D_i \frac{\partial X_i}{\partial r}\right) \tag{7.25}$$

$$= -X_i R_i + S_i + \frac{\partial}{\partial m}\left(\sigma_i \frac{\partial X_i}{\partial m}\right), \tag{7.26}$$

where

$$\sigma_i = (4\pi r^2 \rho)^2 D_i. \tag{7.27}$$

For convection we may obtain an effective D_i from mixing length theory. Suppose there is a composition gradient $\partial X_i/\partial r$ in a convective region. During the turnover of a convective cell of length l, moving at speed v, material of mass fraction $X_i + \Delta X_i$ is exchanged with material of mass fraction X_i in a time l/v. Essentially, in a cell of mass m, a mass $m\Delta X_i$ has moved $-l$ in radius at speed v. This is equivalent to material of mass mX_i moving at the diffusion speed

$-v_i$ and we may write

$$v_i \approx -v\frac{\Delta X_i}{X_i} \approx -vl\frac{1}{X_i}\frac{\partial X_i}{\partial r}. \qquad (7.28)$$

So averaging over convective cells and using Eq. (4.91) we have, for all species,

$$D_i \approx \langle v\delta r\rangle \approx \frac{1}{4}\left(g\frac{\Delta\nabla T}{T}\right)^{\frac{1}{2}}l^2, \qquad (7.29)$$

from mixing length theory of convection. Just as for the heat transport there are a number of different formulations for D_i within mixing length theories.

In the solar convection zone $D_i \approx 2 \times 10^{11}\,\mathrm{cm^2\,s^{-1}}$ so that

$$\tau_{\mathrm{mix}} \approx \frac{R_\odot^2}{D_i} = 2 \times 10^{10}\,\mathrm{s} \approx 800\,\mathrm{yr} \ll \tau_{\mathrm{nuc}}. \qquad (7.30)$$

This time-scale is sufficiently fast that the convective envelope of the Sun is fully mixed, as are most convection zones in stars. So, again, models do not greatly depend on the particular theory of convective transport that is used. This assumption breaks down when the nuclear burning time-scale approaches that of convection, as is the case in the late burning stages of many stars, and then modelling mixing on the correct time-scale becomes important. Other mixing processes, such as rotationally driven or thermohaline mixing, can also be included by calculating appropriate diffusion coefficients.

We require two boundary conditions for each composition equations. These are usually taken to be

$$\frac{\partial X_i}{\partial m} = 0 \qquad (7.31)$$

at the centre and surface of the star. However if a star is accreting material the boundary condition at the surface must be such that material is accreted with the correct composition.

7.2.   Methods of Solution

In constructing detailed stellar structure or evolution models we encounter problems because nonlinear partial differential equations behave unpredictably. In general full physical models require numerical solutions because no analytic solution exists. Each of the equations is written as a numerical difference over finite steps usually in mass and time though the code we use to make the models presented in Chapter 8 has a mesh that moves relative to m and we write $m = m(k)$ where k is the mesh point number.

7.2.1. *Shooting*

When only a single structure model is required, a shooting method can be used to solve the equations. For a given composition as a function of mass we can guess the unknown variables at the surface, $L_r = L$ and $r = R$ when $m = M$, and at the centre, $T = T_c$ and $\rho = \rho_c$, or alternative state variables, when $m = 0$. We then integrate outwards from the centre and inwards from the surface in m towards a chosen mass shell $m = m_{\text{match}}$ in the stellar interior (see Fig. 7.1). On

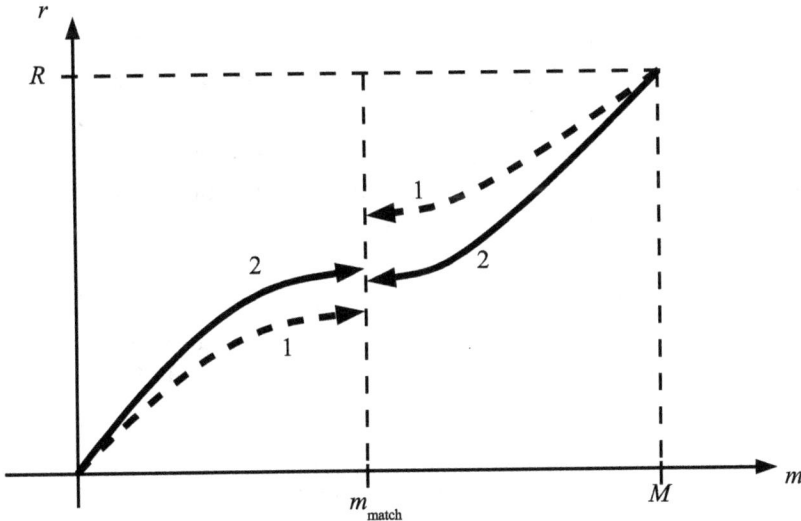

Fig. 7.1. Illustration of the shooting method for the solution of the equations of stellar structure. Variables not fixed by the boundary conditions are guessed at the centre and surface. The equations are numerically integrated to some meeting point in the middle. When they do not meet, the guessed variables at the boundaries are changed so as to decrease the discrepancy at m_{match}. This procedure is iterated until some desired accuracy at m_{match} is reached.

the initial guess the variables and their derivatives calculated from the two integrations do not match at this shell. By making small variations to the guesses, we can find derivatives of the variables at the matching shell with respect to changes in our boundary guesses and then use a Newton–Raphson method to converge on the solution.

Shooting is inefficient when we already have a solution close to the one we want, for instance when we model a homologous set of stars such as the zero-age main sequence or when we evolve a single star in time. It is however useful to create models *ab initio*.

7.2.2.   *Relaxation*

Today nearly all stellar evolution calculations are made with a technique in which an existing model is relaxed into a similar model. The star is divided into spherical shells distributed so as to resolve changes in the variables and minimise these changes from one shell to the next. Typically the number N_{mesh} of shells, or mesh points, needed is somewhere between 100 and 2000 with more required in later phases of evolution when the stellar structure is more complicated. A model consists of each of the variables at each mesh point through the star. For a simple stellar model five variables, say m, r, L, ρ, T plus n_{comp} composition variables[1] are required at each mesh point. Typically abundances of all isotopes that grossly affect stellar structure are given separate variables. A minimal set to model to the end of helium burning might be ^{1}H, ^{3}He, ^{4}He, ^{12}C and ^{16}O, with the abundance of ^{14}N calculated from the initial CNO abundances and assumptions about its destruction during helium burning. More compositions must be added if stars are to

---

[1] It is possible to work with only a single composition variable, the meaning of which is changed depending which phase of burning the particular shell is in or approaching. Up until hydrogen exhaustion it can be taken to be the hydrogen abundance X. The metallicity Z can be assumed fixed, with a fixed combination of isotopes and the helium abundance $Y = 1 - X - Z$. Once hydrogen is exhausted the composition variable becomes Y and the abundances of carbon and oxygen calculated assuming that they are produced by helium burning in a constant ratio. With fast computers with large memories this is no longer necessary.

be evolved further or whenever other reactions, such as deuterium and lithium burning in pre-mainsequence stars, affect the stellar structure. Suppose the mesh is fixed in m. The total number of variables v_i required to model the star at any given time is then

$$N_V = (4 + n_{comp})N_{mesh}. \qquad (7.32)$$

There must therefore be a set of $N_E = N_V$ difference equations $\{E_i\}$, including those for the boundary conditions. These consist of the four Eqs. (7.2)–(7.5) between each neighbouring pair of the $N_{mesh} - 1$ shells and n_{comp} equations (of the form 7.26) for each set of three neighbouring shells, $N_{mesh} - 2$ sets, plus the boundary conditions. The equations give us

$$4(N_{mesh} - 1) + n_{comp}(N_{mesh} - 2) = N_V - 4 - 2n_{comp} \qquad (7.33)$$

difference equations. The two boundary conditions for each composition variable provide a further $2n_{comp}$ equations and at the centre of the star we have two boundary conditions from $r = 0$, $L_r = 0$ and at the surface two from the atmosphere such as (7.15) and (7.16). It can also be useful to allow the mesh to move relative to the mass coordinate so that mesh points are moved closer together when variables change faster with mass. This is also a convenient way to add or remove material from the star's surface. In this case a mesh spacing equation is evaluated between each $N_{mesh} - 1$ pairs of neighbouring shells and boundary condition (7.17) completes the set. We write all $N_E = N_V$ equations in the form

$$E_i(\{v_j\}) = 0. \qquad (7.34)$$

The solution is found by a Newton–Raphson technique originally developed by Wilets and first applied to stellar structure by Henyey *et al.* (1959). An initial guess, almost exclusively determined from an existing model, is made for the set of N_V variables $\{v_j\}$. This does not solve the equations precisely but gives a set of errors

$$\delta E_i = E_i(\{v_j\}). \qquad (7.35)$$

A matrix **A**, of derivatives of each error with respect to each variable, with elements

$$A_{ij} = \frac{\partial E_i}{\partial v_j} \qquad (7.36)$$

is calculated either by evaluation of analytic derivatives or by making small changes to each variable and evaluating the corresponding change to δE_i. The change in the variables $\delta V = (\delta v_1, \delta v_2, \ldots, \delta v_{N_V})^\top$ necessary to solve the equations exactly can be estimated by evaluating

$$\delta V = -\mathbf{A}^{-1} \delta E, \qquad (7.37)$$

where $\delta E = (\delta E_1, \delta E_2, \ldots, \delta E_{N_E})^\top$, and iterating until the desired accuracy is achieved. When the equations only depend on neighbouring mesh points the matrix **A** can be constructed so that it is block diagonal[2] and can easily be inverted by Gaussian elimination. The arithmetic complexity of this inversion is $O(N_{\text{mesh}} n_v^3)$ where n_v is the total number of variables required at each mesh point. In general care must be taken in the construction of **A** in order to avoid singularities and conditioning is often required to ensure convergence.

In time the equations are stiff and so time derivatives must be calculated implicitly to maintain stability. Though a first-order scheme of the form

$$y(t + \delta t) - \dot{y}(t + \delta t)\delta t = y(t) \qquad (7.38)$$

is normally used higher order schemes could be employed. A star is essentially evolved quasistatically with the relaxation applied at the end of each time step. However it is often, but not always, efficient to calculate the initial guess for the for the Newton–Raphson iterations explicitly in the form

$$y(t + \delta t) = y(t) + \dot{y}(t)\delta t. \qquad (7.39)$$

---

[2]Modern numerical techniques to invert sparse matrices achieve similar results without the requirement that **A** be block diagonal. So the need for all equations to be locally evaluated can be relaxed.

The rate of evolution with time varies considerably as a star evolves and so time steps must be chosen by means of an algorithm that ensures that no variable anywhere in the star changes too much. The choice of time step should also be made implicitly but in practice it usually suffices to choose the next time step based on changes over the previous one.

7.3. Homology

In general the evolution of a star is non-homologous because nuclear burning is localised. So it is hard to predict the behaviour in advance. Equally *ad hoc* explanations of stellar evolution often prove to be wrong beyond the certain special cases for which they were formulated. Nevertheless there are many aspects of stars and their evolution that can be understood for sequences that either are almost homologous or can be assumed to be close to homologous. To exploit this we begin by scaling all the variables to obtain a dimensionless set. For example for mass, radius, density and pressure we write

$$m = m'M, \tag{7.40}$$
$$r = r'R, \tag{7.41}$$
$$\rho = \rho'\rho_c \quad \text{and} \tag{7.42}$$
$$P = P'P_c, \tag{7.43}$$

where M is the total mass of the star, R is its radius and ρ_c and P_c are its central density and pressure. The mass equation 2.2 can then be written as

$$\frac{dm}{dr} = \frac{M}{R}\frac{dm'}{dr'} = 4\pi r'^2 \rho' R^2 \rho_c \tag{7.44}$$

or equivalently

$$\frac{dm'}{dr'} = 4\pi r'^2 \rho' k(M,t), \tag{7.45}$$

where

$$k(M,t) = \frac{R^3 \rho_c}{M} \tag{7.46}$$

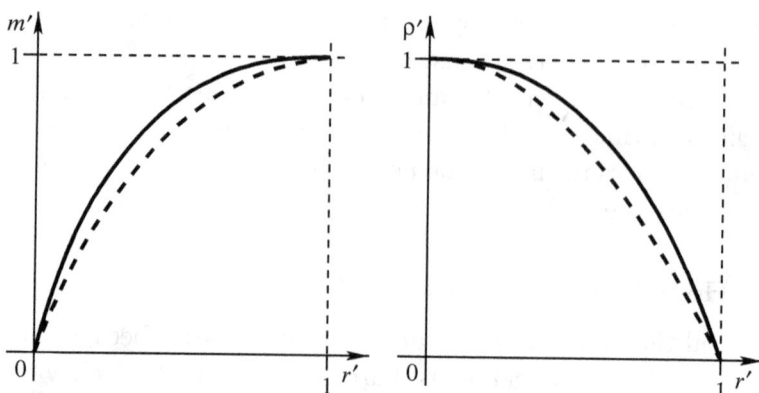

Fig. 7.2. The variation of m' and ρ' with r' in two stars with similar mass. All homologous stars lie on a single curve, such as the solid line, in these plots. Real stars are not quite homologous but a sequence in which the structure does not vary significantly, as for the dashed lines, can be treated as almost homologous.

depends only on the mass and evolutionary state of the star. For stars of uniform composition, as at the zero-age main sequence, k is a function of mass only. Now let us suppose that we have a homologous set of stars in which the dimensionless variables vary in an identical way. As an example Fig. 7.2 shows the dependence of m' and ρ' on r' inside two zero-age main-sequence stars of slightly different mass. The differences in the functional forms $\rho'(r')$ and $m'(r')$ are slight so that we may further argue that $k(M) \approx$ const. Note that this equality is precise for polytropes of the same index n.

This kind of analysis is particularly useful for the zero-age main sequence. By definition such stars have uniform composition, are in thermal equilibrium and are non-rotating. A sequence with the same uniform composition can be expected to have a structure that depends only on mass. The misstyled Vogt–Russell theorem asserts that this is indeed the case but has never been proved. Also in real stars such a state is never perfectly reached because minor reactions in the pp-chain that destroy deuterium and lithium take place in the cores of stars before hydrogen fusion takes over from gravitational binding energy as the major source of luminosity. At masses above that of the Sun equilibrium among the CNO elements begins during this pre-mainsequence evolution too. Nevertheless the variations in composition are minor and local homologous sequences do provide a useful description of the main sequence. So let us assume $k = \text{const}$ and thence that

$$\rho_c \propto \frac{M}{R^3}. \tag{7.47}$$

Similarly, just as the mass equation

$$\frac{dm}{dr} = 4\pi r^2 \rho \quad \Rightarrow \quad \frac{M}{R} \propto R^2 \rho_c, \tag{7.48}$$

we can deduce that hydrostatic equilibrium Eq. (2.10)

$$\frac{dP}{dr} = -\frac{Gm\rho}{r^2} \quad \Rightarrow \quad \frac{P_c}{R} \propto \frac{M\rho_c}{R^2} \tag{7.49}$$

so that

$$P_c \propto \frac{M^2}{R^4}. \tag{7.50}$$

The heat transport equations (4.35) or (4.95) require a bit more care but energy generation (Eq. 6.2),

$$\frac{dL}{dr} = 4\pi r^2 \rho\epsilon \quad \Rightarrow \quad \frac{L}{R} \propto R^2 \rho_c\epsilon, \tag{7.51}$$

where we may also approximate

$$\epsilon = \epsilon_0 \rho^\nu T^\eta \tag{7.52}$$

as discussed in Sec. 6.3.4.1. Thence Eq. (7.51) becomes

$$\frac{L}{R} \propto R^2 \rho_c^{\nu+1} T_c^\eta, \tag{7.53}$$

where T_c is the central temperature.

7.3.1. *Zero-age solar-like stars*

Now assume that zero-age main-sequence stars with masses close to that of the Sun have a homologous structure. Through most of the star the equation of state is well modelled as an ideal gas with a fixed mean molecular weight so we assume this throughout and in particular at the centre of each star so that we may write

$$P = \frac{\rho \mathcal{R} T}{\mu} \quad \Rightarrow \quad P_c \propto \rho_c T_c. \tag{7.54}$$

This we combine with Eqs. (7.50) and (7.47) to obtain

$$T_c \propto \frac{P_c}{\rho_c} \propto \frac{M}{R}. \tag{7.55}$$

By mass, the convective envelope of the Sun is small so we further assume that the whole star can be reasonably modelled as fully radiative and then the equation of radiative transfer (4.35)

$$\frac{dT}{dr} = -\frac{3\kappa\rho L_r}{16\pi a c r^2 T^3} \quad \Rightarrow \quad \frac{T_c}{R} \propto \frac{\rho_c L \kappa_c}{R^2 T_c^3}. \tag{7.56}$$

The important opacity source is not the electron scattering that dominates at the centre of the Sun but rather a Kramers' form that applies to the bulk of the radiative heat transport and so is most important for the stellar structure. We therefore write

$$\kappa = \kappa_0 \rho T^{-3.5} \quad \Rightarrow \quad \kappa_c \propto \rho_c T_c^{-3.5}, \tag{7.57}$$

with which Eq. (7.56) gives us

$$L \propto \frac{\mathcal{R} T_c^{7.5}}{\rho_c^2}. \tag{7.58}$$

Using Eqs. (7.47) and (7.55) we can eliminate ρ_c and T_c in favour of M and R to arrive at

$$L \propto \frac{M^{5.5}}{R^{0.5}}. \tag{7.59}$$

Finally we recall that the pp-chain dominates hydrogen burning at the centre of the Sun with

$$\epsilon = \epsilon_0 \rho T^4 \tag{7.60}$$

so that the relation (7.53) becomes

$$L \propto R^3 \rho_c^2 T_c^4 \propto \frac{M^6}{R^7}. \qquad (7.61)$$

Combining Eqs. (7.59) and (7.61) we can eliminate either L or R to find

$$R \propto M^{1/13}, \qquad (7.62)$$

$$L \propto M^{71/13} \approx M^{5.5}. \qquad (7.63)$$

We deduce that radius varies only weakly with mass while luminosity increases very strongly for stars similar to the Sun. To determine how this sequence appears in the Hertzsprung–Russell diagram we require the effective temperature T_e. We are careful not to confuse T_e with the central temperature by which we have scaled the whole star. Instead we must use the boundary condition (7.15), from which

$$L \propto R^2 T_e^4. \qquad (7.64)$$

Thence

$$T_e^4 \propto L R^{-2} \propto M^{69/13} \qquad (7.65)$$

and

$$T_e \propto M^{69/52} \propto L^{69/284} \approx L^{1/4}, \qquad (7.66)$$

which can be usefully expressed as

$$\log L \approx 4 \log T_e + \text{const}, \qquad (7.67)$$

consistent with the almost constant radius of these stars. In the H–R diagram this sequence of stars appears a straight line as in Fig. 7.3. However in reality the deepening convective envelopes of stars less massive than the Sun mean they are not well modelled as a homologous sequence and this slope does not fit well with either what is observed or what is found from detailed models.

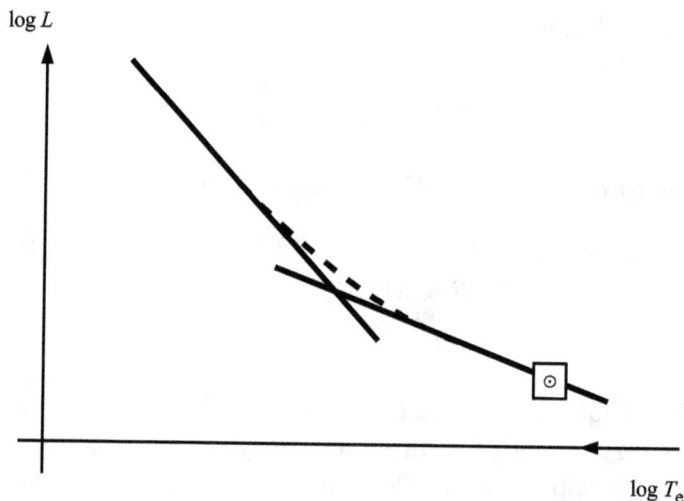

Fig. 7.3. A schematic Hertzsprung–Russell diagram showing homologous sections of the zero-age main sequence. The low-luminosity, low-mass slope through the Sun is shallower than that for more massive stars.

7.3.2. *Higher masses*

As the mass increases so $T_c \propto M/R$ rises. Above $2 \times 10^7 \, \mathrm{K}$ the CNO cycle dominates and

$$\epsilon = \epsilon_0 \rho T^{16}. \tag{7.68}$$

The ideal gas equation of state remains valid until radiation pressure becomes important but this is not until $M \approx 100 \, M_\odot$. Though the cores of these stars become convective their structure is still dominated by radiation transport through most of their mass. However the higher temperatures do mean that electron scattering takes over as the dominant source of opacity and

$$\kappa = \mathrm{const.} \tag{7.69}$$

Eqs. (7.47), (7.50) and (7.55) are all still valid. Radiative transfer now gives us

$$L \propto M^3 \tag{7.70}$$

directly and energy generation yields

$$R \propto M^{15/19}. \tag{7.71}$$

Thence we have

$$L \propto T_{\mathrm{e}}^{76/9} \tag{7.72}$$

or

$$\log L \approx 8.4 \log T_{\mathrm{e}} + \text{const.} \tag{7.73}$$

So the slope of the main sequence steepens compared to solar-like stars (Fig. 7.3). Without convective envelopes these more massive stars are more homologous than the lower mass stars and this is a better representation of their behaviour in the H–R diagram.

7.3.3. *Stellar lifetimes*

For homologous stars the total mass of hydrogen fuel available for burning is just proportional to the mass of the star. The luminosity is a measure of the rate at which this fuel is consumed. So we can find the dependence of main-sequence lifetime on mass, and assume that the total nuclear lifetime of a star is not much longer than this, to estimate that

$$\tau_{\mathrm{nuc}} \propto \frac{M}{L} \propto \begin{cases} M^{-4.5} & \text{solar-like stars,} \\ M^{-2} & \text{massive stars.} \end{cases} \tag{7.74}$$

More massive stars are so much more luminous that they have shorter lifetimes than less massive stars even though they start their lives with more fuel.

7.3.4. *Fully convective stars*

The situation is more complicated when the major energy transport mechanism through a star is convection. A fully convective star behaves as an $n = 3/2$ polytrope and so a homologous sequence does apply well but we must be careful because the surface of a star must radiate luminosity to space. It is this radiative atmosphere that determines the adiabat on which the interior of the star must lie.

Recall that deep within a convective star the temperature gradient is adiabatic (Sec. 4.4.3)

$$\frac{\mathrm{d}T}{\mathrm{d}r} = \nabla_\mathrm{a} \frac{T}{P} \frac{\mathrm{d}P}{\mathrm{d}r} \qquad (7.75)$$

so that the bulk of the star is isentropic with $P = K\rho^\gamma$, where $\gamma = 5/3$ for an ionised mixture and $K = \mathrm{const}$. This is similar to the non-relativistic white dwarfs we modelled in Chapter 3 but, unlike that case, the constant K is not fixed by the equation of state.

The surface boundary condition (7.16) encapsulates the essence of the radiative atmosphere relating the surface pressure P_s to the gravity g. Deep convective envelopes generally exist because hydrogen ionisation drives down ∇_a. They are therefore cool, $L \ll L_\mathrm{Edd}$ and radiation pressure can be neglected. Thus

$$\kappa P_\mathrm{s} = \frac{2}{3} g = \frac{2}{3} \frac{GM}{R^2}, \qquad (7.76)$$

when $T = T_\mathrm{e}$. We also write $\rho = \rho_\mathrm{s}$ at this stellar photosphere. At the typical low temperatures of deep convective envelopes the H$^-$ opacity dominates and

$$\kappa = \kappa_0 \rho^{1/2} T^9 \qquad (7.77)$$

so that, at the surface,

$$\kappa_0 \rho_\mathrm{s}^{1/2} T_\mathrm{e}^9 P_\mathrm{s} = \frac{2}{3} \frac{GM}{R^2}. \qquad (7.78)$$

The material behaves as an ideal gas and also lies on the same adiabat as the stellar interior so that

$$P_\mathrm{s} = \frac{\mathcal{R} \rho_\mathrm{s} T_\mathrm{e}}{\mu} \qquad (7.79)$$

and

$$P_\mathrm{s} = K \rho_\mathrm{s}^{5/3}. \qquad (7.80)$$

From these equations we deduce that

$$\rho_\mathrm{s} \propto T_\mathrm{e}^{3/2} K^{-3/2}, \qquad (7.81)$$
$$P_\mathrm{s} \propto T_\mathrm{e}^{5/2} K^{-3/2} \qquad (7.82)$$

and thence that

$$K \propto \left(M^{-4} R^8 T_{\mathrm{e}}^{49} \right)^{\frac{1}{9}}. \tag{7.83}$$

We can now use the homologous structure to relate the centre of the star to its surface. Equations (7.48) and (7.49) for density and pressure still apply and now we can add

$$P_{\mathrm{c}} = K \rho_{\mathrm{c}}^{5/3} \;\Rightarrow\; \frac{M^2}{R^4} \propto K \frac{M^{\frac{5}{3}}}{R^5} \tag{7.84}$$

which simplifies to

$$K \propto R M^{\frac{1}{3}}, \tag{7.85}$$

as expected for a polytrope of $n = 3/2$. Eliminating K we arrive at

$$T_{\mathrm{e}} \propto M^{1/7} R^{1/49} \quad \text{or} \quad R \propto M^{-7} T_{\mathrm{e}}^{49}. \tag{7.86}$$

The photospheric boundary condition (7.15) gives

$$L \propto R^2 T_{\mathrm{e}}^4 \propto M^{-14} T_{\mathrm{e}}^{102} \tag{7.87}$$

so that, for a fixed mass,

$$\log L = 102 \log T_{\mathrm{e}} + \text{const} \tag{7.88}$$

which, relative to the main sequence, is a near vertical line in the H–R diagram (Fig. 7.4). A sequence of close parallel tracks at increasing temperatures corresponds to a sequence of increasing total stellar masses.

Pre-mainsequence stars, with no central nuclear energy source are fully convective and contract as they radiate their gravitational energy. A pre-mainsequence star descends such a track, known as a Hayashi track, in the H–R diagram until either hydrogen fusion begins or it develops a radiative core. In practice Hayashi tracks are not exactly straight lines because the opacity dependence on temperature and density varies as the star becomes more compact. Note that for a star of fixed mass and radius, in hydrostatic equilibrium, movement to the right, to lower temperature, in the H–R diagram would require a superadiabatic temperature gradient throughout the star and this is not sustainable with efficient convection. For this

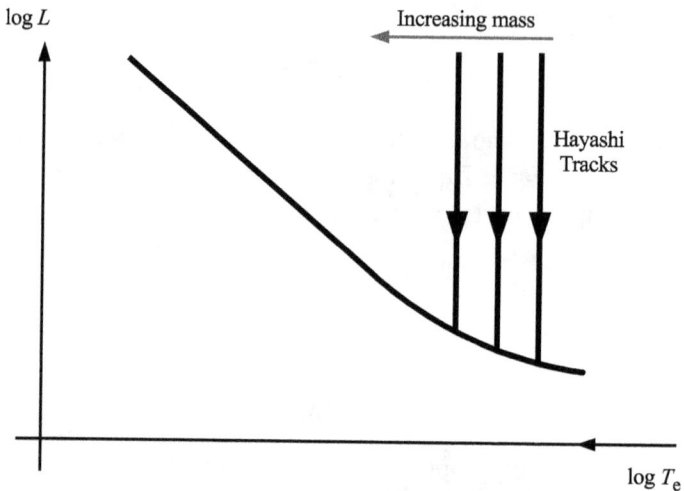

Fig. 7.4. Hayashi tracks of fully convective stars. A star of a given mass cannot lie to the right of its Hayashi track which it follows when fully isentropic. Pre-mainsequence stars descend these tracks as they radiate their gravitational energy and evolve towards the main sequence. Later in their lives stars expand and cool as red giants ascending similar tracks as their burnt out cores grow and their luminosities increase.

reason stars cannot lie to the right of their appropriate Hayashi track which is therefore sometimes referred to as the Hayashi limit.

7.3.5. *Red giants*

Red giants have burnt out degenerate cores, effectively hot white dwarfs at their centres. Nuclear burning proceeds in one or more hot shells in a radiative region of small mass just outside the core and the whole is surrounded by a deep massive convective envelope (Fig. 7.5). The evolutionary track of a giant in the H–R diagram is dominated by the structure of its convective envelope which is very similar to that of a fully convective star. As nuclear burning proceeds material is added to the core which, being degenerate or approaching degeneracy, shrinks. To maintain pressure support the temperature of the core and burning shells rises with a corresponding increase in luminosity. So, as it evolves, a giant ascends a Hayashi track similar to the one it descended at birth.

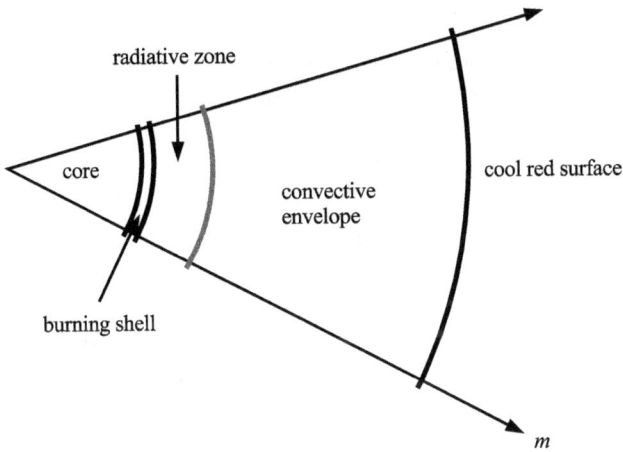

Fig. 7.5. The interior structure of a red giant star showing the fuel-exhausted core, around which is the thin burning shell at the base of a low-mass radiative zone, and the deep convective envelope extending to the stellar surface.

7.4.   Homologous Evolution

Though stars do not remain homologous as they evolve on a nuclear time-scale, it is still possible to obtain some insight into how they might behave at any stage simply by looking at the dependence on mean molecular weight μ which increases as major burning phases proceed. However most arguments based on such analyses prove to be invalid on deeper consideration because the departure from a homologous sequence is too great. An exception is the evolution of fully mixed stars. Very low-mass, less than about $0.3\,M_\odot$, main-sequence stars are fully convective and so do evolve homologously but they do so on such long time-scales that they have not noticeably changed during the lifetime of our Galaxy. More interesting are fully mixed massive stars. Their existence has been postulated to result from mixing driven by instabilities caused by rapid rotation. Many young massive stars are in fact observed to spin rapidly enough for this but whether they can continue to do so for a nuclear time-scale is uncertain. Nevertheless it is instructive to investigate how they might behave if they could.

We add to our equations dependence on μ and the hydrogen mass fraction X. Suppose we have a similar sequence of stars to those we examined in Sec. 7.3.2 for which the material behaves as an ideal gas, the dominant opacity source is electron scattering and hydrogen burning is by the CNO cycle. We now have

$$P = \frac{\rho \mathcal{R} T}{\mu},$$
(7.89)

$$\kappa = \kappa_0 (1 + X) \quad \text{and}$$
(7.90)

$$\epsilon = \epsilon_0 \rho X T^{16}.$$
(7.91)

Equations (7.47) and (7.50) for central density and pressure are unaffected so Eq. (7.55) for the central temperature becomes

$$T_c \propto \mu \frac{M}{R}.$$
(7.92)

Radiative equilibrium Eq. (7.56) gives us

$$\frac{T_c}{R} \propto \frac{(1 + X)\rho_c L}{R^2 T_c^3}$$
(7.93)

so that

$$L \propto \frac{\mu^4 M^3}{1 + X}.$$
(7.94)

From the energy generation Eq. (6.2) we obtain

$$L \propto \mu^{16} X \frac{M^{18}}{R^{19}}.$$
(7.95)

Eliminating L we arrive at

$$R^{19} \propto \mu^{12} M^{15} X (1 + X).$$
(7.96)

Though the CNO elements are rearranged the metallicity Z remains approximately constant so that, for solar metallicity Eq. (3.13),

$$\mu = \frac{4}{2.98 + 5X}.$$
(7.97)

So Eqs. (7.95) and (7.96) become

$$L \propto \frac{M^3}{(1 + X)(5X + 2.98)^4}$$
(7.98)

and

$$R \propto \frac{M^{15}X(1+X)}{(5X+2.98)^{12}}. \tag{7.99}$$

The derivatives

$$\frac{\mathrm{d}L}{\mathrm{d}X} < 0 \quad \text{and} \quad \frac{\mathrm{d}R}{\mathrm{d}X} > 0 \tag{7.100}$$

because X is initially sufficiently less than 1. Subsequently X falls as hydrogen is converted to helium so such a fully mixed star would brighten and shrink as it evolves.

Detailed stellar models of non-rotating main-sequence stars do show an increasing luminosity as they evolve from the main-sequence but an expansion rather than contraction. We shall see in the next chapter that they have a shrinking convective core so that their evolution is far from homologous and it is not surprising that they follow a different course to their, perhaps hypothetical, fully mixed cousins.

7.5. Questions

1. Mesh points in a star are labelled $k = 0, 1, 2, 3, \ldots$ from the centre. The mass at each point i is $m_i = m(x_i)$, where x_i is some independent monotonically increasing variable x through the star. Write down the Taylor expansion for $m(x + \delta x)$ and its first derivative $m'(x + \delta x)$, where δx is small.

The equations of stellar structure are written such that the first derivative $m'_i = m'(x_i)$ can be evaluated at each point $k \geq 1$. The central boundary condition is $m(x_0) = 0$ but the singularity $m = 0$ at the centre prevents the use of a mesh point at $k = 0$. Using the expansion for m to order δx^2 and m' to order δx show that an appropriate central boundary condition is

$$m_1 - \frac{3}{2}m'_1 + \frac{1}{2}m'_2 = 0.$$

2. A set of fully radiative stars has uniform mean molecular weight μ, constant opacity and energy generation rate given by $\epsilon = \epsilon_0 \rho T^{16}$, where ϵ_0 is constant. Radiation pressure is negligible.

Use a simple homology argument to show that for the set of stars

$$L \quad \text{varies as} \quad \mu^4 M^3,$$

$$T_c \quad \text{varies as} \quad \mu^{7/19} M^{4/19} \quad \text{and}$$

$$R \quad \text{varies as} \quad \mu^{12/19} M^{15/19},$$

where M is the stellar mass, L the stellar luminosity, R the stellar radius and T_c the central temperature.

Hence show that the slope of the theoretical main sequence for such a set of stars is

$$\frac{d \log L}{d \log T_e} = \frac{76}{9},$$

where T_e is the effective (surface) temperature.

Consider now two such sets of stars which differ in that one set is composed of pure hydrogen while the other is of pure helium. By considering the ratio of luminosities at fixed effective temperature for the two sets of stars, or otherwise, show that the helium main sequence lies below and to the left of the hydrogen main sequence in the Hertzsprung–Russell diagram.

3. Indicate why in a fully convective star the equation of state may be taken to be $P = KT^{5/2}$ where K is a constant.

Integrations for the atmospheric structure show that $K = Ag^\nu T_e^{-\lambda}$, where A, ν and λ are constants, g is the surface gravity and T_e the effective temperature. Derive a luminosity–mass–radius relation in the form

$$\frac{L}{4\pi\sigma R^2} = C R^\alpha M^\beta,$$

where C, α and β are constants and α and β depend solely on ν and λ.

Show that, when $\nu = 3/4$, T_e is constant. In this case show that the time for a fully convective star to contract to radius R_s radiating its gravitational energy is

$$t = \frac{GM^2}{7\left(4\pi R_s^3 T_e^4\right)\sigma}.$$

[*You may quote any properties of polytropes that you need.*]

4. A white dwarf may be approximated by a two-zone model. A helium interior is composed of a non-relativistic fully degenerate electron gas at constant temperature T_c with equation of state

$$P_e = K\rho^{5/3},$$

where K is constant and P_e is the electron pressure. The very thin outer layers are composed of hydrogen gas in radiative equilibrium obeying an ideal gas equation of state with negligible radiation pressure and with opacity given by Kramers' law in the form $\kappa = \kappa_0 \rho T^{-3.5}$. The transition between the inner and outer zones is defined to be where the pressure given by the ideal gas law in the outer zone is identical to the electron pressure in the inner zone. Show that,

  (i) in the very thin outer layers of the white dwarf,

$$P^2 = \frac{64}{51}\frac{\pi acGM}{\kappa_0 L}\left(\frac{R}{\mu}\right)T^{8.5} \equiv JT^{8.5},$$

where M and L are the mass and luminosity of the white dwarf,

  (ii) the temperature at the transition is

$$T_{\text{tr}} = \left(\frac{R}{\mu}\right)^{10/7} K^{-6/7} J^{-2/7} \quad \text{and}$$

(iii) the luminosity of the white dwarf is

$$L = \frac{64}{51}\frac{\pi acGM}{\kappa_0}\left(\frac{\mu}{R}\right)^4 K^3 T_c^{3.5}.$$

By making plausible numerical estimates, which should be stated, find, to order of magnitude, the temperature in the interior of the white dwarf.

Comment on the source of energy for the white dwarf's luminosity and estimate, in years, the cooling time-scale of the white dwarf. [$K \approx 10^{13}\,\text{dyne cm}^3\,\text{g}^{-5/3}$, $\kappa_0 \approx 4 \times 10^{24}\,\text{cm}^5\,\text{g}^{-2}\,\text{K}^{7/2}$.]

5. Consider a cluster of chemically homogeneous stars. The stellar material is an ideal gas and radiation pressure is negligible. The energy generation, opacity and whether the stellar interiors are convective or radiative depends on a star's mass. There are three distinct mass ranges over the full main sequence, (a) for $M > 2\,M_\odot$ the stars are fully radiative, energy generation is by the CNO cycle and opacity is dominated by electron-scattering, such that $\epsilon = \epsilon_a X \rho T^{13}$ and $\kappa = \kappa_a$, (b) for $0.5 < M/M_\odot < 2$ the stars are fully radiative, energy generation is by the pp-chain and opacity obeys Kramers' law, such that $\epsilon = \epsilon_b X \rho T^7$ and $\kappa = \kappa_b \rho T^{-7/2}$ and (c) for $M < 0.5 M_\odot$ the stars are fully convective, energy generation is by the pp-chain and opacity is dominated by H^- anions, such that $\epsilon = \epsilon_c X \rho T^3$ and $\kappa = \kappa_c \rho^{1/2} T^9$, where κ_a, κ_b, κ_c, ϵ_a, ϵ_b and ϵ_c are constants. Use homology show that, in each case, $L \propto M^\alpha$ with $\alpha = 3$ for case (a), 119/19, for (b) and 113/66 for (c).

Determine the gradient that each section of the main sequence has and sketch them in a single theoretical Hertzsprung–Russell diagram.

For case (c) show that these very low-mass stars evolve with $L \propto X^{17/66} \mu^{103/66}$, if $Z = 0$ and initially $X = 1$, and sketch the track followed by such a star, as it evolves, on your Hertzsprung–Russell diagram.

Why is a similar analysis not suitable for stars that have radiative interiors?

6. A red giant can be modelled by an isothermal degenerate helium core surrounded by a thin hydrogen-burning shell, above which is a radiative region which is itself surrounded by a deep convective envelope. The core is of mass M_1 and radius R_1 related by $M_1^{1/3} R_1 = A = \text{const}$. At the base of the radiative region, just above the core boundary, is the thin hydrogen-burning shell which generates all the luminosity L. The entire radiative envelope has a negligible mass while the convective envelope above it has a significant mass. The opacity obeys $\kappa = \kappa_0 \rho^n / T^m$, with n and m constant. Show that, when radiation pressure is neglected, the relation between P and T

in the radiative region is

$$P = C\left(T^{4+m+n} + T_0^{4+m+n}\right)^{1/(n+1)},$$

where

$$C = \left[\frac{16\pi acGM_1}{3\kappa_0 L}\left(\frac{R}{\mu}\right)^n \frac{(n+1)}{(n+m+4)}\right]^{1/(n+1)}$$

and T_0 is an appropriate constant of integration.

The temperature at the boundary between the radiative zone and the convective envelope is T_b. Show that

$$T_0 = T_b\left\{\left(\frac{\gamma-1}{\gamma}\right)\left(\frac{4+m+n}{n+1}\right) - 1\right\}^{1/(4+n+m)},$$

where γ is the ratio of specific heats throughout the convective zone.

Show that, in regions near the shell, well below the inner boundary of the convective envelope, where $T \gg T_b$ and hence $T \gg T_0$ that the dependence of temperature on radius r is approximately

$$T = \frac{\mu}{R}\frac{GM_1(n+1)}{(4+n+m)r}.$$

Use this to show that, when $n = 1$, $m = 3$ and the energy generation rate is given by $\epsilon = \epsilon_0\rho T^{10}$, with $\epsilon_0 = $ const,

$$L = \frac{4\pi}{13}C^2\epsilon_0\left(\frac{\mu}{R}\right)^2\left(\frac{\mu GM_1}{4R}\right)^{16}\frac{1}{R_1^{13}}.$$

Hence show that the luminosity depends on the core mass as

$$L \propto M_1^{32/3}.$$

What happens to L if mass is removed from the stellar envelope?

Chapter 8

Stellar Evolution

We have described the physical ingredients required to model a star and discussed how these are combined to create models of its structure and evolution. In this chapter we describe full models of the evolution of stars from their births to their deaths either in violent supernovae or gently cooling as white dwarfs. We take as our starting point not our Sun but a star of five times its mass because, with this, we can discuss most of the interesting phases of evolution. Subsequently we shall describe the differences in the evolution of both less and more massive stars.

8.1. Stellar Evolution Models

Figure 8.1 is a Hertzsprung–Russell diagram showing models calculated with the Cambridge STARS code. The zero-age main sequence is plotted along with evolutionary tracks for stars of different masses. Stars close in mass evolve similarly. All these stars have an initial hydrogen mass fraction $X = 0.7$, helium mass fraction $Y = 0.28$ and metallicity $Z = 0.02$. They are not truly zero-age because we have evolved them from the pre-main sequence to ensure that abundances of ^{3}He and the CNO elements are in equilibrium for the structure of the star. Otherwise the star would make a rapid unphysical excursion away from and back to the start of the main sequence while this equilibrium is achieved.

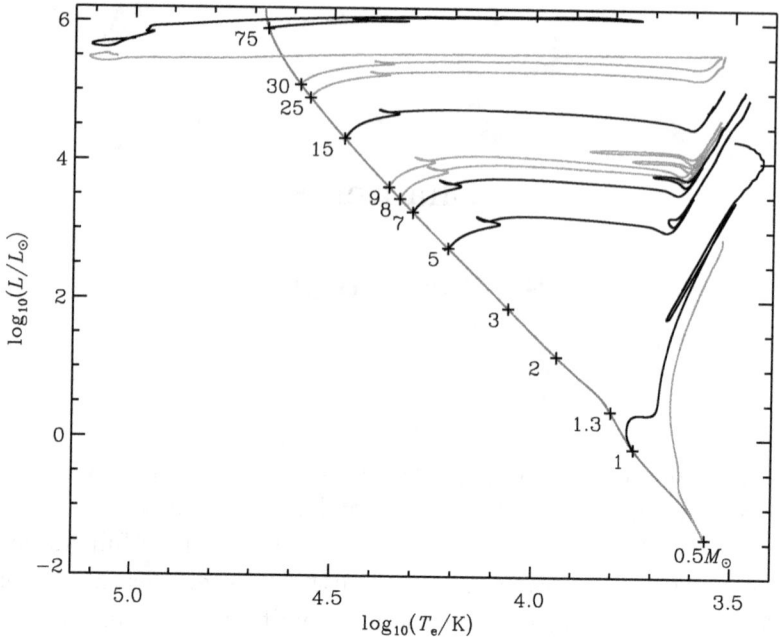

Fig. 8.1. Evolution of Cambridge STARS code stellar models in a Hertzsprung–Russel diagram. The zero-age main sequence is shown with various stars each marked by its initial mass in solar masses. Some of these stars have been evolved and their paths in the H–R diagram are plotted. We examine the models with black tracks in greater detail in this chapter.

On the zero-age main sequence the most important differences between stars of various masses are determined by the temperatures at their centres and photospheres. The central temperature determines the mechanism by which hydrogen burns and the rate at which it does so. The greater temperature sensitivity of the CNO cycle drives the formation of a convective core at higher masses (Sec. 4.4.1). The effective temperature of the photosphere determines the nature of the stellar envelope. If it is low enough, $T_e < 10\,000\,\mathrm{K}$, for neutral hydrogen atoms to exist then a near-surface ionisation layer forces a convective envelope. At higher temperatures the envelope is radiative and the temperature determines the nature of the opacity. In Fig. 8.2 we show the extent of convective zones along our zero-age main sequence. Major structural changes take place between about 0.3

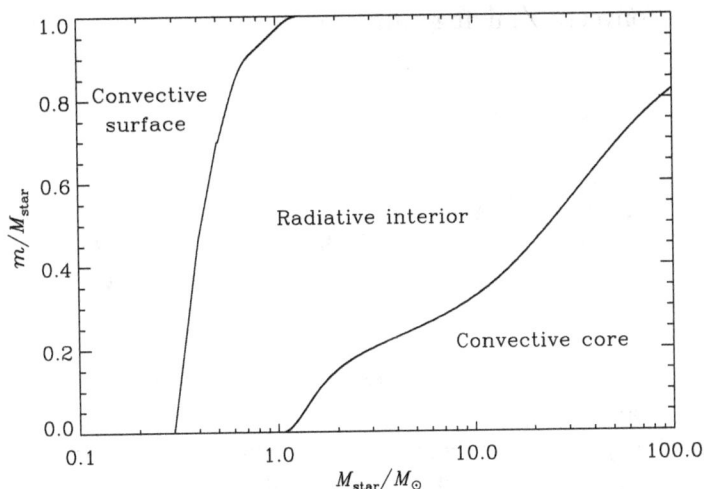

Fig. 8.2. The extent of convective zones by mass m within zero-age main sequence stars of varying total mass M_{star}. Up to 0.3 $M_\odot$ stars are fully convective. Between 0.3 and 1.1 $M_\odot$ they have convective envelopes and above 1.3 $M_\odot$ convective cores.

and 2 $M_\odot$. For $M < 2\,M_\odot$ the pp-chain dominates hydrogen burning so that below about 1.3 $M_\odot$ the core is radiative. For $M < 1.1\,M_\odot$ the envelope is convective and for $M < 0.3\,M_\odot$ it extends to the centre so that the whole star is convective.

8.1.1. *Convective overshooting*

Before proceeding we note that, in making the models we describe, we have included a prescription for extra mixing beyond the convective boundaries determined by the Schwarzschild criterion (4.75). Such prescriptions are usually called convective overshooting and are included to obtain better agreement with various observations. However, being without any real physical basis, convective overshooting is treated with a number of rather different algorithms in various evolution codes and for individual models. Its most important effect is to increase the size of a convective core during nuclear burning. This prolongs the lifetime of burning phases, in particular the main-sequence lifetime of stars with convective cores. In the Cambridge STARS code convective overshooting is included by altering the convective stability criterion by effectively reducing ∇_a. A star is stable to convective mixing only when

$$\nabla_r > \nabla_a - \delta, \tag{8.1}$$

where

$$\delta = \frac{\delta_{ov}}{2.5 + 20\zeta + 16\zeta^2}, \tag{8.2}$$

$$\zeta = \frac{P_r}{P_g}, \tag{8.3}$$

the ratio of radiation to gas pressure which increases with mass and evolution (Sec. 3.5), and δ_{ov} is a constant parameter. The dependence on radiation pressure is used to reduce the effect at higher masses. The parameter $\delta_{ov} = 0.12$ is calibrated to match observations of a number of evolved binary star systems for which masses can be determined independently of the models. There is no change to convective heat transport in the STARS code which is modelled with the commonly used MLT theory devised by Erika Böhm-Vitense (1958) and we take the ratio of the mixing length to pressure scale-height to be $\alpha = 2$.

As well as increasing the lifetime of a star, convective overshooting affects later stages of its evolution because of the larger burnt core at the end of the main sequence. The inclusion of this extra mixing makes the later stages of evolution of a star of a given initial mass similar to those of a higher mass star that, without overshooting, would grow a similar sized core. There is much less effect on stars with radiative cores.

8.2. The Evolution of a 5 $M_\odot$ Star

Stars of intermediate mass, between about 2.3 and 8 $M_\odot$, evolve in a similar way that covers the most important aspects of stellar evolution so we begin by looking at an example in detail. Figure 8.3 shows the track followed by a 5 $M_\odot$ star in the H–R diagram. Important points in its evolution are marked with crosses and discussed in detail below. Real stars of this mass experience important phases of mass loss as giants, particularly as asymptotic giants beyond point 12. In

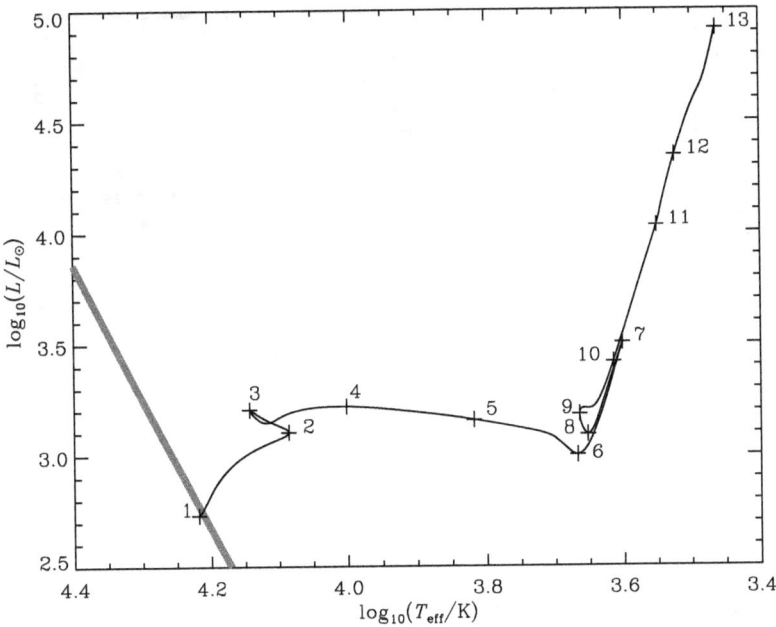

Fig. 8.3. The evolution of a $5\,M_{\odot}$ star in the Hertzsprung–Russell diagram. Numbered crosses indicate the important points which we describe in more detail in the text. The thick grey line is the zero-age main sequence.

order to concentrate on the interior evolution, for this model, we have not included mass loss but return to it in Sec. 8.7.2. Such stars also experience thermal pulses (Sec. 8.3) but modelling these in detail is rather time consuming, particularly given that their number would normally be truncated by the mass loss that we have omitted. So we suppress thermal pulses in this particular model too. In Fig. 8.4 we show the interior evolution of convective zones and burning shells.

(1) The zero-age main sequence at time $t = 0$. The hydrogen mass fraction $X = 0.7$ is uniform throughout the star. Its radius $R = 2.84\,R_{\odot}$ and its luminosity $L = 540\,L_{\odot}$. Hydrogen is burning by the CNO cycle with energy generation rate $\epsilon \propto \rho T^{16}$. At the centre of the star L_r rises rapidly with mass. The radiative temperature gradient $\nabla_r \propto L_r/m$ and exceeds ∇_a so that the core of the star is convective out to $m = 1.2\,M_{\odot}$. The envelope is radiative. As hydrogen is burnt X decreases so, in

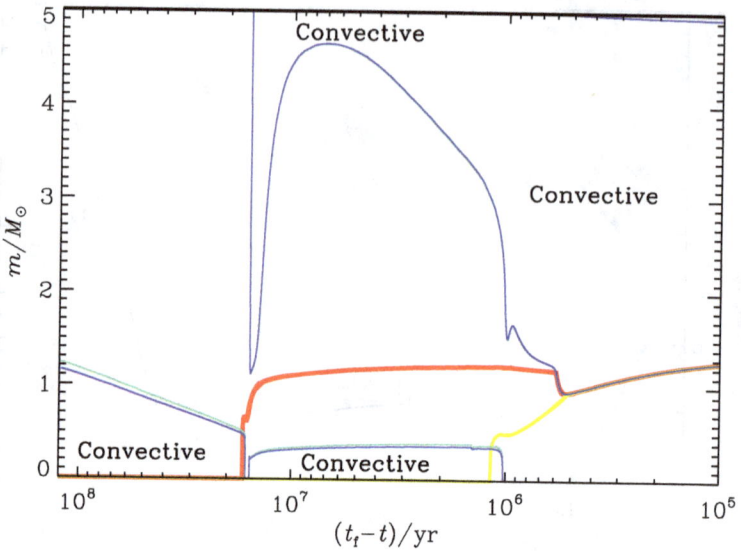

Fig. 8.4. The interior of an evolving $5\,M_\odot$ star. Time is plotted logarithmically and measured backwards from degenerate core carbon ignition. Blue lines are boundaries to fully convective regions which are labelled as such. Cyan lines are semi-convective boundaries such that regions between blue and cyan lines have $\nabla_r \approx \nabla_a$. Initially hydrogen is burning in a shrinking convective core. The red line indicates where in the star the hydrogen abundance X drops below 0.01 and so tracks close to the site of hydrogen burning. A deep convective envelope forms during the brief red giant evolution, and dredges into previously burnt core. Soon afterwards helium ignites and burns in a convective core while the convective envelope recedes back towards the surface. The yellow line is where the helium abundance Y drops below 0.01, tracking the site of helium burning. As helium burning moves to a shell the convective envelope deepens again and dredges past the temporarily extinct hydrogen-burning shell to create the close double shell burning star. Thermal pulses and mass loss are suppressed in this model so the core grows to $1.38\,M_\odot$ when carbon ignites degenerately.

the core, μ rises with time. Because the central temperature T_c is thermostatically controlled the density ρ in the core rises to maintain the pressure P. The luminosity L increases and the radius R of the star slowly rises. The convective core gradually shrinks in mass to $0.5\,M_\odot$ when the central hydrogen is nearly exhausted. As it recedes a composition gradient is left behind (Fig. 8.5).

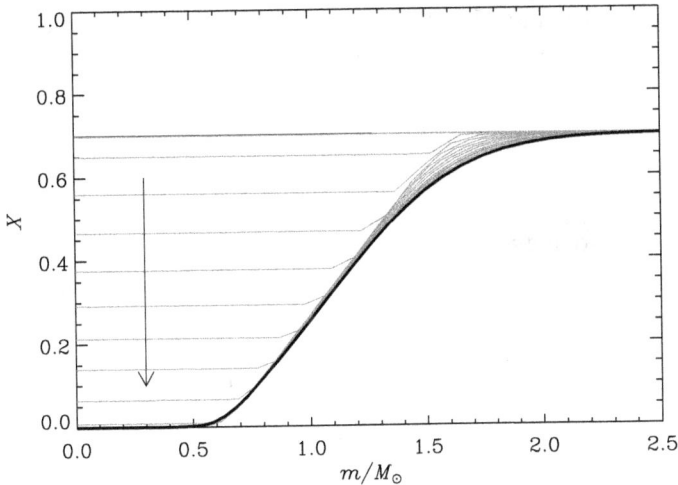

Fig. 8.5. The change in the composition profile of the core of a $5\,M_\odot$ star as it evolves from point (1) to point (2) in Fig. 8.3. At point (1) the hydrogen mass fraction is uniform. The arrow indicates the progression of time as burning proceeds in a shrinking convective core. The solid black line is the profile at point (2) with the thin grey lines showing earlier evolution. There is some numerical diffusion, caused by the moving non-Lagrangian mesh in the STARS code and we see evidence for this at the outer edge of the initial core around $1.75\,M_\odot$. This is an artefact that can be reduced by increasing the resolution of the mesh.

(2) At $t = 1.055 \times 10^8$ yr, $L = 1207\,L_\odot$ and $R = 8.01\,R_\odot$. Core hydrogen is not quite exhausted, $X_c = 0.029$, but it becomes energetically favourable for the star to contract temporarily, generating about one-tenth of its luminosity thermally as it shrinks.

(3) Just 1.5×10^6 yr later hydrogen is exhausted in the core, $X_c = 0$ and burning moves to a shell that proceeds outwards through the star leaving behind a growing isothermal helium-rich, $Y = 0.98$, core.

(4) After another 8×10^5 yr, when $M_c = 0.63\,M_\odot = 0.13\,M_{star}$, it ceases to be possible to support the stellar envelope with an isothermal core, dominated by gas pressure, in thermal equilibrium. This is the Schönberg–Chandrasekhar limit M_{SC} found in the early days of stellar modelling (Schönberg and

Chandrasekhar, 1942). Evolution accelerates as the core contracts on a thermal time-scale.

(5) The star evolves from point 4 to point 6 in only $2.8 \times 10^5 \, \text{yr}$ so that it is rare to find stars in this region of the H–R diagram which is consequently known as the Hertzsprung gap. As the star crosses this gap its core continues to contract with electron degeneracy pressure becoming important. At the same time the stellar envelope expands and its luminosity drops. The photospheric temperature consequently falls and the star becomes a red giant. A deep convective envelope begins to form as hydrogen recombines near the surface. Though stellar models universally demonstrate this erythrogigantism it has not yet been possible to identify precisely which part of stellar physics drives it.

(6) The star has reached its Hayashi limit (Sec. 7.3.4). Between points 6 and 7 the convective envelope deepens and crosses into the composition gradient left behind by the retreating convective core of the main sequence (Fig. 8.6). Helium is dredged to the surface at the expense of hydrogen together

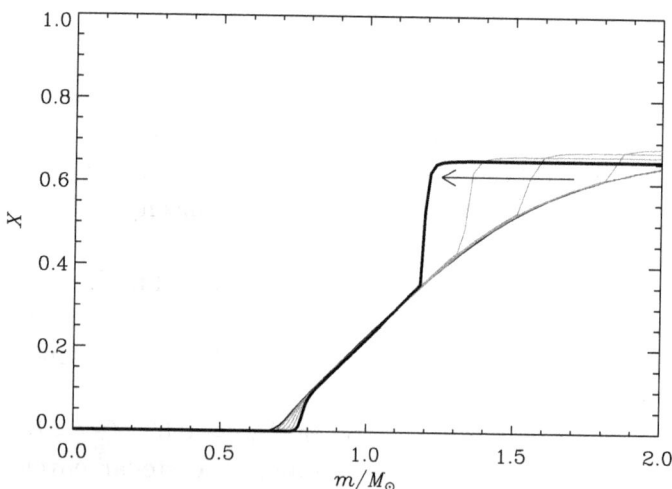

Fig. 8.6. Similar to Fig. 8.5 but now for the change in the hydrogen composition at the first dredge-up by the deepening convective envelope between points 6 and 7. The arrow indicates the progression of the profile with time.

with ^{14}N at the expense of the ^{12}C from which it formed in the CNO cycle. This first dredge up is seen in globular clusters in which the surface compositions of main-sequence stars can be compared with those of red giants. It is evidence that nuclear processing has taken place in the cores of these stars. Thermal equilibrium is regained. Between points 6 and 7 hydrogen shell burning adds mass to the core which contracts and heats up as P_e rises. The luminosity rises as the core temperature T_c rises. This is the red giant branch (RGB) sometimes called the first giant branch (FGB).

(7) After another 3.1×10^5 yr helium burning by the triple-α process ignites when $T_c \approx 1.2 \times 10^8$ K. At its maximum $P_e \approx 0.053\, P_g$ and gas pressure support still dominates so helium ignition is thermostatically controlled and relatively gentle. Again the extreme sensitivity of the burning rate to temperature means that L_r rises rapidly with m in the centre and a convective core forms (medium thickness lines in Fig. 8.7).

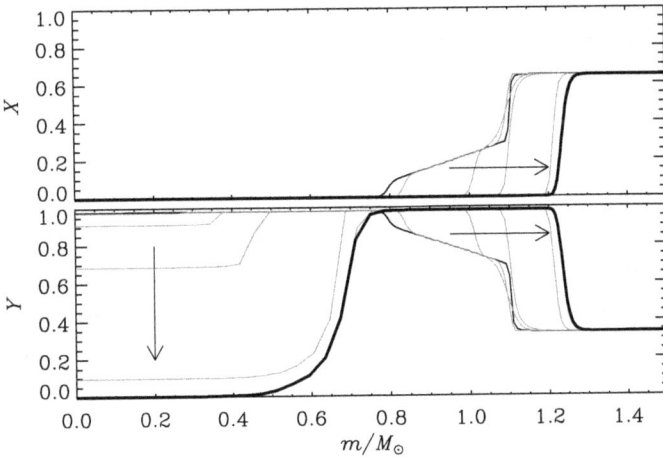

Fig. 8.7. Core helium burning from points 7 (medium thickness lines) to 10 (thickest lines) in Fig. 8.3. The upper panel shows the evolution of the hydrogen abundance profile and the lower the helium. Helium burns in a growing convective core, the outer parts of which are semi-convective, while hydrogen continues to burn out through the star in a shell.

(8) The star reaches another minimum in luminosity after a further 3.1×10^6 yr. At this point the core helium abundance $Y = 0.80$, carbon $X_C = 0.17$ and oxygen $X_O = 0.008$. The central temperature is 1.3×10^8 K. Helium burning by the triple-α process continues in a growing convective core, the outer edge of which becomes progressively more semi-convective (Sec. 4.4.7), avoiding a sharp boundary to the CO core at the end (thick lines in Fig. 8.7). As the burning proceeds the contribution from α captures by ^{12}C to form ^{16}O increases. Meanwhile hydrogen continues to burn in a shell at the edge of the helium core. The star's energy generation remains dominated by hydrogen shell burning throughout most of core helium burning with $L_H/L_{He} \approx 3$.

(9) The extent of the blueward excursion during helium burning is somewhat dependent on both the physics and the numerical methods used to make the models. If the maximum blueward extent of this loop were hotter by 20 to 25% then it would cross the Cepheid instability strip, a near vertical strip in the H–R diagram, about 100 K in width, in which stars pulse on a dynamical time-scale. The over-stable oscillation is driven by the second ionisation of helium

$$\text{He}^+ \rightarrow \text{He}^{++} + e^- \tag{8.4}$$

between about 35 000 and 50 000 K. As the star shrinks it heats up and helium ionises, absorbing energy from the luminosity. The increase in the number of electrons increases the opacity initially trapping the luminosity so the star heats up and ionisation is enhanced. Once helium is fully ionised the high opacity drives expansion, cooling and recombination that in turn leads to a drop in opacity that then enhances cooling and recombination. Because the strip has a narrow temperature range and the luminosity of the blue loop is dictated by the mass of the star the oscillation period

$$\Pi \propto \tau_{\text{dyn}} \propto \sqrt{\frac{R(L, T_{\text{eff}})^3}{GM(L)}} \tag{8.5}$$

is a function of luminosity only. Knowing the distance to nearby Cepheids a period–luminosity relation can be devised and used to estimate the absolute magnitude and hence distance to far-off Cepheids, including those bright enough to be seen in other galaxies. Typically about $5\,M_\odot$ is the, model dependent, lower limit for a star to cross the instability strip. The blue loops of more massive stars extend further to the left. The same instability strip also passes through the horizontal branch in globular clusters where we find the RR Lyrae variables, of lower metallicity, and the main sequence where we find the δ Scuti stars. At the hottest extent of our blue loop helium has been burning in the core for 8×10^6 yr. The helium abundance has dropped to $Y = 0.44$, carbon has risen to $X_C = 0.45$ and oxygen to $X_O = 0.08$. Core helium burning continues for further 7.7×10^6 yr before it is finally exhausted. Notice in Fig. 8.4 that even when $Y < 0.01$ in the centre helium burning is still sufficient to drive convection for a brief period. When core helium burning is about to conclude even a small amount of fresh fuel mixed into the core can be sufficient to reinvigorate the burning. This phenomenon leads to the core breathing pulses seen in some models. They may actually be physical or may be numerical artefacts.

(10) Helium is finally exhausted in the core when $X_C = 0.27$ and $X_O = 0.70$. The entire core helium-burning phase has lasted for 1.4×10^7 yr or about one-seventh of the core-hydrogen-burning lifetime. Helium burning now moves to a shell and hydrogen shell burning briefly dominates the luminosity again. The carbon/oxygen ash from the helium-burning shell is added to the burnt-out core, which becomes increasingly supported by electron degeneracy pressure as it shrinks and heats up. The helium-burning rate rises rapidly. Once again it soon dominates the luminosity of the star but this time to the extent that, outside the burning shell, the star expands and cools. The hydrogen-burning shell is extinguished and this permits the deep convective envelope to dredge in past the composition gradient left by the now extinct burning (Fig. 8.8).

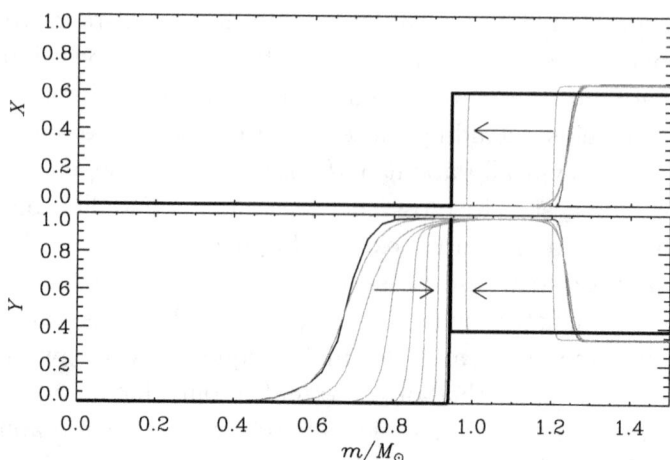

Fig. 8.8. The composition evolution from the end of core helium burning (point 11) to the end of the 2nd dredge-up (point 13). Initially the helium shell rapidly burns outwards, while the hydrogen-burning shell hardly moves. As the CO core grows and shrinks so L_{He} grows. Outside the helium-burning shell the star expands and L_H falls eventually fizzling out by point 12 when the deep convective envelope dredges through the composition jump carrying hydrogen back down to just above the growing helium-burning shell.

(11) This second dredge up begins 4.1×10^5 yr after the end of core helium burning by which time $P_e/P_c = 0.375$. More He and CNO cycled products are dredged to the surface. This time oxygen as well as carbon is depleted in favour of ^{14}N.

(12) Another 8.3×10^4 yr later second dredge up ends when convection brings fresh hydrogen to just above the helium-burning shell, where a hydrogen-burning shell reignites. Electron degeneracy pressure now dominates at the centre where $P_e/P_g = 0.7$ and it is dense and hot enough that neutrino loss cooling (Sec. 6.6) leads to a significant temperature inversion. The core temperature is 1.8×10^8 K some 2.6 times smaller than the maximum temperature in the helium-burning shell. The mass between the two burning shells has been reduced to less than $0.01 \, M_\odot$ and this, coupled with the very strong temperature dependence of the triple-α reaction leads to thermal pulses, driven by the Härm–Schwarzschild instability, that we discuss

in Sec. 8.3. The nature of the pulses and mass loss govern the remaining evolution of real stars.

(13) In the absence of both thermal pulses and mass loss, in this model the two burning shells continue to burn out through the star with the hydrogen burning dominating the luminosity until, when the core mass $M_c = 1.38\,M_\odot$, 1.23×10^8 yr from the zero-age main sequence, carbon ignites degenerately at the centre. In real stars, and suitable models, thermal pulses delay core growth and mass loss removes the envelope long before such carbon ignition. The exposed burning shells cool and the stars evolve to white dwarfs (Sec. 8.8).

8.3. Thermal Pulses

The very high temperature dependence of the triple-α reaction means that the helium-burning shell is very thin, both in radius and mass. Such thin burning shells suffer a thermal instability described by Schwarzschild and Härm (1965). Essentially increased energy production, owing to a small increase in temperature, leads to further rise in the temperature in the burning shell because the energy cannot leak out fast enough and expansion of the very thin shell cannot relieve the pressure needed to support the overlying massive envelope.

Heat transport in the burning shell is by radiation so the radiative transfer (4.35) combined with the mass (2.2) gives us the luminosity as

$$L_r = \frac{64\pi^2 a c r^4 T^3}{3\kappa}\frac{\mathrm{d}T}{\mathrm{d}m}. \tag{8.6}$$

The upper panel of Fig. 8.9 shows an approximation to the temperature profile through the burning shell. We assume no energy generation or loss in the core which therefore remains isothermal out to the base of the burning shell at m_0 at a temperature determined by the burning. The temperature then starts to fall off halfway through the burning shell, the outer edge of which at $m_0 + \Delta m$, is where the burning is negligible. The more temperature sensitive the reaction, the thinner the shell. At the inner edge we have $L_r(m_0) = 0$ and at

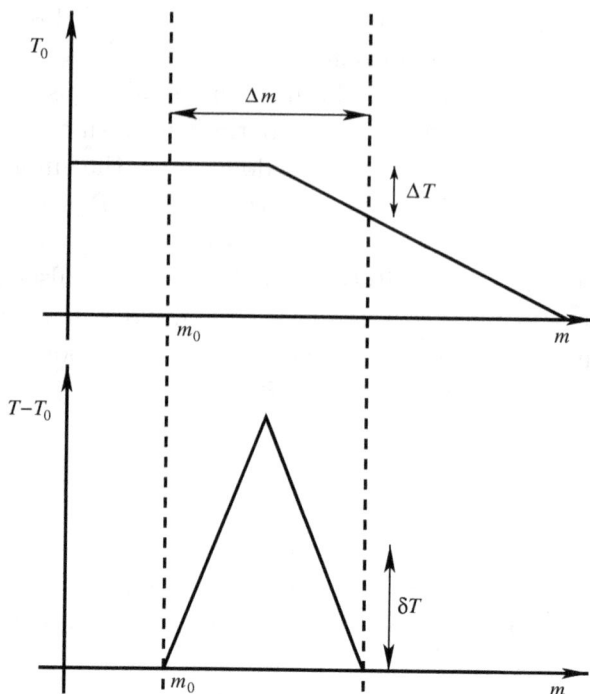

Fig. 8.9. Idealised temperature profile and perturbation to illustrate the Härm–Schwarzschild instability. The burning shell extends in mass from m_0 to $m_0+\Delta m$. The upper panel illustrates the unperturbed temperature profile T_0 which we take to be isothermal throughout the core and into the inner half of the burning shell, where $m < m_0 + \Delta m/2$, and then falling off linearly, dropping by ΔT by $m = m_0+\Delta m$. In the lower panel we introduce a perturbation to the temperature in the shell peaking at $2\delta T$ at $m < m_0 + \Delta m/2$ and vanishing at the edges.

the outer edge

$$L_r(m_0 + \Delta m) = \frac{64\pi^2 a c r^4 T^3}{3\kappa} \frac{2\Delta T}{\Delta m} = L \qquad (8.7)$$

in this unperturbed state. In a quasistatic state the energy generation Eq. (6.3) reduces to

$$\epsilon = \frac{\mathrm{d}L_r}{\mathrm{d}m} \approx \frac{L}{\Delta m}. \qquad (8.8)$$

Now consider introducing a temperature perturbation as illustrated in the lower panel of Fig. 8.9. This leads to a perturbation δL_r in luminosity such that just inside the edges of the shell

$$\delta L_r(m_0) = -\frac{64\pi^2 acr^4 T^3}{3\kappa} \frac{4\delta T}{\Delta m} \qquad (8.9)$$

and

$$\delta L_r(m_0 + \Delta m) = \frac{64\pi^2 acr^4 T^3}{3\kappa} \frac{4\delta T}{\Delta m}. \qquad (8.10)$$

So there is a perturbation to the luminosity gradient of

$$\delta\left(\frac{\mathrm{d}L_r}{\mathrm{d}m}\right) = \frac{\delta L_r(m_0 + \Delta m) - \delta L_r(m_0)}{\Delta m} = 4\frac{L}{\Delta m}\frac{T}{\Delta T}\frac{\delta T}{T}. \qquad (8.11)$$

For an energy generation rate

$$\epsilon \propto T^\nu, \qquad (8.12)$$

the temperature perturbation leads to a perturbation $\delta\epsilon$ in energy generation rate such that

$$\frac{\delta\epsilon}{\epsilon} = \nu\frac{\delta T}{T} \qquad (8.13)$$

and the perturbation to the time-dependent energy generation (6.3) means that

$$T\frac{\partial \delta s}{\partial t} = \delta\left(\epsilon - \frac{\mathrm{d}L_r}{\mathrm{d}m}\right) = \left(\nu - 4\frac{T}{\Delta T}\right)\frac{L}{\Delta m}\frac{\delta T}{T}, \qquad (8.14)$$

where δs is the perturbation to the specific entropy in the burning shell. Whenever $\nu > 4T/\Delta T$, δs must grow. Recall that for hydrogen burning by the CNO cycle $\nu \approx 14$ while for helium burning $\nu \approx 40$ so that for the latter a fractional temperature difference $\Delta T/T$ of only 10% across the shell is sufficient to trap entropy. The time-scale for the thermal pulse $\tau_{\mathrm{TP}} = c_V T\Delta m/L$ is the Kelvin–Helmholtz time-scale for the unstable burning shell and it is much shorter than for the entire star. For an ideal gas with $\gamma = 5/3$ entropy changes, as in

Eq. (6.22), according to

$$\delta s = c_V T \delta \log_e \frac{P}{\rho^{5/3}} = c_V T \left(\frac{\delta P}{P} - \frac{5 \delta \rho}{3\rho} \right) = c_V T \left(\frac{\delta T}{T} - \frac{2\delta \rho}{3\rho} \right),$$

(8.15)

because

$$P = \rho \frac{\mathcal{R}T}{\mu},$$

(8.16)

means that

$$\frac{\delta P}{P} = \frac{\delta \rho}{\rho} + \frac{\delta T}{T},$$

(8.17)

but $\delta T/T$ is small so the burning shell responds by expanding on a time-scale τ_{TP}. Now suppose the burning shell extends from $r_0(m_0)$ to $r = r_0(m_0) + \Delta r$ in radius so that

$$\Delta m = 4\pi r_0^2 \rho \Delta r.$$

(8.18)

We assume the inner radius of the shell is fixed and let its outer radius increase by $\delta r = \delta \Delta r$ then the density changes according to

$$\frac{\delta \rho}{\rho} = -\frac{\delta r}{\Delta r}.$$

(8.19)

Suppose the outer layers of the star expand homologously so that, because

$$\frac{\mathrm{d}P}{\mathrm{d}m} = -\frac{Gm}{4\pi r^4},$$

(8.20)

we have

$$\frac{\delta P}{P} = -\frac{4\delta r}{r_0}.$$

(8.21)

Combining Eqs. (8.17), (8.19) and (8.21) we arrive at

$$\frac{\delta T}{T} = \left(\frac{r_0}{\Delta r} - 4 \right) \frac{\delta r}{r_0}.$$

(8.22)

So for $\Delta r < r_0/4$ the temperature in the burning shell rises further as the shell expands. This positive feedback removes the usual thermostatic control and the reaction rate rises rapidly on the Kelvin–Helmholtz time-scale of the thin shell τ_{TP}.

Because the helium-burning shell is unstable it burns faster than the hydrogen-burning shell can supply it with fresh fuel. So it cannot burn continuously and instead such stars experience a series of thermal pulses, two of which are illustrated in Fig. 8.10. The cycle can be broken down into three parts.

(i) **Interpulse:** The helium-burning shell is extinguished while the hydrogen shell burns outwards depositing a mantle of helium on the carbon–oxygen core. We set $t = 0$ at the start of an

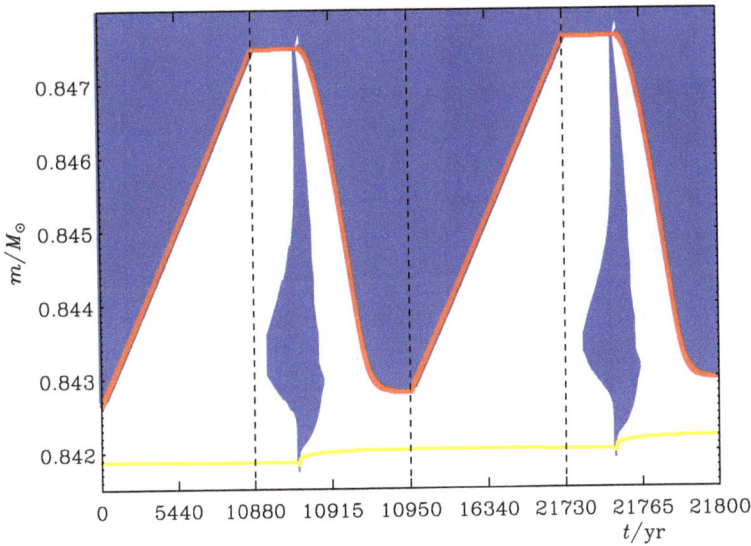

Fig. 8.10. The detailed structure of the intershell region during two thermal pulses. The red line is where $X = 0.01$. During the interpulse and on phase it is the location of hydrogen burning. During the power down phase and dredge up it is at the base of the deep convective envelope. The yellow line, where $Y = 0.01$, is the location of the helium-burning shell. Blue regions are convection zones. Time t is measured from the start of the first interpulse period and is shown at higher resolution during the pulse and dredge up than during the interpulse.

interpulse that lasts about 10^4 yr after which time the helium-burning shell ignites.

(ii) **On phase:** The Härm–Schwarzschild instability leads to thermonuclear runaway. L_{He} reaches 10^6 to $10^9\, L_\odot$ over a burning period that lasts about 30 yr with most of the helium consumed within a few years around the peak of the burning. The rapid release of energy drives an intershell convective region which quickly removes ^{12}C from the burning shell. Very little ^{16}O is formed and the intershell region is typically left with a composition of $Y \approx 0.25$ and $X_C \approx 0.75$. The star expands from the inside out, absorbing the excess luminosity so that, at the surface, L remains relatively constant. The hydrogen-burning shell cools and is extinguished.

(iii) **Power down phase:** As its fuel runs out the helium burning shell cools over a period of about 100 yr. The intershell convection ceases and the surface convective envelope deepens through the extinct hydrogen-burning shell into the intershell region. Thence it dredges the products of helium burning to the stellar surface in what is known as a third dredge up event. In each pulse ^{12}C and s-process elements (recall Sec. 6.5) are mixed throughout the deep convective envelope. Eventually the base of the convective envelope carries fresh hydrogen fuel close to the now extinct helium-burning shell, where the temperature is high enough to reignite shell hydrogen burning and start the next interpulse period.

8.3.1. *Carbon stars*

As thermal pulses proceed, third dredge up slowly increases the carbon content of the deep convective envelope and so too the atmosphere of the star. When the number of carbon atoms exceeds the number of oxygen atoms the chemistry of the atmosphere alters drastically and the star becomes a carbon star. Almost all the oxygen in the atmosphere is then combined with carbon in CO molecules so that none is available to form the typical oxides, such as TiO, seen in the spectra of oxygen-rich giants. Instead excess carbon now combines with nitrogen as CN, molecular bands of which become prominent in the star's spectrum. Before carbon dominates oxygen, zirconium produced in the s-process can displace the prominence of TiO by ZrO in the spectrum and so the star evolves from spectral type M, through S and then to C as a true carbon star. In more massive AGB stars, those whose main-sequence progenitors had masses in excess of about $4\,M_\odot$, the base of the surface convection zone can reach temperatures at which partial CNO burning of hydrogen, hot bottom burning, converts ^{12}C to ^{14}N throughout the envelope. As the C/O ratio falls these stars cease to be carbon stars again as long as hot bottom burning remains faster than dredge up. The nitrogen produced is called primary because it is predominantly formed from the carbon freshly synthesised by helium burning rather than carbon or oxygen that was present when the star formed. Nitrogen produced by the CNO cycle from carbon or oxygen already present in the material from which the star formed is known as secondary. Being very luminous giants, AGB stars tend to have strong stellar winds so much of the processed and dredged material is returned to the interstellar medium. AGB stars are probably the major source of nitrogen and s-process elements in the Universe as well as the source of about one-third of the carbon.

Because of third dredge-up the CO cores of AGB stars do not grow as fast as they would if hydrogen were burnt smoothly to helium and on to carbon and oxygen. During the interpulse period the hydrogen shell advances by ΔM_{H} but during third dredge up it is pushed back by $\lambda \Delta M_{\mathrm{H}}$. The dredge up fraction λ is not well known and varies from model to model as well as star to star. Typically $0.1 < \lambda < 0.9$ with λ usually increasing as the pulses proceed. The actual averaged growth rate of the core is then $(1-\lambda)\Delta M_{\mathrm{H}}/\tau_{\mathrm{interpulse}}$, where $\tau_{\mathrm{interpulse}}$ is the time between pulses, the duration of the pulse itself being insignificant. Because it is the competition between mass loss in stellar winds and core growth (Sec. 8.7.2) that determines the end points of stellar evolution, for a given mass-loss rate, deeper third dredge-up results in a lower mass CO white dwarf descending from an AGB star of a given initial mass.

8.3.2. *Quantitative problems*

Though thermal pulses are now qualitatively well understood their quantitative modelling still has some way to go before agreement is reached within the international community of stellar modellers. Problems arise because the time-scales for nuclear burning, convective mixing and structural adjustment are much more similar during a thermal pulse than during other phases of a star's evolution. In particular there is an interesting problem concerning the source of neutrons for the s-process (Sec. 6.5). In the more massive AGB stars with hotter helium burning the neutrons are easily explained by alpha captures on to ^{22}Ne during helium burning, ^{22}Ne$(\alpha, n)^{25}$Mg. The ^{22}Ne is itself a product of the helium burning of secondary ^{14}N created during earlier CNO burning of hydrogen, ^{14}N$(\alpha, \gamma)^{18}$F$(, e^+\nu_e)^{18}$O$(\alpha, \gamma)^{22}$Ne. The cooler helium burning in the more common, lower-mass stars is unable to activate this source of neutrons yet technetium is still seen in their spectra. Recall that technetium has no stable isotope but is produced by the s-process, the path of which passes through ^{99}Tc and ^{100}Tc. The stabler isotope, ^{99}Tc, has a half-life of only 2.1×10^5 yr, much shorter than the lifetime of these stars. So its presence indicates ongoing s-process and hence a source of neutrons. The most likely source is the reaction ^{13}C$(\alpha, n)^{16}$O which can proceed even during the interpulse phase. Only a small quantity of ^{13}C is left after full CNO burning and so, to produce sufficient neutrons to account for observed s-process yields, it is postulated that incomplete CNO cycling of intershell ^{12}C takes place in a pocket at the base of the convective envelope where protons are mixed in so as to reach a number density similar to that of ^{12}C. The nature of this mixing process has yet to be established. Convective overshooting, rotation-driven mixing and breaking gravity waves are all under consideration. Nucleosynthesis models tend to simply include an artificial ^{13}C pocket calibrated in size so as to reproduce observed s-process distributions in stellar spectra and meteoritic grains.

8.4.   The Evolution of a $1\,M_\odot$ Star

Though our Sun will evolve through similar phases to a $5\,M_\odot$ star, there are significant differences which we now consider in more detail. The evolutionary track of a $1\,M_\odot$ star is illustrated in Fig. 8.11 in which we have labelled equivalent evolutionary stages with the same numbers as for the $5\,M_\odot$ star.

(1) At $t = 0$ the radius $R = 0.89R_\odot$ and the luminosity $L = 0.67L_\odot$. At about $t = 4.8\times10^9$ yr when $R \approx R_\odot$, $L \approx L_\odot$ and $X_c = 0.36$, this star is similar but not identical to the Sun. Its luminosity can be adjusted by varying the initial hydrogen abundance X and, because it has a surface convection zone, the mixing length

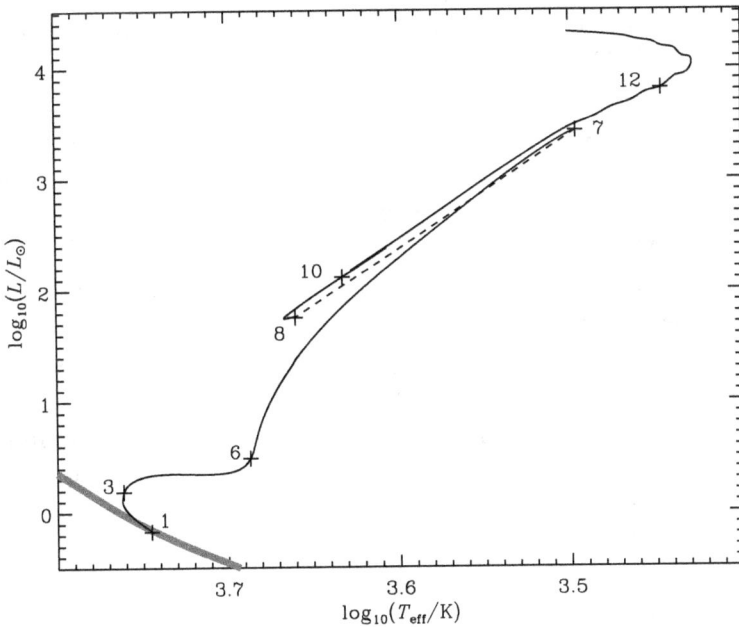

Fig. 8.11. Evolution of a $1\,M_\odot$ star in the Hertzsprung–Russel diagram. Numbered crosses indicate the phases of evolution, corresponding to those in the $5\,M_\odot$ star in Fig. 8.3 and discussed in the text. The thick grey line is the zero-age main sequence.

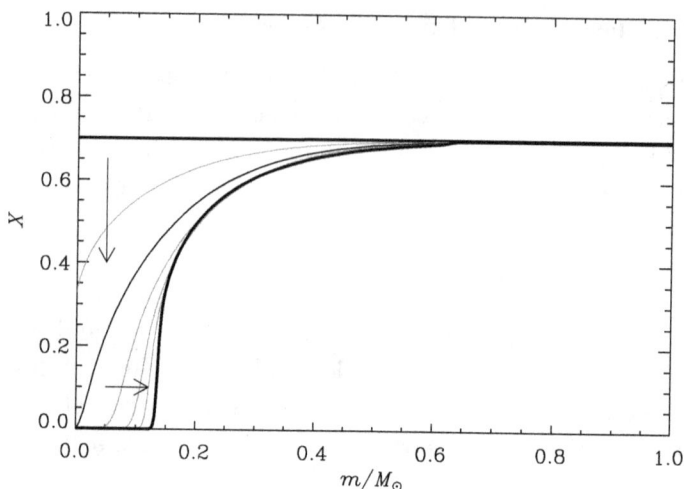

Fig. 8.12. The evolution of the core hydrogen profile between phases (1) and (6) in Fig. 8.11. The thin solid black lines are at (1), (3) and (6). Arrows indicate the progression of burning with time.

can be chosen to fix the radius. This is equivalent to choosing the adiabat on which the majority of the convective envelope lies and it is this that determines the radius (Sec. 4.4.2). Unlike the $5\,M_\odot$ star, as it starts to evolve a $1\,M_\odot$ star grows hotter and more luminous because it has a radiative rather than a convective core (Fig. 8.12). This is itself a consequence of the cooler hydrogen burning by the less temperature-sensitive pp-chains rather than the CNO cycle. At this point the Sun is growing at about one inch per year, while rising in luminosity by about one part in 10^{10}. So, undetectably slowly but surely, the Solar System habitable zone is moving further from the Sun. Simply balancing the energy arriving with the energy leaving the Earth as a radiating black body indicates that the Earth's surface temperature $T_\oplus \propto L^{1/4}$ rises by about $1\,\mathrm{K}$ in $3 \times 10^7\,\mathrm{yr}$ so the habitable zone will have moved outside the Earth's orbit long before the Sun exhausts its core hydrogen fuel some $5 \times 10^9\,\mathrm{yr}$ from now.

(3) As $X_\mathrm{c} \to 0$ shell burning begins smoothly through the composition profile left by nuclear burning in the radiative zone. So

the hook at the end of the main sequence caused by thermal readjustment in the $5\,M_\odot$ star is not seen.

(6) The star begins to ascend the giant branch. At $t = 1.18 \times 10^{10}$ yr the core mass $M_c = 0.14 M_\odot$ and at the centre $P_e = 0.5P$, already significantly more degenerate than the helium core of our $5\,M_\odot$ star.

(7) At $t = 1.27 \times 10^{10}$ yr the electron degeneracy pressure $P_e = 0.81P$ dominates in the centre of the core of total mass $M_c = 0.472\,M_\odot$. The central temperature has risen to $T_c = 7.7 \times 10^7$ K and helium ignites. By then this $1\,M_\odot$ star has already existed as a red giant for over seven times as long as the entire lifetime of the $5\,M_\odot$ star. Pressure in the very degenerate core depends only on density (Sec. 3.7) so when the triple-α reaction begins the temperature rises without thermostatic control and nuclear reactions, aided by the extreme temperature dependence (Sec. 6.3.5.1), run away until the degeneracy can be lifted. This is the helium flash that marks the onset of helium burning in all low-mass stars with $M < 2\,M_\odot$ depending on composition and choice of convective overshooting. Energy generation peaks at about $10^{12}\,\mathrm{erg\,g^{-1}\,s^{-1}}$ so that the core luminosity reaches $L_c \approx 10^{46}\,\mathrm{erg\,s^{-1}}$, about that of an entire galaxy, but this lasts for only a few seconds. Many stellar evolution codes cannot properly model the flash itself but break down and resume only when hydrostatic and thermostatic burning is restored. Typically models show a series of convective burning flashes through the core, often beginning off-centre but eventually merging to form a single convective helium-burning core.

(8) Helium burning settles down within a convective core, Fig. 8.13. In contrast to the $5\,M_\odot$ and more massive stars there is no expansion to the blue.

(12) Unlike high-mass AGB stars this $1\,M_\odot$ star does not experience a second dredge up at the end of core helium burning. Instead the more rapidly burning helium shell simply catches up with the hydrogen-burning shell, Fig. 8.14. Thermal pulses begin when $M_c = 0.63\,M_\odot$. Observations indicate that third dredge

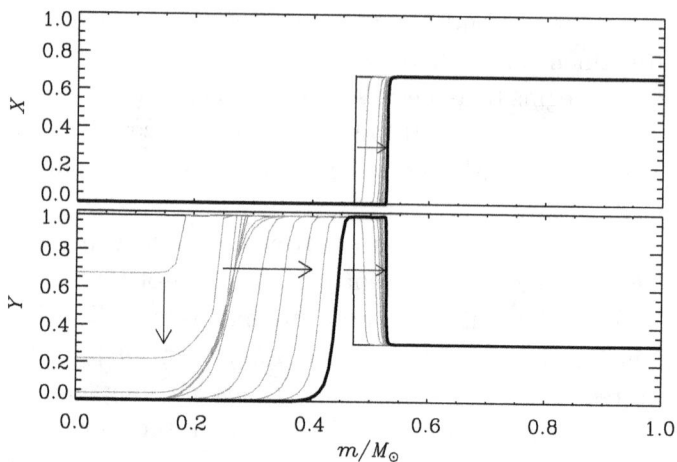

Fig. 8.13. The evolution of the hydrogen X and helium Y composition profiles during core helium burning. The behaviour is similar to that of the $5\,M_\odot$ star once helium is burning gently. However there is no second dredge up. The thin intershell is created by the helium-burning shell catching up with the hydrogen-burning shell without extinction.

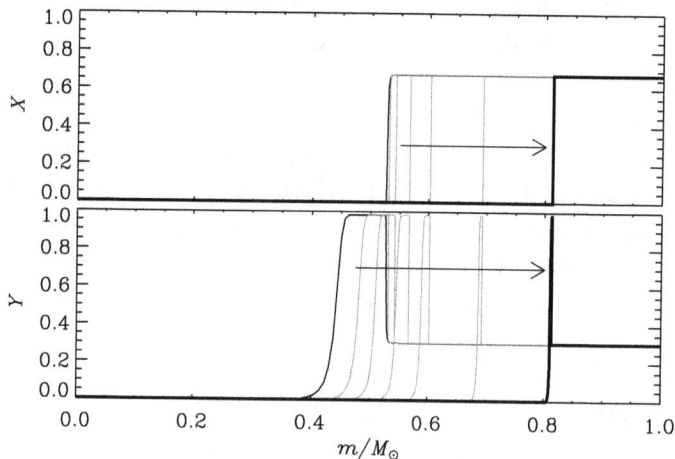

Fig. 8.14. Evolution of the hydrogen and helium composition profiles following from Fig. 8.13 at the onset of helium shell burning. There is no second dredge-up. Instead the more rapidly burning helium-burning shell catches up with the hydrogen-burning shell to form the thin intershell that drives thermal pulses.

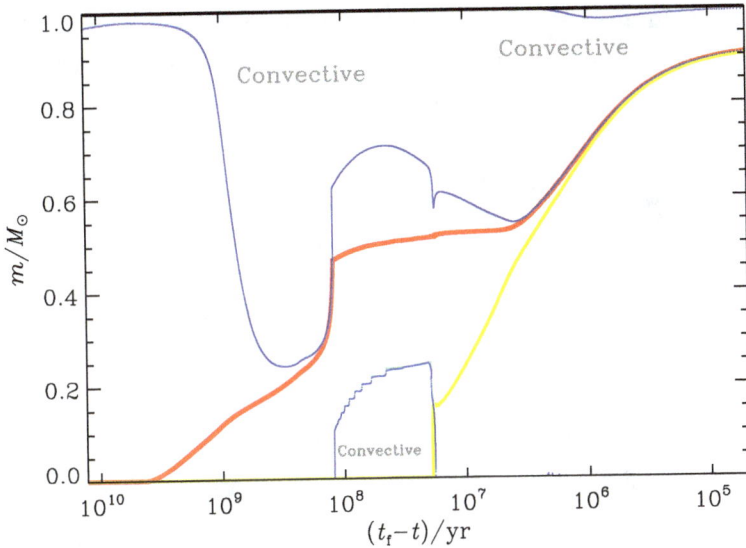

Fig. 8.15. The interior evolution of a $1M_\odot$ star. The red line at $X = 0.01$ is the location of the hydrogen-burning shell and yellow line at $Y = 0.01$ is the helium-burning shell. Blue lines are the convective boundaries to the deep giant envelope and the helium-burning core.

up begins soon afterwards but some models are unable to produce it and none can do so at sufficiently low core masses to match observed luminosities. Temperatures at the base of the convective envelope never rise enough for hot-bottom burning so once this AGB star becomes a carbon star it remains so.

We have again suppressed the thermal pulses to allow core growth in the absence of mass loss. Shell burning proceeds until almost the entire star, $M_c = 0.92\,M_\odot$, is converted to carbon and oxygen. Temperatures are too low for carbon ignition and the star cools to a CO white dwarf. In real stars stellar winds compete with slower core growth to remove the hydrogen envelope and determine the final mass of the CO white dwarf remnant. The maximum radius reached on the AGB depends upon this final core mass. Though it is likely to exceed one astronomical unit, mass loss in the case of the Sun means that the Earth will have migrated outwards. Whether or not Earth is ultimately swallowed by the Sun depends on how much mass is lost as well as the strength of dissipation of tides the planets will have raised on the Sun.

The complete internal evolution of the $1M_\odot$ star appears in Fig. 8.15 which shows how the burning shells and convective boundaries move within the star as it evolves. The helium core appears to grow in bursts because the growth is slowed even when a small quantity of extra fuel is mixed in. As for core breathing pulses, it is not clear whether this is a numerical artefact.

8.5.   The Evolution of a $7\,M_\odot$ Star

At higher initial masses evolution is complicated by higher burning temperatures and consequently more advanced nuclear burning phases. Figure 8.16 shows the internal evolution of a $7\,M_\odot$ star. The early evolution is very similar to that of a $5\,M_\odot$ star except that time-scales are shorter and core masses are larger. At the zero-age

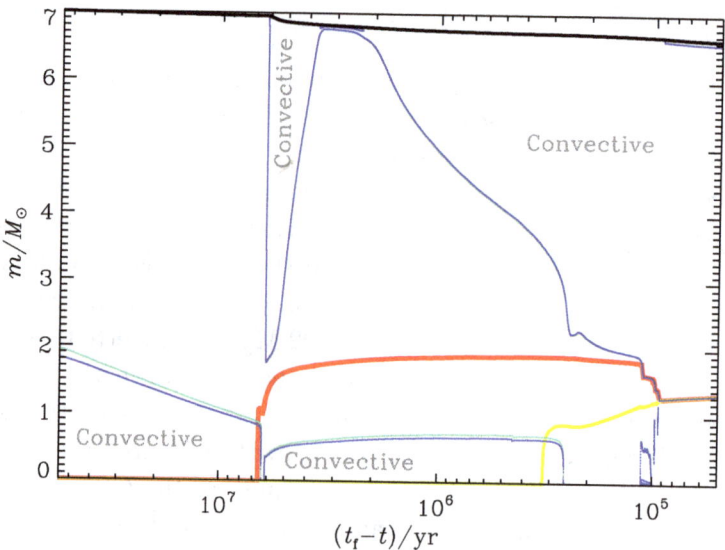

Fig. 8.16.   The interior evolution of a $7\,M_\odot$ star. Early stages are very similar to those of a $5\,M_\odot$ star but about 10^5 yr before the end of the model we see the convective onset of carbon burning.

main sequence the convective core extends to $2\,M_\odot$. This falls to $0.8\,M_\odot$ by the end of the main-sequence evolution after $4.9 \times 10^7\,\mathrm{yr}$. The red giant, core-helium burning and early AGB phases proceed as for the $5\,M_\odot$ star but then, before any second dredge up, at an age of $5.5 \times 10^7\,\mathrm{yr}$ when the hydrogen-burning shell is at $1.9\,M_\odot$ and the helium-burning shell at $1.2\,M_\odot$, conditions are reached for carbon ignition. Like the helium flash in low-mass stars this is a degenerate ignition. It is off-centre because neutrino cooling in the dense core has created a temperature inversion. In a series of flashes carbon is converted to neon throughout the core by which time second dredge up has brought the hydrogen-burning shell to just outside the helium-burning shell and thermal pulses can begin. Such thermally pulsing stars, with ONeMg cores are known as super-AGB stars. Their interpulse periods tend to be shorter and the pulses weaker. If stellar winds remove the hydrogen envelope quickly enough an ONeMg white dwarf remnant is left. However if the mass of the degenerate core can reach $1.37\,M_\odot$, at which point the central density $\rho \approx 10^{9.7}\,\mathrm{g\,cm^{-3}}$, electron captures on to $^{24}\mathrm{Mg}$ and $^{20}\mathrm{Ne}$ begin to remove the electron degeneracy pressure and the core collapses to a neutron star. Most of the gravitational energy escapes in neutrinos but if the surrounding material is sufficiently dense some energy can be trapped and drive a supernova explosion. The existence of these electron capture supernovae has yet to be established but it is likely that very little core material is ejected, making them nickel poor and so relatively faint. They probably experience weaker kicks than the normal core-collapse supernovae (Sec. 8.9.1). Supernova 2008S may be an example.

As initial mass increases the carbon ignition becomes less degenerate, eventually beginning smoothly at the centre of the star. Which stars become ONeMg white dwarfs or electron capture supernovae depends critically on the interplay between core growth and mass loss. The minimum mass depends on how large a CO core is left at the end of core helium burning and this in turn depends critically on the treatment of semi-convection and convective overshooting at the core boundary. This minimum mass may be as low as $6\,M_\odot$ or as high as $8\,M_\odot$. The upper mass limit for progenitors of ONeMg white dwarfs tends to be about $1\,M_\odot$ larger irrespective of assumptions made during core helium burning.

8.6. The Evolution of Stars More Massive Than $8\,M_\odot$

For the models presented here a star with initial mass above $8\,M_\odot$ develops an iron-group element core with $M > M_{Ch}$ at which point it explodes in a supernova driven by the gravitational collapse of its core which can no longer be supported by electron degeneracy pressure. Without mass loss the lifetime of our $8\,M_\odot$ star is 41 Myr and this falls to 3.3 Myr for our $100\,M_\odot$ star. At higher masses the main-sequence luminosity of a star tends to be directly proportional to its mass so that the lifetimes of such massive stars are similar.

Otherwise stars between 8 and $30\,M_\odot$ evolve much as the $7\,M_\odot$ star, except for the aforementioned fact that carbon ignition becomes gentler in the less degenerate cores as mass increases and they go on to successively burn neon, oxygen and silicon (Sec. 6.2.3) to eventually grow an iron core in nuclear statistical equilibrium (Sec. 6.4.1). Approximate ranges for the time-scales of various burning phases are listed in Table 8.1 along with typical ignition temperatures. Above about $30\,M_\odot$, core helium burning begins as the star crosses the Hertzsprung gap before it has established itself as a giant.

It is however mass loss that really determines the late stages of massive star evolution. Broadly there are stars that retain their hydrogen envelopes up to the formation of an iron core and those

Table 8.1. Fusion temperatures and time-scales in massive stars (Woosley, Heger and Weaver, 2002).

Core burning phase	T_c/K	Time-scale
H → He	3×10^7	3 to 40 Myr
He → C&O	2×10^8	0.5 to 2 Myr
C → O, Ne&Mg	8×10^8	1 kyr
Ne → O&Mg	1.6×10^9	1 to 1 000 yr
O&Mg → Si	2×10^9	1 yr
Si → Ni&Fe	3.3×10^9	1 to 30 d

that lose it first, becoming naked helium stars. The more massive a star, the more luminous it is and the stronger the stellar winds, it can drive. As illustrations we discuss the evolution of 15 and 25 $M_\odot$ stars as typical examples that end their lives as red giants of some kind and stars of 40 and 75 $M_\odot$ as typical of those that end their lives as naked helium stars. Again we stress that the particular evolution we describe depends critically on the uncertain mass loss history for these stars (Sec. 8.7.2).

Figure 8.17 shows the evolutionary tracks for typical 15 and 25 $M_\odot$ stars. Phases of evolution are numbered as for the 5 $M_\odot$ star though some no longer apply and later additional burning phases are identified. Figure 8.18 shows the internal evolution of the 15 $M_\odot$ star.

(1) The zero-age main sequence.
(2) The terminal-age main sequence, thermal time-scale shrinkage begins.

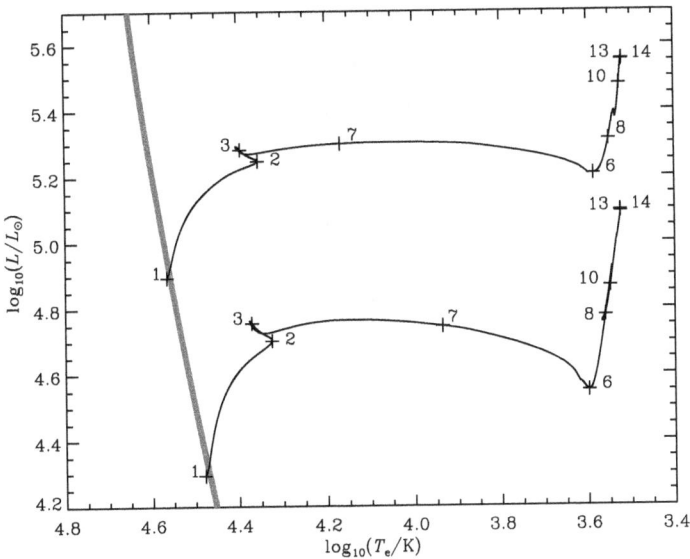

Fig. 8.17. Evolution of 15 (lower L track) and 25 $M_\odot$ stars in a Hertzsprung–Russel. Numbered points indicate phases of evolution. The thick grey line is the zero-age main sequence.

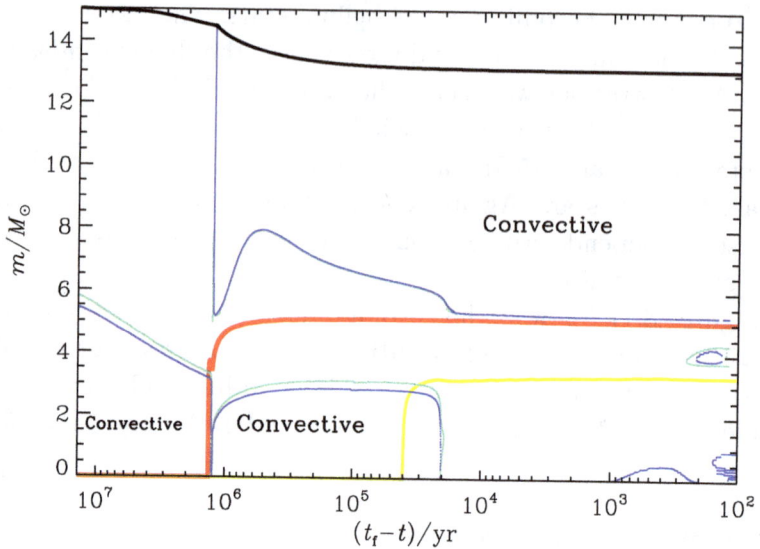

Fig. 8.18. The interior evolution of a 15 $M_\odot$ star. Again the red line, at $X = 0.01$, is the location of hydrogen burning, the yellow line, at $Y = 0.01$, is the helium-burning shell and the blue and cyan lines are convective and semi-convective boundaries. The black line is the surface of the star.

(3) Core hydrogen exhaustion $X_c = 0$.

(7) Helium ignites gently in the Hertzsprung gap.

(6) The Hayashi limit. A deep convective envelope forms and the first dredge up occurs.

(8) Helium burning begins in earnest in a growing convective core.

(10) Core helium exhaustion.

(13) Core carbon ignition. Carbon burns outwards through the core in a series of convective shells.

(14) Carbon burning is soon followed by neon, oxygen and silicon burning to leave a core of iron-group elements. Once the iron-group core is formed and its mass exceeds the Chandrasekhar mass it collapses to a neutron star (Sec. 8.9.1). If enough mass remains the remnant can collapse to a black hole.

Stars more massive than $25\,M_\odot$ or so experience little or no evolution as a red giant because mass loss is sufficiently strong to remove the hydrogen envelope before they reach the Hayashi limit.

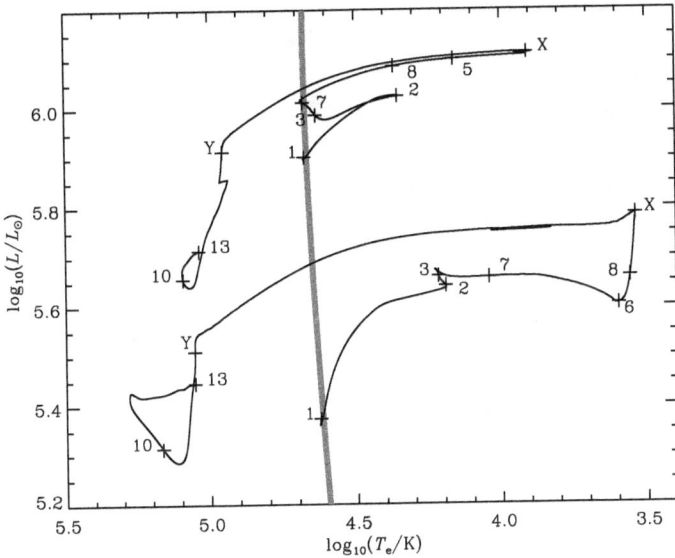

Fig. 8.19. Evolution of (lower L track) 40 and 75 $M_\odot$ stars in the Hertzsprung–Russel diagram. Numbered points again indicate particular phases of evolution and the thick grey line is the zero-age main sequence.

Figure 8.19 shows typical tracks followed by 40 and 75 $M_\odot$ stars with mass loss. Convective core helium burning begins in the Hertzsprung gap, just before a short red giant phase for the 40 $M_\odot$ star and somewhat earlier for the 75 $M_\odot$ star. Details of the internal evolution are shown for the 40 $M_\odot$ star in Fig. 8.20. There are again some notably different evolutionary phases.

(6) The Hayashi limit is reached and the deep convective envelope leads to first dredge up.

(X) Wind mass loss removes enough of the hydrogen-rich envelope that the star can no longer support a giant structure. The envelope collapses on to the core as mass loss continues.

(Y) All hydrogen from the star has been lost and it becomes a naked-helium star that is observed as a Wolf–Rayet star.

In such massive stars both neon and oxygen burn rapidly within a few years or less. Silicon burning then lasts only a matter of days. The lack of a hydrogen envelope does not have much effect on this nuclear

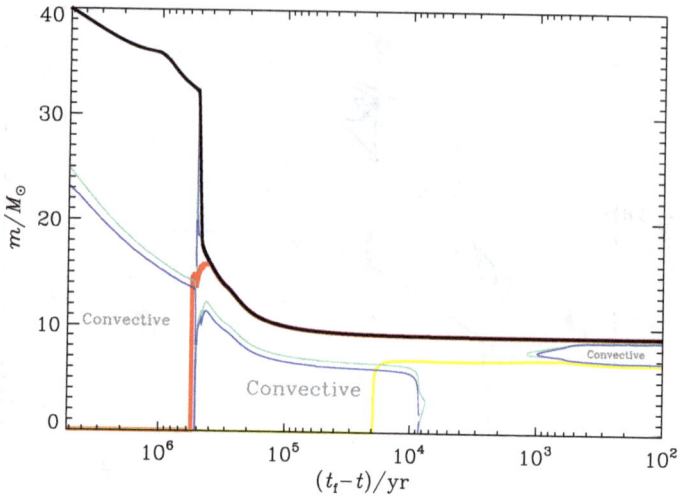

Fig. 8.20. The internal evolution of a 40 $M_\odot$ star. Lines have their now familiar meanings. Note the considerable mass loss experienced by this star during its relatively brief life time.

evolution of the core. Towards the end of their evolution these stars cool and develop convective envelopes. Exactly where in the H–R diagram they finally explode as supernovae is rather uncertain and varies from model to model with metallicity and mass loss history. There is however consensus that they do explode as blue stars some ten or more times hotter than they would as red supergiants with hydrogen-rich envelopes.

8.7. Further Complications and Uncertainties

We have described typical evolution of stars of various masses with compositions similar to the Sun. Even for these stars there are a great many uncertainties, particularly in the later stages of their evolution. Not least is mass loss. Better understood is the effect on internal evolution of different compositions. The abundances of metals affect the stellar opacity. Stars of lower metallicity tend to be smaller and hotter at the same mass and so evolve more quickly. The abundances of the CNO elements also affect the rate at which hydrogen burning proceeds. Rapid rotation can have an important effect if it can drive extra mixing in stars that leads to more hydrogen being mixed into the core on the main sequence and hence a longer lifetime, as well as the generation of anomalous surface abundances.

8.7.1. *Initial metallicity*

So far we have examined stars that have a metallicity of $Z = 0.02$ with an elemental distribution similar to the Sun's. We have not tried to perfectly match the Sun partly because its actual composition still remains uncertain (Chapter 1). The Sun is generally considered to be a metal-rich or population I star. Population II stars, with only about a twentieth the metal content of the Sun, have been known for some time, particularly in globular clusters. Recently population III has been used to refer to primordial stars formed with negligible metal abundances in the very early Universe. As an alternative to a mass fraction of metals Z, metallicity can be expressed relative to the Sun for a particular element, most usually iron, in the form [Fe/H], the decimal logarithm of ratio of the number of iron nuclei to hydrogen nuclei in the star divided by the same ratio in the Sun. So when the fraction of iron is directly proportional to the fraction of all metals

$$[\text{Fe/H}] = \log_{10}(Z/Z_\odot) - \log_{10}(X/X_\odot). \qquad (8.23)$$

To construct our models we assume that Y rises in proportion to Z as stars add to the big bang nucleosynthesis so that, specifically $Y = 0.25 + 1.5Z$. With $Z_\odot = 0.02$, $X_\odot = 0.7$ a typical population II metallicity of $Z = 0.001$ corresponds to [Fe/H] $= -1.336$. The old globular cluster M13 has [Fe/H] ≈ -1.6 while 47 Tuc has [Fe/H] ≈ -0.8. A number of extremely metal poor stars have now been identified in the Galactic field with [Fe/H] as low as -5.

Metallicity can also alter late stages of stellar evolution by affecting mass-loss rates. Winds from low-metallicity stars are even more poorly understood than those from population I stars but it is generally believed, particularly for hot stars, that mass loss weakens as metallicity falls. So the typical mass of a star that loses its envelope at a particular phase of evolution rises as metallicity falls.

In the extreme case of zero-metallicity population III stars hydrogen burning must begin only by the proton–proton chains. The weaker temperature dependence of the proton–proton reaction means that the cores of massive stars at zero-metallicity reach much higher temperatures in order to provide the pressure gradient necessary to support their envelopes. Above about $12\,M_\odot$ their cores are hot enough for helium burning to coexist with hydrogen burning. Carbon production raises the total CNO mass fraction to about 10^{-7} whence hydrogen burning can proceed by the CNO cycle. Such high temperature burning with very low abundances requires careful modelling of the later stages of evolution of these stars.

8.7.2. *Mass loss and stellar winds*

We have alluded to how mass loss plays an increasingly important role as stars evolve. Without it any star with a mass above the Chandrasekhar limit potentially undergoes core collapse. However if there is any carbon left in the core it can undergo a cold degenerate ignition at $1.38\,M_\odot$ before the core can collapse. Below $1.38\,M_\odot$ stars would end their lives as white dwarfs. The Sun is currently losing mass at only $10^{-14}\,M_\odot\,\mathrm{yr}^{-1}$ and, were it to continue at this rate, would lose negligible mass over its entire lifetime. Nevertheless even this weak wind has been responsible for slowing the spin of the Sun to a rate at which it has minimal effect on its evolution. Once it evolves to a giant the mass-loss rate is expected to increase significantly, and more so as it ascends the asymptotic giant branch, to such an extent that the expected remnant of the Sun is a carbon/oxygen white dwarf of about $0.6\,M_\odot$ well below M_{Ch}.

Stellar winds fall into two broad categories, of radiatively or coronally driven, depending on the evolutionary state, particularly the temperature and luminosity of the star.

(1) The main sequence:

> (i) Low effective temperature — the wind is thermally driven from a hot corona. In the Sun temperatures in the sparse corona surrounding the star rise to $10^6\,\mathrm{K}$ from a photosphere at less than $6000\,\mathrm{K}$. Heating of the corona is not understood but is potentially the result of microflares generated by the reconnection of magnetic loops. In turn the Sun's magnetic

field is thought to be maintained by a dynamo, probably in the tachocline, the thin layer between its convective envelope and radiative core, where differential rotation peaks. This dynamo also maintains a weak poloidal magnetic field that enables even the minimal wind to carry off substantial angular momentum, as if it were corotating with the Sun out to an Alfvén radius, where the kinetic energy in the wind exceeds the magnetic energy, at about $50\,R_\odot$. Such magnetic braking appears to be responsible for the slow rotation of all stars with convective envelopes.

(ii) High effective temperatures — the wind is driven by radiation pressure via the opacity of partially ionised metals. Iron tends to dominate because of its high abundance and cross-section. At solar metallicity the luminosity of the star exceeds the effective Eddington luminosity of metals for stars above about $20 M_\odot$ so radiatively driven winds become very important at higher masses.

(2) Post-main sequence:

(i) Low effective temperatures — the winds of giant stars are similarly thought to be thermally and magnetically driven. The potential well from which the wind must escape becomes increasingly shallower as the star grows and loses mass. High on the AGB large-scale oscillations, associated with Mira variables, can enhance mass loss and winds can be accelerated by radiation pressure on dust grains that form in the cool stellar environs.

(ii) High effective temperatures — once hydrogen-rich envelopes have been removed, stars shrink and their effective temperatures rise. Radiation driven winds are again important but the stars are both smaller and more luminous than on the main sequence and there is competition between increased driving and a deeper potential well from which to escape. These hot evolved stars can develop dense, optically thick fast winds and fall into the class of Wolf–Rayet stars, defined observationally as stars with broad emission lines of helium, carbon, nitrogen, silicon or oxygen but no, or very weak, hydrogen lines. They have effective temperatures between about 3×10^4 and 2×10^5 K. The broadening of the lines is due to Doppler shifts of around $1\,000\,\mathrm{km\,s^{-1}}$ with the absorption at various depths in the expanding atmosphere.

For now mass-loss rates included in stellar models remain substantially at the whim of the modeller and, though winds can be modelled, there is little quantitative agreement between different approaches. For population I stars we have observations of current mass-loss rates from individual stars. However there is little reason to assume that mass loss is not highly variable on time-scales short compared with the life times of even the shortest phases of stellar evolution. By examination of stellar remnants, particularly in clusters with a well-defined main-sequence turn-off masses and ages, we can estimate the mass lost integrated over the whole nuclear lifetime.

A simple formula, due to Reimers (1975), is often used in the absence of anything more sophisticated or believable. Physically Reimers' formula sets the mass-loss rate, by dimensional analysis, proportional to the luminosity and inversely proportional to the surface gravitational potential of the star, such that

$$\dot{M} = -4 \times 10^{-13} \eta \left(\frac{R}{R_\odot} \right) \left(\frac{L}{L_\odot} \right) \left(\frac{M_\odot}{M} \right) \mathrm{M_\odot\,yr^{-1}}. \qquad (8.24)$$

with the magnitude fixed by the winds from red supergiants, estimated from P Cygni profiles (Sec. 5.9), and the parameter η is of order unity and chosen by the modeller. More recent calibrations suggest that $\eta \approx 0.5$ for red giant branch stars. With this formula we estimate that mass loss can drive stellar evolution when the time-scale

$$\tau_{\dot{M}} \approx \frac{M}{\dot{M}} < \tau_{\mathrm{nuc}}. \qquad (8.25)$$

This is the case say when $R > 200\,R_\odot$ and $L > 200\,L_\odot$ typical of a giant. In cool supergiants and AGB stars, Mira pulsations and dust driving are found to amplify the mass-loss by a factor of 100 or more.

For completeness we shall summarise the mass-loss rates we have used in some of the stellar models so far presented but it must be borne in mind that this is our particular choice for these models and this is far from universally used or accepted. For main-sequence evolution the theoretically based but observationally tested rates developed by Vink, de Koter and Lamers (2001) predict mass-loss rates between 10^{-6} and $10^{-4}\,M_\odot\,\mathrm{yr}^{-1}$ for massive stars with radiative envelopes. Importantly these rates have a metallicity dependence because the winds are driven by radiation pressure on iron lines, for which $L > L_{\mathrm{Edd}}$, and so more iron gives stronger driving. At solar metallicity this dependence is approximately $\dot{M} \propto (Z/Z_\odot)^{0.5}$ because many of the atomic lines are saturated. The dependence steepens, approaching $\dot{M} \propto Z$ at very low metallicities, because driving of other lines, particularly those of CNO elements, becomes important.

Post main-sequence mass loss is even less well understood. All the rates used are empirically based. For giants we use the rates of de Jager, Nieuwenhuijzen and van der Hucht (1988). These are typically 10^{-6} to $5 \times 10^{-5}\,M_\odot\,\mathrm{yr}^{-1}$ for stars of 7 to $40\,M_\odot$. They can, as we do, also be scaled with metallicity but, given that the driving mechanism is unknown nor is its dependence on metallicity. Stellar populations created without scaled mass-loss rates predict fewer red supergiants at lower metallicities than are observed.

When massive stars lose their hydrogen envelopes to become naked helium stars we use the empirical mass-loss rates of Nugis and Lamers (2000). High luminosities and effective temperatures give rise to rates in excess of $10^{-5}\,M_\odot\,\mathrm{yr}^{-1}$. Such winds are optically thick with velocities of more than $1\,000\,\mathrm{km\,s}^{-1}$ and give rise to the very broad emission lines seen in Wolf–Rayet spectra. These strong winds are also thought to be driven by radiation pressure on iron and so are expected to scale with metallicity.

Stars with initial main-sequence masses $M > 50\,M_\odot$ appear to never become red giants of any kind and so must lose their hydrogen-rich envelopes before or soon after leaving the main sequence. This loss probably occurs in a luminous blue variable (LBV) phase of evolution. A prototypical example is the bright star η Carina which underwent a dramatic outburst, visible to the naked eye, in 1883. It is estimated to have ejected around $10\,M_\odot$ in this one mass-loss event. Less dramatic shorter time-scale variation is typically seen in other massive stars. It is not known exactly why these stars are so unstable. Their luminosity is close to the Eddington luminosity so the outbursts may be caused by continuum opacity driven winds. Binary companions or strange mode oscillations in the envelope may also play a significant role. Whatever drives the mass loss the end result is that the stars likely become Wolf–Rayets once their envelopes are sufficiently reduced.

8.7.3. *Stellar rotation*

Spots moving across the Sun's surface indicate that it rotates roughly once every 25 d at its equator and once every 31 d at its poles. Red and blue giants in the lower spectral classes reveal that they are rotating by a Doppler broadening of spectral lines. For otherwise naturally narrow spectral lines the full width of a rotationally broadened line measures $2v\sin i$, where $v = R\Omega$ is the equatorial velocity of a star with spin period $P_{\text{spin}} = 2\pi/\Omega$ and i is the inclination of the spin axis to our line of sight to the star. Some nearby stars, such as Vega, spin sufficiently fast that they are noticeably oblate because, in a frame rotating with the star, there is a centrifugal force that effectively weakens the gravity at the equator relative to the poles. The lower effective gravity at the equator makes it cooler than the poles and rotation drives meridional circulation within the star that can lead to important extra mixing whenever the time-scale on which it operates is shorter than the nuclear time-scale.

In Question 2 of this chapter you are invited to examine a rotating star using similar ideas to those developed for binary systems in the next chapter. Though we do not examine the details of the circulation flows that question reveals that rotation adds perturbations of order $\Omega^2/4\pi G\rho \approx \Omega^2/\Omega^2_{\text{crit}}$, where $\Omega_{\text{crit}} = \sqrt{GM/R^3}$ is the critical angular velocity at which the centrifugal force balances gravity at the surface of a spherical star. The circulation is required to maintain thermal balance in the star so the time-scale for mixing can be approximated by

$$\tau_{\text{merid}} \approx \tau_{\text{KH}} \left(\frac{\Omega_{\text{crit}}}{\Omega}\right)^2, \tag{8.26}$$

where τ_{KH} is the Kelvin–Helmholtz, or thermal, time-scale of the star. Comparing this with the nuclear time-scale we find

$$\frac{\tau_{\text{merid}}}{\tau_{\text{nuc}}} \approx 0.003 \left(\frac{\Omega_{\text{c}}}{\Omega}\right)^2 \frac{M}{M_\odot} \frac{R_\odot}{R}. \tag{8.27}$$

The Sun has $\Omega/\Omega_{\text{crit}} \approx 10^{-5}$ so that $\tau_{\text{merid}}/\tau_{\text{nuc}} = 3 \times 10^7$ and its rotation has a negligible effect on its evolution.

Hotter stars with radiative envelopes can often spin much faster. When $\Omega \to \Omega_{\text{crit}}$, τ_{merid} exceeds τ_{nuc} whenever $M > 20\,M_\odot$ or so. Such stars might well be homogeneously mixed. Indeed in some cases surface nitrogen is seen to be enhanced at the expense of carbon and oxygen, and this may be evidence of rotationally driven mixing from the convective core, where the CNO cycle operates, to the stellar surface. Not all nitrogen enhanced stars are rapidly rotating now but could well have been in the past. Hydrogen from the envelope is also mixed into the core and this prolongs the main-sequence lifetime. In the extreme case of complete mixing a star could evolve homologously, gradually decreasing the hydrogen content throughout the whole star (Sec. 7.4).

The interaction between rotation, mixing, magnetic fields and braking remains uncertain and is an active area of research. During formation even mildly rotating clouds of gas must lose significant angular momentum just to collapse to the size of a star. It is therefore likely that most stars begin their lives spinning relatively rapidly and that they, except for some young massive stars, no longer do so suggests that spin down is efficient for all stars, particularly those with convective envelopes like the Sun. Stars can also be spun up or down by tidal interactions in binary systems or spun up close to Ω_{crit} if accreting through a disc (Chapter 9).

8.8. Naked Helium Stars, White Dwarfs and Wolf–Rayet Stars

As a star evolves beyond the main-sequence mass loss becomes an increasingly important driver of its evolution. The final fate of a star is determined by competition between mass loss and nuclear burning. The deep convective envelope of a red giant responds to loss of mass by expanding slightly and cooling but, until it is almost completely removed, the underlying structure does not change qualitatively. When the total mass in the envelope is reduced to only 0.02 to 0.05 $M_\odot$ it becomes energetically favourable for the remaining material to collapse down on to the burning shell releasing energy on a thermal time-scale. The burning shell responds more slowly so the star moves horizontally from the red to the blue in a H–R diagram. The remaining fuel is consumed by the burning shell, possibly in a few flashes as the shrinking envelope becomes degenerate, and eventually extinguishes.

Figure 8.21 shows the tracks followed by 0.5, 5 and 30 $M_\odot$ stars in a H–R diagram with typical mass loss. The lowest mass stars lose their hydrogen envelopes during their first ascent of the giant branch. Their helium cores are degenerate and of low mass. We must have $M_c < 0.472\,M_\odot$, so that the helium flash is not reached before the hydrogen envelope is lost and actually M_c must be lower still to avoid subsequent helium ignition that would lead to a horizontal branch star. The stars move horizontally to the blue, at almost constant luminosity, as the remaining hydrogen envelope collapses on to the core. They then turn back to the red evolving along a white dwarf cooling track at a radius dependent only on the total mass of the core. For very low-mass white dwarfs the amount of hydrogen envelope that survives affects the radius and so there is a small dispersion around the cooling track. Intermediate mass stars, illustrated by the 5 $M_\odot$ star, evolve through core helium burning and ascend the asymptotic giant branch before losing their hydrogen envelope in a similar way to leave a CO white dwarf. If they have reached the thermally-pulsing AGB they may undergo a final pulse and move back to the red from any point in their excursion to the blue, a late thermal pulse, or even the early cooling stages, a very

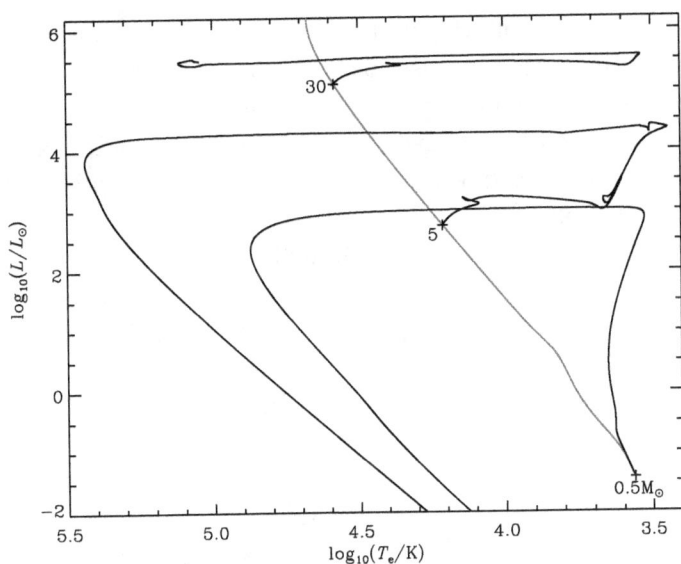

Fig. 8.21. Tracks followed by 0.5, 5 and 30 $M_\odot$ stars with typical mass loss in a H–R diagram. The stars leave the red giant branch and move horizontally to the blue when their hydrogen envelopes are lost. For the 0.5 $M_\odot$ star this occurs on the first ascent of the giant branch and the remnant is a helium white dwarf. For the 5 $M_\odot$ star it is on the asymptotic giant branch and the remnant is a carbon oxygen white dwarf. In both cases the stars converge to white dwarf cooling tracks once the burning shells are extinguished. The 30 $M_\odot$ star loses its envelope on its first ascent of the giant branch but its helium core is sufficiently massive to ignite the triple-α reaction and it becomes a naked helium star.

late thermal pulse. There may also be hydrogen shell flashes as the envelope becomes degenerate while collapsing on to the core.

Yet more massive stars again lose their hydrogen envelopes on their first ascent of the giant branch but for these the helium core is sufficiently massive to go on to ignite non-degenerately. At lower masses the burning is short-lived and consumes only the central parts of the core before extinguishing. The star then cools to a mixed white dwarf with a carbon/oxygen core and helium envelope. At higher masses convective helium burning can be established and the core continues to evolve as a naked helium-burning star. This is illustrated by our 30 $M_\odot$ star in Fig. 8.21. If the exposed core is massive enough nuclear burning can progress all the way to iron whence core collapse results in a supernova. Naked helium stars are often observed as Wolf–Rayet stars when their high luminosities and surface temperatures give rise to strong winds, $|\dot{M}| > 10^{-5}\,M_\odot\,\mathrm{yr}^{-1}$. In binary systems (Chapter 9) efficient mass loss, driven by Roche-lobe overflow or common envelope evolution can expose white dwarfs and naked helium stars from a much wider range of progenitors.

The evolution of naked helium stars in Fig. 8.22 was calculated by artificially generating a zero-age helium main sequence for which $X = 0$, $Y = 0.98$ and $Z = 0.02$. The minimum mass for helium ignition is around $0.32\,M_\odot$ when ignited thermally. Stars less mass than about $2\,M_\odot$ evolve to a helium-giant branch and thence typically to a CO white dwarf following mass loss. More massive stars end their lives as core-collapse supernovae when their processed cores exceed M_{Ch}. Once again mass loss makes the final fate of the stars uncertain. If a CO core reaches $1.38\,M_\odot$ while cold enough for degenerate carbon ignition an explosive thermonuclear runaway similar to a type Ia supernova (Sec. 9.9.4) but with a helium-rich envelope can be set off. Alternatively warm carbon ignition leads to

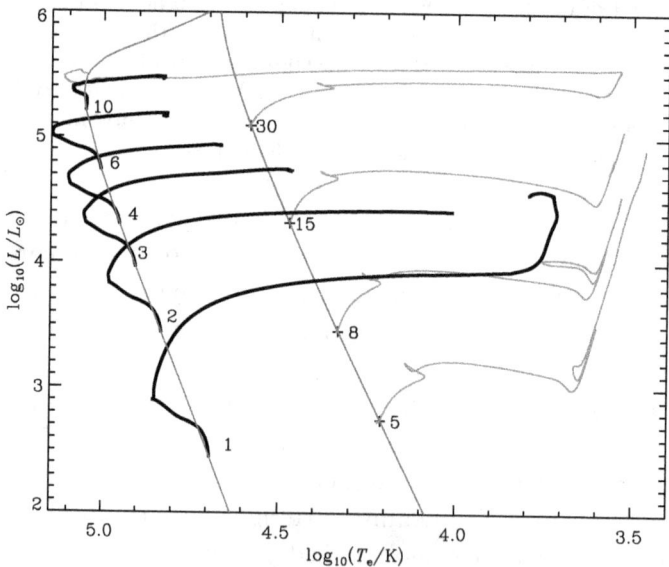

Fig. 8.22. H–R diagram for naked helium stars, black, overlaid on hydrogen stars, grey. The helium main sequence lies blueward of the hydrogen zero-age main sequence and stars of similar masses are more luminous.

more gentle burning to an oxygen/neon/magnesium core which itself collapses when electrons are captured by magnesium and neon at slightly lower masses around $1.37\,M_\odot$.

The horizontal branches seen in the H–R diagrams of globular clusters arise because even a thin hydrogen-rich layer on top of an otherwise naked helium star lowers the effective temperature. The horizontal branch stars are typically red giants stripped of their envelopes close to the point of the core helium flash. This leaves a helium star of just under $0.5\,M_\odot$. Whatever the mechanism, variation in the surface hydrogen from none to about $0.05\,M_\odot$ determines the temperature of the stars, all of which have similar luminosities. The result is a horizontal branch extending from the naked helium star main sequence to the red clump where the normal core-helium-burning stars with hydrogen envelopes are located.

Recall that Wolf–Rayet stars are identified observationally by their broad emission lines that are explained, in theory, by a rapidly expanding optically thick wind with no or very little hydrogen. Their precise descriptions depend on which emission lines they display, with four major types arranged in order of envelope removal, WNL→WNE→WC→WO. The WNL (sometimes WNH) stars have little H and are N rich. The WNE stars have no H and are He and N rich. The WC stars are He poor and C rich. The WO stars have no He and are O rich. These spectral types are, in practice, determined by a mixture of surface temperature as well as stellar composition and deconvolving these contributions is difficult and remains the subject of current research. Because their winds are optically thick there is no single stellar photosphere that can be called the stellar surface and their winds and spectra must be modelled in detail that includes hydrodynamic and non-LTE effects.

8.9. The Deaths of Massive Stars

The ends of the lives of massive stars remain uncertain mostly because of their uncertain mass loss. Nevertheless some stars do end their lives as supernovae. The classification of supernovae, like much in astronomy, is historically arcane. Most important is the spectral evolution of the supernova light. First supernovae are divided according to whether their spectra show hydrogen lines, type II, or not, type I. The type I supernovae are subdivided into the type Ia which show silicon lines and the type Ib/c which show no silicon. Both type II and type Ib/c supernovae result from core collapse while type Ia are a thermonuclear explosion of a carbon-rich cold degenerate core. In all cases much of the supernova light is generated by radioactive decay of nickel-56 to iron-56 and the energy released is determined by nuclear statistical equilibrium (Sec. 6.4.1) in the final moments of a star's active life.

8.9.1.   *Core collapse*

Once no further energy can be extracted by nuclear burning an iron-group element core with $M_c > M_{Ch}$ remains. As this cools it cannot be supported by electron degeneracy pressure but rather collapses on its dynamical time-scale, of about 10 ms, to a neutron star of mass $1.4\,M_\odot$ and radius of about 10 km, when nuclear densities of $\rho \approx 10^{14}\,\mathrm{g\,cm^{-3}}$ are reached and pressure support can be provided by neutron degeneracy. At this point the equation of state stiffens and the gravitational energy must be released. The stellar core has collapsed from a radius of about 1000 km so the total energy liberated

$$E_{\mathrm{sn}} \approx GM_c^2 \left(\frac{1}{R_{\mathrm{NS}}} - \frac{1}{R_{\mathrm{WD}}} \right) \approx 5 \times 10^{45}\,\mathrm{J}. \qquad (8.28)$$

Compare this to the binding energy of the envelope of a $10\,M_\odot$ star,

$$E_{\mathrm{bind}} = \int_{M_{\mathrm{core}}}^{M} \frac{Gm\,dm}{r} \approx 10^{44}\,\mathrm{J} \ll E_{\mathrm{sn}}. \qquad (8.29)$$

However much of this energy is released in neutrinos as electrons combine with protons to form neutrons once $\rho > 10^{12}\,\mathrm{g\,cm^{-3}}$. Many of these neutrinos are simply lost because the mean free path λ_ν of neutrinos is large. However the material in the collapsing core is extremely dense and the neutrino interaction cross-sections are proportional to the atomic mass of the nuclei. In the iron-rich material at the outer edge of the collapsing core $\lambda_\nu \approx 100\,\mathrm{km}$ and so a significant fraction of neutrinos can be trapped and their energy thermalised. The theory of the explosion is still not fully understood and, except in very particular cases, numerical stellar evolution models that take into account our current understanding of the physics still struggle to reach the final explosion which is usually modelled independently with a hydrodynamic code. How the energy is trapped turns out not to be important for the reaction of the envelope and hence we can still produce realistic models of expanding supernovae and match their light curves and spectra in detail.

8.10. Supernova Spectra and Light Curves

The light curves of supernovae can be well modelled by conservation of energy once it has been released at the inner parts of the stellar envelope around the collapsing core whether thermally or mechanically. Either way the envelope rapidly loses hydrostatic equilibrium and expands supersonically. The expanding ejecta cool and emit radiation, often at sufficient luminosity outshine the supernova's host galaxy for a few months. A similar luminosity is provided by the radioactive decay of the cooling iron-group elements that formed in nuclear statistical equilibrium.

Type Ia supernovae are intrinsically the brightest and also seem to be remarkable standard candles in that their peak luminosity varies by only $\pm 60\%$. A correlation between this peak luminosity and the light curve shape reduces this spread to $\pm 15\%$ which, coupled with their brightness makes them useful tools to measure the early Universe. The radioactive decay of ^{56}Ni to ^{54}Co is apparent in their light curves and it is now well established that they originate from the thermonuclear explosion of about $1\,M_\odot$ of degenerate carbon and oxygen. We return to them again in Sec. 9.9.4. Type Ib/c supernovae are thought to arise when the naked helium core of a star that has been stripped of its hydrogen envelope exceeds the Chandrasekhar mass and collapses.

The nature of the hydrogen envelope in a type II supernova leads to an interesting variety of light curves and spectra. Initially the light curve is dominated by the thermal emission of the ejecta. When the hydrogen envelope is massive, as is typical for a red supergiant progenitor, as the ejecta expands in radius, the photosphere, which lies close to where hydrogen atoms recombine, moves inwards in mass. So the photosphere remains at almost constant radius and this gives rise to a long plateau of constant luminosity in the supernova light curve. These are the most common, type IIP, supernovae. Though typical not all type II supernova light curves show such a plateau. The light curves of type IIL supernovae show a linear decay with time which can be explained by a substantially less massive hydrogen envelope in which the expansion does not keep pace with the photosphere. With almost no hydrogen envelope the type IIb supernovae initially show prominent hydrogen lines that fade to leave something looking more like a type Ib. Thus core-collapse supernovae conform to a sequence of types, IIP, IIL, IIb, Ib and Ic as the envelope of the star is progressively removed before the explosion. By looking at well resolved pre- and post-supernova images of nearby galaxies progenitors, such as 2003 gd and 2008 bk, of type IIP supernovae have been confirmed to be red supergiants (Smartt, 2015). As yet insufficient progenitors of the other core-collapse types have been identified to confirm our expectations.

Because the light curves of the type I and eventually the type II supernovae are dominated by radioactive decay it is possible to make a reasonable estimate of the amount of ^{56}Ni ejected. The decay to ^{56}Co is by electron capture which leaves an excited nucleus that rapidly emits a γ-ray,

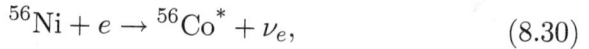

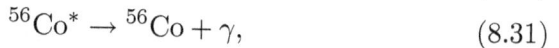

$$^{56}\text{Ni} + e \rightarrow {}^{56}\text{Co}^* + \nu_e, \tag{8.30}$$

$$^{56}\text{Co}^* \rightarrow {}^{56}\text{Co} + \gamma, \tag{8.31}$$

with a half-life of 6.1 d and $Q_\gamma = 1.75\,\text{MeV}$. The decay to ^{56}Fe is either by another electron capture or, 19% of the time, by spontaneous decay with the emission of a positron, the kinetic energy of which is thermalised and itself annihilates with an electron,

$$^{56}\text{Co} + e \rightarrow {}^{56}\text{Fe}^* + \nu_e \tag{8.32}$$

or

$$^{56}\text{Co} \rightarrow {}^{56}\text{Fe}^* + e^+ + \nu_e, \tag{8.33}$$

$$^{56}\text{Fe}^* \rightarrow {}^{56}\text{Fe} + \gamma, \tag{8.34}$$

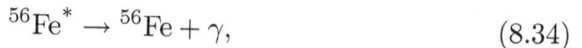

with a combined half-life of 77.12 d and $Q = 3.73\,\text{MeV}$.

The cobalt decay is observed at late times but the nickel decay is often swamped by the emission of thermal energy deposited in the envelope. A full model of the supernova light curve combines the radioactive decay and conservation of the envelope's thermal energy but it is usually possible to make a simple estimate of the original ^{56}Ni mass from the late time evolution which is dominated by the decay of ^{56}Co. The luminosity is proportional to the rate of decay so

$$L \propto \frac{\mathrm{d}N_{\mathrm{Co}}}{\mathrm{d}t} = \lambda N_{\mathrm{Co}}, \tag{8.35}$$

where N_{Co} is the number of cobalt nuclei remaining and $\lambda = (\log_e 2)/\tau_{1/2}$ for a half-life $\tau_{1/2}$. The rate of change of the bolometric magnitude of the supernova confirms that we are seeing the cobalt decay and the total luminosity extrapolated back to the peak gives us the estimate of the original nickel mass.

8.11. Evolution Summary

In this chapter we have discussed the evolution of stars in detail. We can have confidence in the general scheme but precise quantitative details are somewhat dependent on uncertain physical processes such as mass-loss rates, convection and stellar rotation. There are three broad mass ranges over which evolution follows the same general phases, with a few quirks which we do not repeat here. The precise mass ranges for each type are uncertain.

(1) Low-mass stars, between about 0.08 and $0.8\,M_\odot$, only burn hydrogen to helium on the main sequence. They evolve to red giants and then lose their envelopes to end their lives as helium white dwarfs.

(2) Intermediate-mass stars, between about 0.8 and $8\,M_\odot$, ignite helium and form a carbon–oxygen core. They become double shell burning AGB stars. After second dredge up, in the more massive stars, or when the helium-burning shell catches up with the hydrogen, in the less massive, the proximity of the burning shells leads to thermal pulses. The envelope is lost and the stars end their lives as carbon–oxygen white dwarfs. At the top end of

this range core carbon burning can begin before the envelope is lost and leave an oxygen–neon white dwarf.

(3) Massive stars, above $8\,M_\odot$, progress their nuclear burning further to eventually form an iron-group core which must collapse once its mass exceeds the Chandrasekhar limit. They end their lives as supernovae, the observational characteristics of which depend on how much of the hydrogen envelope is lost before core-collapse. Their final state is a neutron star or black hole.

8.12. Questions

1. A red giant of total mass M has luminosity

$$L = CM_c^6,$$

where C is a constant and M_c is its core mass. Explain why, to a good approximation the rate of core growth

$$\dot{M_c} = AL,$$

where A is a constant.

The giant is losing mass in a stellar wind at a rate such that

$$\dot{M} = -B\frac{RL}{M},$$

where R is the radius of the giant and is related to its luminosity and mass according to

$$R = DLM^{-1/3},$$

where D is a constant. The initial mass of the giant is M_0 and its initial core mass is M_{c0}. Show that it leaves a white dwarf remnant of mass M_{WD} such that

$$M_0^{7/3} - M_{WD}^{7/3} = \frac{BDC}{3A}\left(M_{WD}^7 - M_{c0}^7\right),$$

as long as M_{WD} is sufficiently small.

2. In a frame in which all the material is corotating with angular velocity $\mathbf{\Omega}$ show that the equation of hydrostatic equilibrium in a

star can be written as

$$\nabla P = -\rho \nabla \phi,$$

where P is the pressure, ρ is the density and $\phi(\boldsymbol{r})$ is a combined gravitational and centrifugal potential which satisfies

$$\nabla^2 \phi = 4\pi G \rho - 2\Omega^2.$$

Show that P and ρ must be constant on equipotential surfaces. Hence deduce that $\nabla^2 \phi$, but not necessarily $|\nabla \phi|$, is constant on equipotential surfaces.

Argue that, for a star of uniform composition, temperature T is also constant on equipotential surfaces.

The star is in radiative equilibrium with heat flux

$$\boldsymbol{F} = -\chi \nabla T = -\chi \frac{\mathrm{d}T}{\mathrm{d}\phi} \nabla \phi,$$

where χ is the conductivity which is related to the opacity $\kappa(\rho, T)$ by

$$\chi = \frac{4acT^3}{3\kappa\rho},$$

a is the radiation constant and c is the speed of light. Show that the effective temperature on the surface of the star

$$T_{\mathrm{e}} \propto g^{1/4},$$

where g is the magnitude of the effective gravitational acceleration and sketch the cross-section of a rapidly spinning star and indicate where it is hottest.

Why is it not in general possible for the energy balance to be given simply by

$$\nabla \cdot \boldsymbol{F} = \rho \epsilon,$$

where $\epsilon(\rho, T)$ is the energy generation rate per unit mass?

Now suppose that there is a steady circulation velocity field $\boldsymbol{v}(\boldsymbol{r})$ so that the energy balance is given instead by

$$\rho T \frac{\mathrm{D}s}{\mathrm{D}t} = \rho \boldsymbol{v} \cdot T \nabla s = \rho \epsilon - \nabla \cdot \boldsymbol{F},$$

where $s(\rho, T)$ is the specific entropy. Use continuity and the thermodynamic relation

$$T \, \mathrm{d}s = \mathrm{d}h - \frac{1}{\rho} \mathrm{d}P,$$

where $h(\rho, T)$ is the specific enthalpy, to show that

$$\int_S \boldsymbol{F} \cdot \mathrm{d}\boldsymbol{S} = \int_V \rho \epsilon \, \mathrm{d}V,$$

where S is an equipotential surface enclosing a volume V.

Hence show that the radiative gradient is given by

$$\frac{\mathrm{d} \log T}{\mathrm{d} \log P} = \frac{3 \kappa P L}{16 \pi a c G m T^4} \left(1 - \frac{\Omega^2 V}{2 \pi G M}\right)^{-1},$$

where L is the rate of energy generation in V and m is the mass in V.

3. The light curve of a type Ia supernova is powered by the decay of ^{56}Ni to ^{56}Co with a half-life of 6.1 d. The energy per unit mass generated by this reaction is 3.1×10^{16} erg g^{-1}. Given that at least $1 \, M_\odot$ of ^{56}Ni is formed in a type Ia supernova estimate the initial luminosity of the explosion.

The ^{56}Co subsequently further decays to ^{56}Fe with half-life of 77.1 d and an energy per unit mass of 6.4×10^{16} erg g^{-1}. Estimate the time since the explosion when this second decay begins to dominate the emission from the supernova.

In comparison most core-collapse supernovae produce approximately 0.01 to 0.1 $M_\odot$ of nickel-56. Estimate how many magnitudes fainter they are at late times than the thermonuclear type Ia events.

4. Most nuclear reactions throughout a star's life proceed so that the ratio of the mean number of protons to the mean number of neutrons is given by $\langle Z \rangle / \langle N \rangle = 1$. However during helium-burning nitrogen-14 is converted to neon-22 by the reactions

$^{14}N(\alpha,\gamma)^{18}F(,e^+\nu_e)^{18}O(\alpha,\gamma)^{22}Ne$. Neon-22 nuclei have $\langle Z\rangle/\langle N\rangle = 5/6$. Nitrogen becomes the most abundant metal after hydrogen burning via the CNO cycle. Assume it contributes the entire metallicity of a star at that point and that its progeny, ^{22}Ne, is the only element that contributes excess neutrons in a carbon/oxygen white dwarf and everything other than ^{22}Ne is converted to ^{12}C and ^{16}O in equal amounts by mass. Show that

$$\frac{\langle Z\rangle}{\langle N\rangle} = \frac{11-Z}{11+Z},$$

where Z is the metallicity mass fraction.

The progenitor of a type Ia supernova is a carbon–oxygen white dwarf. Describe the light curve of a type Ia supernova and how it is powered.

If a supernova with a progenitor of zero metallicity forms $1M_\odot$ of nickel, calculate to two significant figures how much nickel a solar metallicity ($Z = 0.02$) type Ia supernova produces when the equilibrium is reached at the same temperature and density and that the only products are ^{56}Ni and ^{54}Fe.

If the luminosity of a type Ia supernova is assumed to be a standard candle and calibrated at solar metallicity, show that the error in the distance to an extremely low metallicity type Ia supernova is about 3%.

[*Note* $(Q(56,28) - Q(54,26))/k = -2.7 \times 10^{10}\ K\ and\ \sqrt{550/523} \approx 1.03$.]

Chapter 9

Binary Stars

In our Galaxy only a minority of stars are as isolated from other stars as is our own Sun. In our case the Sun rotates slowly and the angular momentum of the solar system lies in the orbits of the planets. For many systems the angular momentum is shared between two or more objects of sufficient mass to become stars. William Herschel (1803) introduced the name binary star for a gravitationally bound system of two stars. For the majority of such systems the physics is simply Newton's gravity and the description of their motions is a generalised form of Kepler's laws for planetary motion. Thus the scientific study of binary stars is in some respects older than that of the evolution of single stars. However if the two stars in a binary system are close enough their individual evolution can influence that of their companion. In the extreme case, material at the surface of one star can end up more attracted to its companion than itself giving birth to a number of interesting exotic objects, particularly when the companion is a dense degenerate object, a white dwarf, a neutron star of even a black hole. In this chapter we explore the physics governing binary star orbital evolution and interaction and look in more detail at some examples of exotic binary stars.

Historically the concept of double stars has existed since ancient times. Written records in Ptolemy's Almagest, dating from the second century AD, report several double stars almost certainly recorded by Hipparchos almost three centuries earlier. These however are not among Herschel's binary stars but are in fact chance superpositions of relatively bright stars on the celestial sphere. Indeed it was generally believed that this was the case for all double stars even when the advent of telescopes meant that some apparently single stars could be resolved into two. Galileo resolved the components of Mizar soon after he turned his telescope to the celestial sphere. Other telescopic double stars were soon found by all who looked. This paradigm remained unchallenged until the Revd John Michell (1767) considered the expected incidence of binary stars if all stars were simply placed at random on the celestial sphere. He correctly determined "... that it is highly probable in particular, and next to a certainty in general, that such double stars, &c. as appear to consist of two or more stars placed very near together, do really consist of stars placed near together, and under the influence of some general law, ..." Speaking of clusters, and binary stars by inference, he concluded that "they form a kind of systems, ..., to whatever cause this may be owing, whether to their mutual gravitation, or to some other law or appointment of the Creator." — a statement that still well sums up our understanding of the physics of the formation of binary stars. Herschel clung to the idea of random placement because he remained keen to use the movement of a relatively close component of a double star relative to its more distant companion as a measure of parallax, an idea that had been around at least since Galileo and still seemed only to need better instrumentation. It was such observations that finally led him to the confirmation of the existence of true binary stars.

9.1. Numbers

The relative numbers of stars in binary systems is usually expressed either as the fraction of systems that are single, binary or multiple or as the number of stars that find themselves in such systems. Table 9.1 lists the multiplicity identified amongst bright stars. The majority of the systems listed as multiple are triple but the number includes multiples of up to seven stars, found in two systems. Often the ratio of the total number of stars to the total number of systems is referred to as the typical multiplicity. For bright systems this amounts to 1.52.

Table 9.1. The multiplicity of stars brighter than sixth magnitude.

	Single	Binary	Triple or higher multiples
systems	59%	31%	9%
stars	39%	41%	20%

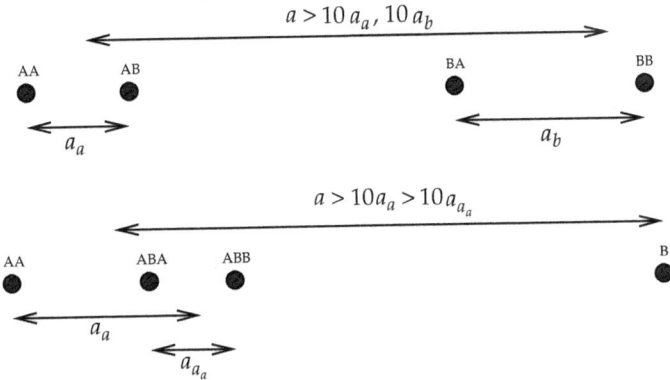

Fig. 9.1. Two possible configurations of quadruple systems with long lifetimes. The convention of labelling the stars in a binary as A and B is extended through the hierarchy. Separations of binary systems within the hierarchy must be typically a factor ten larger when moving from one level up to the next for long-term survival.

Because fainter companions are missing from the observations it is estimated that the true multiplicity is about 2 and from this it is often said that half the stars in the Galaxy have companions. There is also a tendency for more massive stars to have companions. Almost all O stars are in binary systems and most M stars appear to be single. Many systems are very wide and the presence of a companion is hardly felt. However we estimate that at least half of those stars with companions interact at some point in their lives, the closest by rapid mass transfer and the widest at least by weak tides.

In order to be reasonably stable over the stellar lifetimes multiple systems must be hierarchical and so can be considered as a sequence of binary stars within binary stars. For instance quadruple systems (Fig. 9.1) can take essentially two forms. Either they have, like

Capella, two pairs of stars both orbiting one another or, like ξ Tau, a very close binary in a wider orbit with a third star and then this triple system in a yet wider orbit with the fourth star. Typically the separations of nested pairs are factors of ten to a thousand or so smaller. However long-term stability is often achievable when this ratio is as low as three or four if the orbits are coplanar but no multiple system is indefinitely stable. There are certainly a few instances where interactions between three stars in a system could be important but for the most part interaction is between pairs of stars and we can concentrate on just interacting binary stars.

9.2. Observed Binary Stars

Binary stars fall into three classes according to the type of observations possible. Visual binaries tend to have long periods and are near enough to us that their orbits that can be seen in the motion of the component stars relative to the fixed background of distant stars and galaxies. When only one component of the pair can be seen executing its elliptical motion, the system is called an astrometric binary star. To determine the orbit precise observations are required and the stellar parallax and proper motion must be subtracted carefully. Even amongst the bright stars visible to the naked eye there are just over a hundred doubly bright visual binary stars in which the two components can be telescopically separated from the ground. Astrometric satellites such as *Hipparcos*, at the end of the twentieth century and *Gaia*, launched on 19th December 2013, are revolutionising our data. *Gaia* is expected to find some fifty million visual binary stars and a further fifty million astrometric systems. For spectroscopic binaries changes in the radial velocities of the stars are measured by the Doppler shifts of their spectral lines. If these shifts can be seen in the spectra of both components of a binary star the system is classified as SB2. It is SB1 if only one set of moving spectral lines are seen. Cross correlation of the entire spectrum with that of a similar stationary star either computationally or with a machine that passes the light through an appropriate mask can give extremely accurate measurements even when there is significant light pollution near the telescope. Photometric binaries, identified by variations in their magnitude, are of two types. In eclipsing systems one star periodically passes in front of the other. Ellipsoidal binaries show sinusoidal like variations in luminosity because the stars are tidally distorted by their companions. Eclipsing SB2s are the best available test of theoretical stellar evolution models beyond the Sun because they allow direct measurements of the masses and radii of both components.

9.3. Orbits

Before we consider the interaction of stars, which comes about because of their extended size, we shall treat them as point masses in order to calculate their underlying orbital motion. As a consequence of Newton's laws of motion and gravity the orbits of binary stars (Fig. 9.2) obey a generalised form of Kepler's laws for planetary motion. First the orbits are conic sections and bound orbits are ellipses. The diagram shows the semi-major axis a, the semi-minor axis b and the semi-latus rectum l. These are related to the eccentricity e by

$$l = a(1 - e^2) \tag{9.1}$$

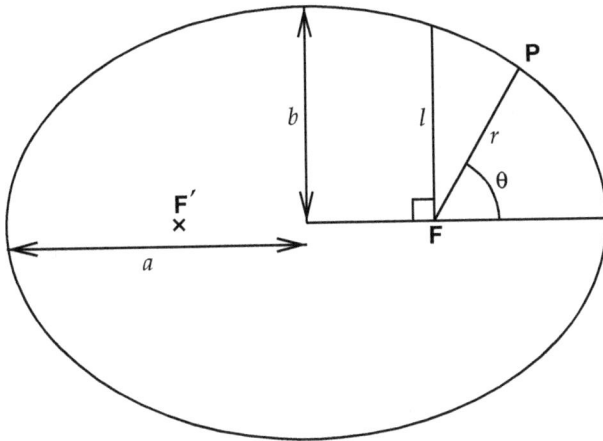

Fig. 9.2. Stars in a bound binary star follow elliptical orbits. One star is at the focus F and the other orbits at P around the ellipse.

and

$$e^2 = 1 - b^2/a^2. \tag{9.2}$$

A general point on the ellipse is given parametrically by

$$r = \frac{l}{1 + e \cos \theta}, \tag{9.3}$$

where r is the distance from the primary focus F and θ is the angle from the line joining the stars at periastron to the line joining F to P.

Kepler's second law states that the line connecting the two bodies sweeps out equal areas in equal times. If one body is considered fixed at F while the other orbits at P this is equivalent to

$$\frac{1}{2}r^2\dot{\theta} = \frac{\pi a^2 (1 - e^2)^{1/2}}{P}, \tag{9.4}$$

where the numerator is the area of the ellipse and the denominator P is the orbital period of the binary, the time taken for a complete orbit. We shall see that this follows from conservation of angular momentum.

Kepler's third law states that the period and separation are related by

$$\left(\frac{P}{2\pi}\right)^2 = \frac{a^3}{G(M_1 + M_2)}, \tag{9.5}$$

where G is Newton's gravitational constant and M_1 and M_2 are the masses of the two stars.[1] In the case of the Solar System the masses of the planets themselves are negligible compared to the mass of the Sun and the total mass $M_1 + M_2$ can be replaced by $M_\odot$ to

---

[1] It is often convenient to use units appropriate to the Earth's orbit around the Sun and write $(M/M_\odot)(P/\text{yr})^2 = (a/\text{AU})^3$, where M is the total mass of the binary system and $\text{AU} \approx 215\,R_\odot$ is one astronomical unit, the mean distance of the Earth from the Sun.

recover Kepler's original third law which, we shall see, follows from the inverse square law of gravity.

9.3.1. *Newton's laws*

Consider the binary system in the centre of momentum frame (Fig. 9.3). Let star 1 be the more massive star and star 2 its companion. Let star 1 have mass M_1 and be at position vector $\mathbf{r}_1$ relative to the centre of momentum. Similarly star 2 has mass M_2 and position vector $\mathbf{r}_2$. Define the total mass $M = M_1 + M_2$ and the position vector of star 1 relative to star 2 to be $\mathbf{r} = \mathbf{r}_1 - \mathbf{r}_2$. Let $\hat{\mathbf{r}}_1$, $\hat{\mathbf{r}}_2$ and $\hat{\mathbf{r}}$ be unit vectors in the directions of $\mathbf{r}_1$, $\mathbf{r}_2$ and $\mathbf{r}$ so that $\hat{\mathbf{r}}_2 = -\hat{\mathbf{r}}_1 = -\hat{\mathbf{r}}$. By the definition of the centre of momentum,

$$\mathbf{r}_1 = \frac{M_2}{M}\mathbf{r} \quad \text{and} \quad \mathbf{r}_2 = -\frac{M_1}{M}\mathbf{r} \qquad (9.6)$$

and the rates of change of these vectors with respect to time are related by

$$\dot{\mathbf{r}} = \dot{\mathbf{r}}_1 - \dot{\mathbf{r}}_2 \quad \text{and} \quad \ddot{\mathbf{r}} = \ddot{\mathbf{r}}_1 - \ddot{\mathbf{r}}_2. \qquad (9.7)$$

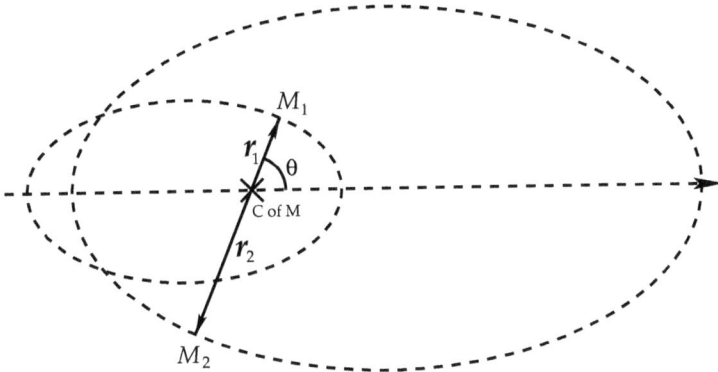

Fig. 9.3. Each star orbits the centre of momentum (C of M) of the system. Let the stars be at $\mathbf{r}_1$ and $\mathbf{r}_2$ relative to this centre and let θ be the angle measured from the line through the centre of momentum and the periastron position of the more massive star of mass M_1 at $\mathbf{r}_1$.

We apply Newton's law of gravity to star 1 to obtain

$$M_1 \ddot{\boldsymbol{r}}_1 = -\frac{GM_1M_2}{r^2}\hat{\boldsymbol{r}}_1, \qquad (9.8)$$

where $r = |\boldsymbol{r}|$ and similarly for star 2

$$M_2 \ddot{\boldsymbol{r}}_2 = -\frac{GM_1M_2}{r^2}\hat{\boldsymbol{r}}_2. \qquad (9.9)$$

These combined give us

$$\ddot{\boldsymbol{r}} = -\frac{GM}{r^2}\hat{\boldsymbol{r}}. \qquad (9.10)$$

9.3.2. *Angular momentum of the orbit*

The total angular momentum is given by

$$\boldsymbol{J} = M_1\boldsymbol{r}_1 \times \dot{\boldsymbol{r}}_1 + M_2\boldsymbol{r}_2 \times \dot{\boldsymbol{r}}_2 = \frac{M_1M_2}{M}\boldsymbol{r} \times \dot{\boldsymbol{r}}, \qquad (9.11)$$

and we define a reduced mass $\mu = M_1M_2/M$ and a specific angular momentum $\boldsymbol{h} = \boldsymbol{r} \times \dot{\boldsymbol{r}}$ so that

$$\boldsymbol{J} = \mu\boldsymbol{h}. \qquad (9.12)$$

Because $\dot{\boldsymbol{r}} \times \dot{\boldsymbol{r}} = 0$ the rate of change of $\boldsymbol{h}$ is

$$\dot{\boldsymbol{h}} = \boldsymbol{r} \times \ddot{\boldsymbol{r}} = -\frac{GM}{r^3}\boldsymbol{r} \times \boldsymbol{r} = \boldsymbol{0}, \qquad (9.13)$$

by Eq. (9.10). Thus the total orbital angular momentum $\boldsymbol{J}$ is conserved. This is a consequence only of the fact that gravity is a central force. The component of the velocity $\dot{\boldsymbol{r}}$ perpendicular to $\boldsymbol{r}$ is simply $r\dot{\theta}$ so that

$$|\boldsymbol{h}| = h = r^2\dot{\theta} = \text{const} \qquad (9.14)$$

is twice the rate at which area is swept out by the line joining the two stars. This is Kepler's second law.

9.3.3. *Energy*

Combining the kinetic energy of the two stars with their gravitational binding energy we have total energy

$$E = \frac{1}{2}M_1\left|\dot{\boldsymbol{r}}_1\right|^2 + \frac{1}{2}M_2\left|\dot{\boldsymbol{r}}_2\right|^2 - \frac{GM_1M_2}{r} \tag{9.15}$$

$$= \frac{1}{2}\mu\dot{\boldsymbol{r}}\cdot\dot{\boldsymbol{r}} - \frac{GM\mu}{r}. \tag{9.16}$$

Its rate of change is

$$\dot{E} = \mu\dot{\boldsymbol{r}}\cdot\ddot{\boldsymbol{r}} + \frac{GM\mu}{r^2}\dot{r} \tag{9.17}$$

$$= -\frac{GM\mu}{r^3}\dot{\boldsymbol{r}}\cdot\boldsymbol{r} + \frac{GM\mu}{r^2}\dot{r}, \tag{9.18}$$

again by Eq. (9.10). Now, with $\dot{r} = d|\boldsymbol{r}|/dt \neq |\dot{\boldsymbol{r}}|$,

$$\frac{d}{dt}(r^2) = 2r\dot{r} = 2\boldsymbol{r}\cdot\dot{\boldsymbol{r}} \tag{9.19}$$

so that $\dot{E} = 0$ and the total orbital energy is conserved, this time as a consequence of the inverse square law for the force of gravity.

9.3.4. *The Laplace–Runge–Lenz vector*

We also define a vector $\boldsymbol{e}$, the Laplace–Runge–Lenz vector, by

$$GM\boldsymbol{e} = \dot{\boldsymbol{r}} \times \boldsymbol{h} - \frac{GM}{r}\boldsymbol{r}. \tag{9.20}$$

It will turn out that $\boldsymbol{e}$ has magnitude e equal to the eccentricity of the orbit and is parallel to $\boldsymbol{r}$ when the stars are at periastron. The rate of change of $\boldsymbol{e}$, by Eq. (9.19), is

$$GM\dot{\boldsymbol{e}} = \ddot{\boldsymbol{r}} \times \boldsymbol{h} + \dot{\boldsymbol{r}} \times \dot{\boldsymbol{h}} + \frac{GM}{r^3}(\boldsymbol{r}\cdot\dot{\boldsymbol{r}})\boldsymbol{r} - \frac{GM}{r}\dot{\boldsymbol{r}} = 0 \tag{9.21}$$

because $\dot{\boldsymbol{h}} = 0$ and, by Eq. (9.10) and the definition of $\boldsymbol{h}$,

$$\ddot{\boldsymbol{r}} \times \boldsymbol{h} = -\frac{GM}{r^3}\boldsymbol{r} \times (\boldsymbol{r} \times \dot{\boldsymbol{r}}) \tag{9.22}$$

$$= -\frac{GM}{r^3}(\boldsymbol{r}\cdot\dot{\boldsymbol{r}})\boldsymbol{r} + \frac{GM}{r}\dot{\boldsymbol{r}}. \tag{9.23}$$

So e is conserved.

Now let ϕ be the angle between $\boldsymbol{r}$ and $\boldsymbol{e}$ so that

$$GM\boldsymbol{r} \cdot \boldsymbol{e} = GMre\cos\phi = \boldsymbol{r} \cdot (\dot{\boldsymbol{r}} \times \boldsymbol{h}) - GMr \qquad (9.24)$$

$$= \boldsymbol{r} \cdot (\dot{\boldsymbol{r}} \times (\boldsymbol{r} \times \dot{\boldsymbol{r}})) - GMr \qquad (9.25)$$

$$= h^2 - GMr. \qquad (9.26)$$

so

$$r\,(1 + e\cos\phi) = \frac{h^2}{GM} = l = \text{const} \qquad (9.27)$$

and if $\boldsymbol{r}$ is parallel to $\boldsymbol{e}$ at periastron we may set $\phi = \theta$ and recover Eq. (9.3) and hence Kepler's first law. We note here too that the conserved orbital angular momentum determines the semi-latus rectum l.

9.3.5.   *Orbital energy and Kepler's third law*

We take the dot product of Eq. (9.21) with itself to obtain

$$(GM)^2\,e^2 = (\dot{\boldsymbol{r}} \times \boldsymbol{h}) \cdot (\dot{\boldsymbol{r}} \times \boldsymbol{h}) - 2\frac{GM}{r}\boldsymbol{r} \cdot \dot{\boldsymbol{r}} \times \boldsymbol{h} + (GM)^2. \qquad (9.28)$$

Expanding the vector triple product and noting that $\dot{\boldsymbol{r}} \cdot \boldsymbol{h} = 0$, we may rewrite

$$(\dot{\boldsymbol{r}} \times \boldsymbol{h}) \cdot (\dot{\boldsymbol{r}} \times \boldsymbol{h}) = [(\dot{\boldsymbol{r}} \times \boldsymbol{h}) \times \dot{\boldsymbol{r}}] \cdot \boldsymbol{h} = \dot{r}^2 h^2 - (\dot{\boldsymbol{r}} \cdot \boldsymbol{h})^2 = \dot{r}^2 h^2. \qquad (9.29)$$

Finally

$$\boldsymbol{r} \cdot \dot{\boldsymbol{r}} \times \boldsymbol{h} = \boldsymbol{r} \times \dot{\boldsymbol{r}} \cdot \boldsymbol{h} = h^2 \qquad (9.30)$$

so that

$$(GM)^2\,(1 - e^2) = -2\frac{Eh^2}{\mu} \qquad (9.31)$$

which, together with Eqs. (9.1) and (9.27), gives us

$$E = -\frac{GM\mu}{2a} = -\frac{GM_1 M_2}{2a}. \qquad (9.32)$$

Thus the total orbital energy E fixes the semi-major axis a of the orbit.

The area of the ellipse is πab and so, with Kepler's second law (9.14) the orbital period

$$P = \frac{\pi ab}{\frac{1}{2}r^2\dot{\theta}} \qquad (9.33)$$

and thence

$$\left(\frac{P}{2\pi}\right)^2 = \frac{a^2 b^2}{h^2} = \frac{a^3 l}{h^2} = \frac{a^3}{GM} \qquad (9.34)$$

which is Kepler's third law. Hence energy and mass fix P.

9.4. Orbital Elements

Seven quantities determine the orbit of a binary system in space. Three of these, the orbital period P, the semi-major axis a and the eccentricity, are intrinsic to the binary star. The first two combine to give the total mass of the system through Kepler's third law. They also determine the total orbital energy E. Adding the eccentricity determines the orbital angular momentum J. The remaining four observable elements are extrinsic. They determine the orientation of the orbit in time and space. Most important is the inclination i, the angle between the orbital angular momentum vector and the line of sight to the system. If the stars are to eclipse then $i = 90°$. The second angle Ω is that between the line of nodes and a fixed direction on the celestial sphere. Consider a plane perpendicular to the line of sight that passes through the centre of the ellipse of the orbit. The line of nodes is defined as the intersection of the plane of the orbit and this plane. The third angle ω, the longitude of periastron, is the angle between the line of nodes and the direction to periastron $\mathbf{e}$. Finally a time of periastron passage T fixes the orbit in time. For the physical evolution of the system we are only interested in the intrinsic elements. Different sets of elements can be measured for the different observable types of binary system.

9.4.1. *Visual binary stars*

When the orbit can be seen P, e, i, Ω, ω and T are directly measurable. If the distance to system is known by parallax or by knowing the absolute luminosity of one of the stars the semi-major axis can also be measured. Kepler's third law then tells us the total mass M of the system. If in addition the absolute orbit of both stars can be seen against a background of distant objects then the two masses M_1 and M_2 can be found.

9.4.2. *Spectroscopic binary stars*

The radial velocity of one or both stars can be measured by Doppler shift. Consider first a circular orbit in which star 1 moves with speed v_1 at a_1 from the centre of momentum. Gravity provides the centripetal force so that

$$\frac{M_1 v_1^2}{a_1} = \frac{GM_1 M_2}{a^2}. \qquad (9.35)$$

Noting that $P = 2\pi a_1/v_1$ and combining this with Kepler's third law (9.6) we have

$$v_1^3 = 2\pi \frac{GM_2^3}{PM^2}. \qquad (9.36)$$

It is the projection of the velocity in the direction of the line of sight that is measured and this depends on both the position in the orbit and the inclination of the system. It has a maximum of

$$K_1 = v_1 \sin i \qquad (9.37)$$

and we define a mass function

$$F_1 = \frac{M_2^3 \sin^3 i}{M^2} = \frac{P}{2\pi G} K_1^3 \qquad (9.38)$$

constructed from the observed quantities. It follows that the mass M_2 of the companion to the observed star always has a mass greater than F_1. Often $M_2 \gg F_1$ and it is only useful if we have some other information on M_1. However it is with measurements of such mass functions that we have been able to confirm the existence of black holes in some close binary systems because only an object of mass F_1 that is smaller than its Schwarzschild radius could physically fit within the separation. When radial velocities of both stars can be measured we have the mass ratio

$$q = \frac{M_1}{M_2} = \frac{K_2}{K_1}. \tag{9.39}$$

If in addition the inclination of the orbit i can be found, either because the system is also eclipsing or has a visual orbit, both M_1 and M_2 can be found without the need for a parallax to obtain the distance. When the orbit is eccentric it is possible to determine the eccentricity e and the longitude of periastron ω from the shape of the radial velocity curve plotted against time because the stars move faster at periastron than at apastron when the orbit is eccentric. A slightly more complex analysis shows that the mass function for an eccentric binary can be written as

$$F_1 = \frac{M_2^3 \sin^3 i}{M^2} = \frac{P\left(1 - e^2\right)^{3/2}}{2\pi G} K_1^3. \tag{9.40}$$

9.4.3. *Eclipsing binary stars*

Eclipses occur when one star passes in front of its companion temporarily cutting out some of its light. Figure 9.4 shows how the combined luminosity varies with time for the case where a small hot star is totally eclipsed by a large cool star. This is the primary eclipse of greatest depth d_1. There is a secondary eclipse, of depth d_2, when the hot star passes in front of the cool star. The depth of the secondary eclipse is smaller even if the cool star is actually brighter because the small hot star only obscures its own surface area of the cool star and that area radiates less energy because it is cooler. For simple black bodies the depths of the eclipses are

$$d_1 = 4\pi\sigma R_1^2 T_1^4 \quad \text{and} \quad d_2 = 4\pi\sigma R_1^2 T_2^4. \tag{9.41}$$

So the ratio of the eclipse depths tells us the ratio of surface temperatures. For a circular orbit in which the relative velocity of the stars v is constant the time from the start of the eclipse to its maximum depth is $2R_1/v$ while the time from the start of eclipse to the end of the maximum depth is $2R_2/v$. Thus the eclipse timing

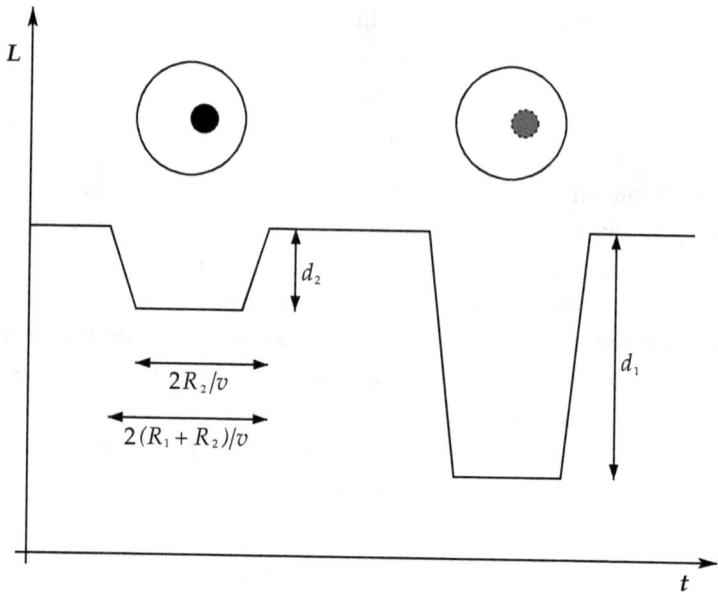

Fig. 9.4. The eclipse light curve for two stars with a total eclipse. In this case star 2 is larger in radius but cooler than star 1. The primary eclipse, the greatest reduction in luminosity L by d_1 occurs when the hotter star 1 is eclipsed by the cooler star 2 and the secondary eclipse of depth d_2 when star 2 is eclipsed by star 1. The time taken for the eclipses to reach the greatest depth is the time for star 1 to cross the edge of star 2 while the eclipse lasts for the time between when the leading edge of star 1 first crosses the edge of star 2 to when its trailing edge crosses the other side.

gives the ratio of the stellar radii. If the binary system is not circular the asymmetry of the eclipse and variations in timing between the two eclipses give the eccentricity. The sharp corners on the light curve shown in Fig. 9.4 indicate an inclination of $i = 90°$. If the eclipse of star 1 by star 2 is only partial the depth is reduced and the corners rounded off. Thus the shape of the light curve can be used to determine the inclination i rather precisely.

If, as well as eclipsing, a binary system has measurable radial velocity variations of both stars we can measure the two masses, because we now also know i, and hence the absolute dimensions of the system. We then have measurements of the absolute luminosities, the masses, the radii and the effective temperatures of both components along with the distance to the system. These double lined eclipsing binary stars are excellent tests of theories of stellar evolution. For now only about two hundred such pairs are well-measured and only about twenty of these contain one or, more commonly, two giants. However they do demonstrate excellent agreement with the theory of the structure of main-sequence stars over a wide mass range. Conversely the properties of eclipsing SB2s can be used to provide a dynamical measure of the distance to the system given its apparent luminosity. This has proved a useful measure of the distance to nearby galaxies in which binary systems can be separated from the surrounding stars.

9.5. Tides

So far we have thought of both stars as point masses. This is a good approximation when they are well separated but when they are closer the finite size of the stars becomes important and there are tidal interactions and even mass transfer between the two. First assume that star 2 remains sufficiently small to still be considered a point mass. Let star 1 have a radius R and spin with angular velocity Ω about an axis with moment of inertia I. In practice the star is likely to rotate differentially and the moment of inertia depends on the amount of distortion but to simplify our discussion we shall assume that it rotates as a solid body and that the moment of inertia is simply that of a sphere. To proceed we must calculate the distortion, and its variation with time, of the extended star 1 owing to the presence of the point mass star 2. We can then calculate the gravitational potential felt by the point mass star 2 owing to the distorted star 1. Because the potential is no longer simply proportional to $1/r$ the quantities h and e are no longer conserved. Their rates of change can be calculated given the potential felt by star 2. For an isolated binary star the total angular moment, $H = J + I\Omega$, is conserved. When energy is dissipated we also have $\dot{E} < 0$. In the absence of any dissipation J and e can only precess around H. These apsidal motions only affect extrinsic orbital elements and it is the effects of dissipation on the intrinsic elements that most interest us.

9.6.   Tidal Equilibrium

Though angular momentum can be lost in stellar winds and gravitational radiation we shall first consider cases in which the binary system is mechanically isolated and its total angular momentum is conserved. Because the stars are luminous they can radiate orbital and spin energy if it is converted to heat by tides or any other process. Stable equilibrium is then a minimum in the energy for a given angular momentum.

9.6.1.   *Circularisation*

First consider the case where $|\boldsymbol{J}| \gg I|\boldsymbol{\Omega}|$ so that the spin of the stars can be ignored and $\boldsymbol{J}$ is conserved. The energy can be written in terms of the angular momentum and eccentricity as

$$E = -\frac{GM_1M_2}{2h^2}GM\left(1 - e^2\right) \qquad (9.42)$$

from which we can see that

$$\left(\frac{\partial E}{\partial e}\right)_J \propto 2e \quad \text{and} \quad \left(\frac{\partial^2 E}{\partial e^2}\right)_J > 0 \quad \text{when} \quad e = 0. \qquad (9.43)$$

Thus a circular orbit is the most stable, least energetic, configuration for a given angular momentum.

9.6.2.   *Synchronization*

Now consider a circular binary system in which the spin axis of star 1 is aligned with the orbital axis so $\boldsymbol{\Omega}$ is parallel to $\boldsymbol{J}$. Write $\Omega = |\boldsymbol{\Omega}|$ and let $\omega = 2\pi/P$ be the orbital angular velocity. The magnitude of the specific angular momentum is then $h = |\boldsymbol{h}| = a^2\omega$ for an orbital separation a and we may write the magnitude of the total angular momentum

$$H = |\boldsymbol{H}| = I\Omega + \mu h = \text{const.} \qquad (9.44)$$

Conservation of $\boldsymbol{H}$ means that

$$\dot{H} = I\dot{\Omega} + \mu\dot{h} = 0. \qquad (9.45)$$

Differentiation of the logarithm of Kepler's third law gives

$$\frac{3\dot{a}}{a} = -\frac{2\dot{\omega}}{\omega} \tag{9.46}$$

so that

$$\dot{h} = 2a\dot{a}\omega + a^2\dot{\omega} = -\frac{1}{3}a^2\dot{\omega}. \tag{9.47}$$

Thence

$$I\dot{\Omega} = \frac{1}{3}\mu a^2 \dot{\omega} \tag{9.48}$$

and so Ω and ω either both increase or both decrease together. If the star spins up angular momentum is removed from the orbit which consequently shrinks and acquires a shorter period. The total energy is

$$E = \frac{1}{2}I\Omega^2 - \frac{GM_1 M_2}{2a} \tag{9.49}$$

so that, again with Kepler's third law,

$$\dot{E} = I\Omega\dot{\Omega} - \frac{1}{3}\mu a^2 \omega\dot{\omega} \tag{9.50}$$

$$= (\Omega - \omega)\, I\dot{\Omega}. \tag{9.51}$$

At equilibrium $\dot{E} = 0$ so $\Omega = \omega$ and the spin of star 1 is synchronized with its orbital angular velocity. Differentiating again

$$\ddot{E} = I(\dot{\Omega} - \dot{\omega})\dot{\Omega} + (\Omega - \omega)\, I\ddot{\Omega}. \tag{9.52}$$

The equilibrium is stable if and only if $\ddot{E} > 0$ when $\dot{E} = 0$ and this requires

$$\dot{\Omega}^2 > \dot{\omega}\dot{\Omega} \tag{9.53}$$

and, recalling that $\dot{\omega}$ and $\dot{\Omega}$ always have the same sign, the equilibrium is stable if and only if

$$|\dot{\Omega}| > |\dot{\omega}| \tag{9.54}$$

or, by Eq. (9.48), if and only if

$$\frac{1}{3}\mu a^2 > I. \tag{9.55}$$

That is the effective moment of inertia of the orbit is greater than thrice the moment of inertia of the star. At equilibrium let $\omega = \Omega = \Omega_{eq}$ so that the total angular momentum can be written as

$$H = I\Omega_{eq} + \mu a^2 \Omega_{eq} \tag{9.56}$$

but this has a minimum when

$$I = \frac{1}{3}\mu a_{eq}^2, \tag{9.57}$$

where a_{eq} is the separation at equilibrium, and there can be no equilibrium if

$$H < H_{crit} = 4I\Omega_{eq}. \tag{9.58}$$

Consequently, if the system does not have enough angular momentum the stars can ultimately spiral together. This instability was identified by George Darwin (1879) who was one of the first to investigate the operation of tides at the end of the nineteenth century. Figure 9.5 illustrates the situation. While the star spins faster than its orbit, $\Omega > \omega$, tides act to spin down the star and so transfer angular momentum to the orbit which in turn expands with consequent drop in ω. Similarly while $\Omega < \omega$ the star spins up while the orbit shrinks. Most binary systems evolve towards the stable branch of the equilibrium. We do not really expect stars to spin so much faster than their orbits as to lie in the unstable region to the right of the diagram. However it is quite possible for some close systems, particularly those with extreme mass ratios, to have insufficient angular momentum to reach equilibrium at all. A number of contact systems fall into this category as do many of the close orbiting exoplanets, including 51 Peg the first to be found. The fact that we observe such systems is indicative of the rather long time-scale on which tidal dissipation is taking place.

9.6.3. *The tidal mechanism*

Figure 9.6 shows the action of dissipative tides on the expanded envelope of star 1 which orbits a point mass star 2. The potential at

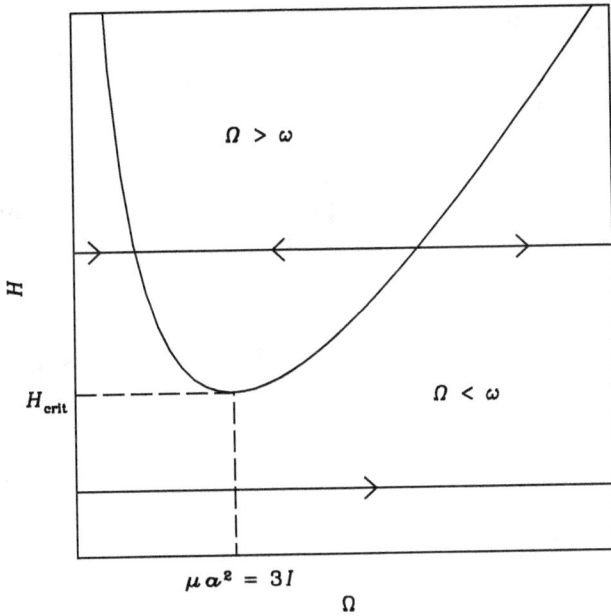

Fig. 9.5. The total angular momentum H of a binary system with angular velocity ω with one expanded star spinning at Ω. The solid curve is the tidal equilibrium. Horizontal lines are the direction of evolution when not in equilibrium. Though most detached and semi-detached binaries lie in regions where they can evolve to stable equilibrium, some contact systems and many close orbiting planetary systems are ultimately unstable.

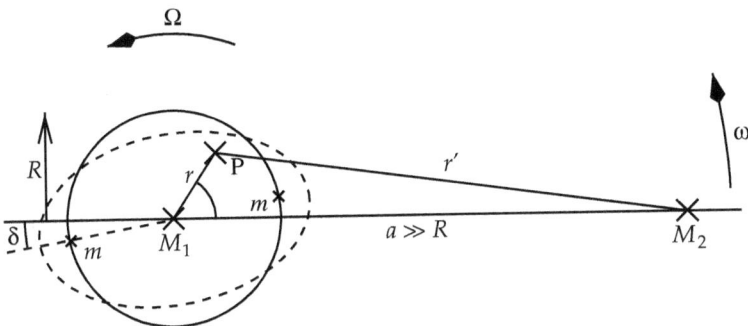

Fig. 9.6. The tidal potential of star 2 distorts star 1. If, as here, the star is spinning faster than the orbit ($\Omega > \omega$) viscosity drags the tidal bulges ahead of the orbit and dissipates energy. The force between star 2 and the two bulges provides a torque that transfers angular momentum from star 1 to the orbit.

a point P owing to star 2 can be expanded as

$$\Phi_2 = \frac{GM_2}{r'} = \frac{GM_2}{\sqrt{a^2 + r^2 - 2ar\cos\theta}} = \frac{GM_2}{a} \sum_{n=0}^{\infty} \left(\frac{r}{a}\right)^n P_n(\cos\theta),$$

(9.59)

where P_n is the nth Legendre polynomial. The $n = 0$ and $n = 1$ terms balance the orbital motion of the material of star 1. Of most interest for the evolution of the system is the $n = 2$ term because this leads to both transfer of angular momentum between star 1 and the orbit and dissipation of energy. Star 1 is distorted as illustrated by the dashed curve in Fig. 9.6. To calculate the shape of the distortion, surfaces of constant pressure and density can be assumed to lie on surfaces of constant potential in the frame in which star 1 is not rotating. These potential surfaces take into account the centrifugal potential of star 1 combined with the potential of star 2 expanded to the $n = 2$ term. Some manipulation[2] leads to a simplified picture in which the distorted star can be treated as a rigid combination of three point masses, M_1 at its centre and two masses of $m \ll M_2$ placed at the ends of a diameter on the surface of star 1. We may write

$$m = \frac{1}{2}kM_2 \left(\frac{R}{a}\right)^2,$$

(9.60)

where k is an apsidal constant that depends only on the structure of the unperturbed star 1 and typically lies between 0.01 and 0.1. When star 1 rotates synchronously with the orbit, the distortion rotates with the star and the three-point masses simply lie on a line through the centres of the two stars. There is no torque on either star but the change in gravitational potential leads to precession of the periastron direction or apsidal motion. However when the star is not rotating synchronously with the orbit the distortion must flow around it. Viscosity leads to a lag so that the tidal bulges are either dragged ahead of or fall behind the line joining the stars. The

---

[2]The interested reader can consult the appendices of *Evolutionary Processes in Binary and Multiple Stars* by Peter Eggleton (2001) for a detailed discussion.

situation illustrated in Fig. 9.6 has star 1 spinning faster than the orbit so that the two bulges lie on a line ahead of that joining the centres of the two stars by a tidal lag angle δ. The gravitational force between the two bulges and star 2 now provides a torque

$$T_{\text{tid}} = -\frac{R^5}{a^6} k G M_2^2 \delta \tag{9.61}$$

for small δ. The size of the tidal lag angle depends on the rate of working of viscosity within the tidal bulge and the difference between the spin of the star and the orbital angular velocity. It can be conveniently be written in terms of a tidal lag time τ_{lag} such that

$$\delta = (\Omega - \omega)\, \tau_{\text{lag}}. \tag{9.62}$$

The calculation of this tidal lag time is involved but may again be evaluated in terms of the unperturbed structure of star 1 when the perturbations are small. We want to relate it to the time-scale τ_{damp} on which tidal perturbations are damped. Consider the energy E_{tid} temporarily stored in the tidal bulge of mass m which rises and falls with a typical velocity v. The rate of change of energy in the tidal bulge as it moves around the star is then

$$\frac{E_{\text{tid}}}{\tau_{\text{damp}}} \approx \frac{mv^2}{\tau_{\text{damp}}} \approx Fv, \tag{9.63}$$

where

$$F \approx m\frac{dg}{dz} z, \tag{9.64}$$

where g is the gravitational acceleration at the surface and $z \ll R$ is the height reached, is the force the bulge must overcome to rise. It takes τ_{lag} for the bulge to rise so $z \approx v\tau_{\text{lag}}$. The gravitational acceleration is $g = GM_1/(R+z)^2$ so

$$F \approx m\frac{GM_1}{R^3} z \tag{9.65}$$

and

$$\tau_{\text{lag}} \approx \frac{R^3}{GM_1} \frac{1}{\tau_{\text{damp}}}. \tag{9.66}$$

Thus for a circular orbit the torque gives the rate of change of angular velocity by

$$I\dot{\Omega} = -\frac{3R^5}{a^6} k M_2^2 G \left(\Omega - \omega\right) \tau_{\text{lag}} \qquad (9.67)$$

and writing the moment of inertia $I = M_1 \left(r_g R\right)^2$, so that r_g is the ratio of the star's radius of gyration to its total radius, we arrive at

$$\dot{\Omega} = -\frac{3k}{\tau_{\text{damp}}} \frac{1}{q^2} \frac{1}{r_g^2} \left(\frac{R}{a}\right)^6 \left(\Omega - \omega\right), \qquad (9.68)$$

where $q = M_1/M_2$ is the mass ratio. Angular momentum of the system is conserved (Eq. 9.48) so this leads to

$$\frac{\dot{a}}{a} = \frac{6k}{\tau_{\text{damp}}} \frac{1}{q} \left(1 + \frac{1}{q}\right) \left(\frac{R}{a}\right)^8 \left(\frac{\Omega}{\omega} - 1\right) \qquad (9.69)$$

and the orbit decays if $\omega > \Omega$.

To obtain a circularisation rate the torque and the radial acceleration must both be integrated over the orbit. When eccentricity e is small

$$\frac{\dot{e}}{e} = -\frac{27k}{\tau_{\text{damp}}} \frac{1}{q} \left(1 + \frac{1}{q}\right) \left(\frac{R}{a}\right)^8 . \qquad (9.70)$$

Note that changes to the orbital separation and eccentricity that alter the energy of the system, with moment of inertia proportional to a^2 take place at a rate proportional to $(R/a)^8$ while synchronization, which requires change to the spin energy of the star, with moment of inertia proportional to R^2, takes place at a rate proportional to $(R/a)^6$. Alignment of the stellar spin with the orbital axis proceeds at a similar rate to synchronization, proportional to $(R/a)^6$.

9.6.4. *Time-scales*

The time-scale τ_{damp} depends on the energy dissipation mechanism. In convective parts of a star the bulk motion provides a natural bulk viscosity which can be estimated from mixing length theory. The damping time-scale is approximately the convective turnover time-scale and, because of the strong dependence on the ratio of stellar radius to binary separation, a small error in the time-scale is of little consequence for stars that grow by several factors of ten in radius over their evolution. Thus for stars with convective envelopes we may estimate

$$\tau_{\text{damp}} \approx \tau_{\text{conv}} \tag{9.71}$$

$$\approx \left(\frac{M_{\text{env}} R_{\text{env}} R}{3L} \right)^{\frac{1}{3}} \tag{9.72}$$

$$\approx \begin{cases} 22\,\text{d} & \text{for the Sun,} \\ 1\,\text{yr} & \text{for giants,} \end{cases} \tag{9.73}$$

where M_{env} is the mass in the convective envelope and R_{env} is its depth. Again because of the strong dependence on R/a, convective cores do not play much part in damping tides. Instead stars with radiative envelopes must rely on alternative dissipation mechanisms which are not well modelled. Typically tidal disturbances in the surface become evanescent and radiate energy but at such a rate that tidal damping can be as much as ten thousand times weaker than in convective envelopes.

For convective stars the typical circularisation time-scale is

$$\tau_{\text{circ}} \approx \frac{2q^2}{1+q} \left(\frac{a}{R} \right)^8 \text{yr} \tag{9.74}$$

and synchronization time-scale

$$\tau_{\text{sync}} \approx q^2 \left(\frac{a}{R} \right)^6 \text{yr.} \tag{9.75}$$

In Sec. 9.7 we define the Roche-lobe radius R_{L} to be the maximum radius a star can reach before mass at its surface is more attracted to its companion. When $q \approx 1$, $R_{\text{L}} \approx \frac{1}{3}a$ so

$$\tau_{\text{circ}} \approx \begin{cases} 2000\,\text{yr} & \text{when } R = R_{\text{L}}, \\ 6 \times 10^5\,\text{yr} & \text{when } R = \frac{1}{2}R_{\text{L}}. \end{cases} \tag{9.76}$$

These are short compared with the nuclear evolution time-scales of stars,

$$\tau_{\text{nuc}} \approx \begin{cases} 10^{10}\,\text{yr} & \text{for a } 1\,M_\odot \text{ star,} \\ 10^7\,\text{yr} & \text{for a } 30\,M_\odot \text{ star,} \end{cases} \tag{9.77}$$

or indeed the tenth of this time these stars spend as giants with deep convective envelopes. Similarly synchronization time-scales are

$$\tau_{\text{sync}} \approx q^2 \left(\frac{R}{a}\right)^6 \text{yr} \tag{9.78}$$

$$\approx \begin{cases} 300\,\text{yr,} & R = R_{\text{L}}, \\ 2 \times 10^4\,\text{yr,} & R = \frac{1}{2}R_{\text{L}} \end{cases} \tag{9.79}$$

when $q = 1$. Synchronization is usually faster circularisation so stars tend to pseudosynchronize with their orbits at periastron. For $e > 0.2$ we find

$$\Omega \approx 0.8\,\omega_{\text{periastron}}. \tag{9.80}$$

In general tides are strong enough that we may expect alignment, synchronization and circularisation before the Roche-lobe overflow we describe in the next section.

9.7. Mass Transfer

When two stars are very close so that $R \approx a$ the higher order terms in the expansion of the tidal potential cannot be ignored and instead we use the fact that by the time the radius of either star gets large enough tides have already circularised the orbit and synchronized the spin of the star. We work in a frame rotating with the star at Ω and illustrated in Fig. 9.7. Suppose all the material is stationary except for a test particle at P with position vector $\boldsymbol{x}$ relative to the centre of momentum of the system and moving at $\dot{\boldsymbol{x}}$ in the rotating frame. Then in an inertial frame the velocity of P is

$$\boldsymbol{v} = \dot{\boldsymbol{x}} + \Omega \times \boldsymbol{x} \tag{9.81}$$

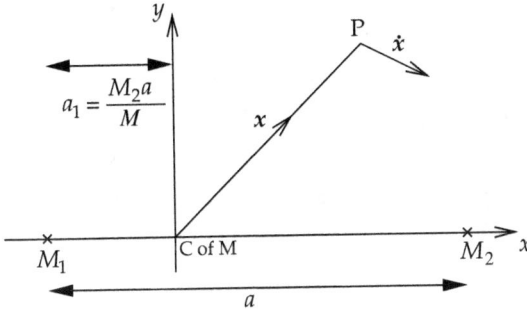

Fig. 9.7. Coordinates rotating with the binary system centred on its centre of mass with the z-axis perpendicular to the orbital plane.

and its acceleration is

$$a = \ddot{x} + 2\Omega \times \dot{x} + \Omega \times (\Omega \times x). \tag{9.82}$$

The first term is familiar as the Coriolis force and the second as the centrifugal force. The Euler momentum equation in the inertial frame is

$$\rho a = -\nabla P - \rho \nabla \phi_G, \tag{9.83}$$

where ρ is the density, P is the pressure and

$$\nabla^2 \phi_G = 4\pi G\rho \tag{9.84}$$

is the gravitational potential. In corotation $\dot{x}$ and $\ddot{x}$ vanish and so, aligning the z-axis with Ω, we may write

$$\Omega \times (\Omega \times r) = -\nabla \phi_\Omega, \tag{9.85}$$

with

$$\phi_\Omega = -\frac{1}{2}\Omega^2 s^2, \tag{9.86}$$

where s is the distance from the z-axis. With a combined potential $\Phi = \phi_G + \phi_\Omega$ the Euler equation reduces to

$$\frac{1}{\rho}\nabla P + \nabla \Phi = 0. \tag{9.87}$$

Therefore gradients of P and Φ are parallel so surfaces of constant pressure are surfaces of constant Φ. Taking the curl of Eq. (9.87) we obtain $\nabla P \times \nabla \rho = 0$ and so surfaces of constant ρ are also surfaces of constant Φ and hence, for uniform composition, all state variables are constant on surfaces of constant Φ. We may approximate our distorted star by a one-dimensional model as long as we now think of surfaces of constant potential labelled by some mean radius. In particular the surface of the star, if defined as $\rho = 0$, is a surface of constant Φ. Stars are centrally condensed so to a good approximation ϕ_G is just the gravitational potential of two point masses at the centres of the stars. In Cartesian coordinates with star 1 at the origin and star 2 at $(a, 0, 0)$ we have

$$\Phi = \frac{-GM_1}{\sqrt{x^2 + y^2 + z^2}} + \frac{-GM_2}{\sqrt{(x-a)^2 + y^2 + z^2}}$$
$$- \frac{1}{2} \frac{GM}{a^3} \left[\left(x - \frac{a}{1+q} \right)^2 + y^2 \right], \qquad (9.88)$$

a function of the mass ratio $q = M_1/M_2$, GM and a. Moreover if we scale all lengths by the separation $x \rightarrow x/a$ the shape of the equipotential surfaces is a function of q only. We plot them for $q = 2$ in Fig. 9.8.

Corotating material in hydrostatic equilibrium fills up to an equipotential surface. When the stars are small compared to their separation their equipotential surfaces are spheres. Far from the binary equipotential surfaces are again spheres. Of interest to us are the two innermost critical surfaces on which the lines meet at stationary Lagrangian points. Moving outwards from the centres of the stars the first, meeting at the inner Lagrangian point L_1 determines the maximum size of either star before material at its surface becomes more attracted to its companion at L_1. The second opens to the right, in Fig. 9.8, at the L_2 point and determines the maximum size of a compound star, or contact binary, around the two orbiting masses. The three other stationary points are also shown but are not of interest to us because, beyond the surface through the L_2 point, there is nothing to keep the material corotating and Eq. (9.87) is no longer valid.

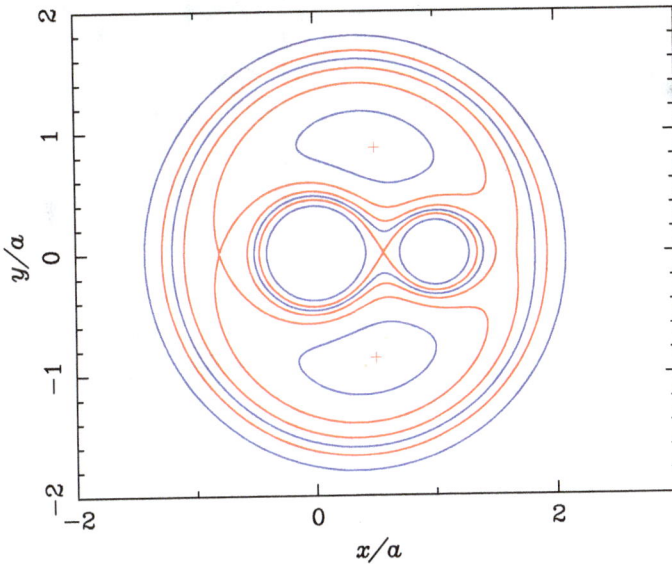

Fig. 9.8. Equipotential surfaces intersecting the x–y plane. Solid red lines pass through Lagrangian points where $\nabla\Phi = 0$.

Figure 9.9 shows the variation of the potential Φ along the x-axis and illustrates how stars fill their equipotential surfaces to form three different classes of binary star. In a wide binary system both stars have radii small compared to their separation and the system is detached. As either star grows it gradually distorts until it fills the critical potential surface that crosses at the inner Lagrangian L_1 point between the two stars. This equipotential around the star is its Roche lobe. If the star grows any larger material at L_1 is more attracted to its companion than to itself and flows from it to the other star in a stream. This is known as Roche-lobe overflow and the system is said to be semi-detached. Algols (Sec. 9.9.1) and cataclysmic variable stars (Sec. 9.9.2) are in this state. If the second star expands so that it too overfills its Roche lobe the two stars can exist in equilibrium in contact. Such systems appear to be common but do not last long. Material and heat are transferred between them until the mass ratio becomes large and Darwin's tidal instability shrinks the orbit and merges the two stars.

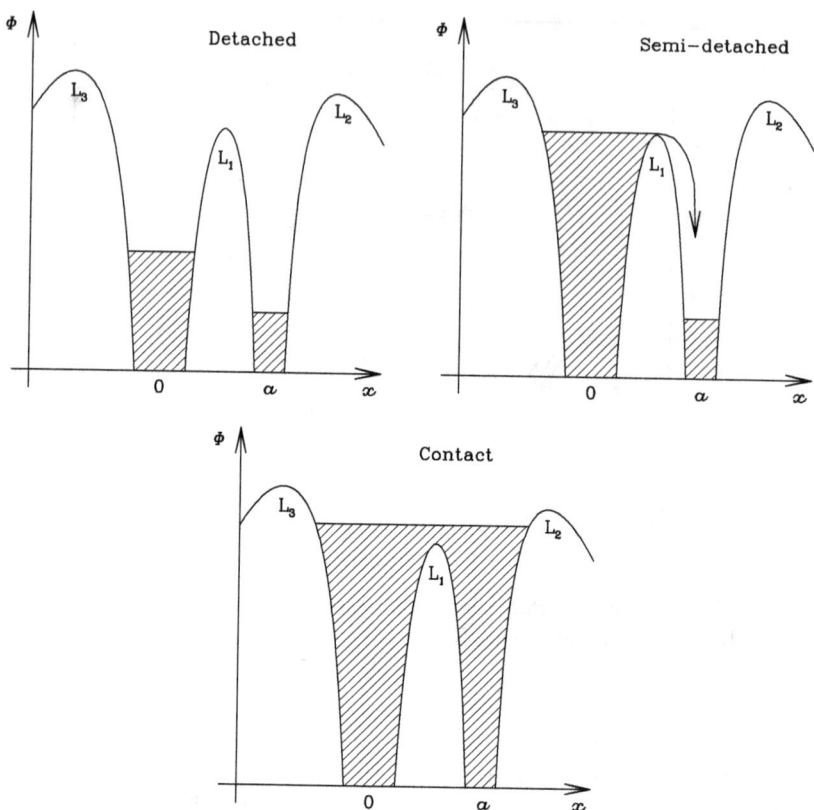

Fig. 9.9. The potential along the x-axis in Fig. 9.8. Three binary star configurations, representing detached, semi-detached and contact systems, are shown.

Calculation shows that the surface through the L_1 point is in fact close to spherical. When the mass ratio $q = 1$ the ratio of the difference in extent between the x and z directions is only 5% of the diameter. This rises to only 10% when $q = 10$ and tends to one-third as $q \to \infty$. This allows us to reasonably approximate the components of a binary system with one-dimensional spherical models even when they fill their Roche lobes. We define the Roche-lobe radius R_L to be the radius of a sphere with the same volume as the Roche lobe,

$$V_L = \frac{4}{3}\pi R_L^3. \qquad (9.89)$$

The volume can be evaluated numerically and various simple fits to R_L have been deduced. Peter Eggleton (1983) fitted the Roche-lobe radius of star 1 by

$$\frac{R_L}{a} = \frac{0.49q^{2/3}}{0.6q^{2/3} + \log_e\left(1 + q^{1/3}\right)}.$$ (9.90)

This is accurate to better than 1% over the whole range $0 < q < \infty$. It is the preferred form for numerical work. For analytic work a formula deduced by Bodan Paczyński (1971)

$$\frac{R_L}{a} = 0.462 \left(\frac{M_1}{M}\right)^{\frac{1}{3}},$$ (9.91)

which is accurate to better than 3% for $0 < q < 0.8$, is much more useful.

9.7.1. *Mass transfer rate*

When a star overfills its Roche lobe, material at the L_1 point accelerates towards the companion while far from L_1 material at the surface of star 1 is almost stationary on an equipotential surface (Fig. 9.10). The expansion is similar to flow through a rocket nozzle formed by the equipotential surface. Close to the L_1 point the stream passes through a sonic surface where it has a typical width W and is flowing at the sound speed c_s. The mass transfer rate is then the flux through this surface,

$$\dot{M}_1 = -(\rho c_s W^2)_{L_1}.$$ (9.92)

By considering a streamline along the surface of the star we may write

$$\frac{1}{2}c_s^2 \approx \Delta\Phi(x_1, W),$$ (9.93)

where

$$\Delta\Phi(x, y) = \Phi(x_1, y, 0) - \Phi(x_1, 0, 0) \approx \frac{1}{2}\left(\frac{\partial^2\Phi}{\partial y^2}\right)_{L_1} y^2$$ (9.94)

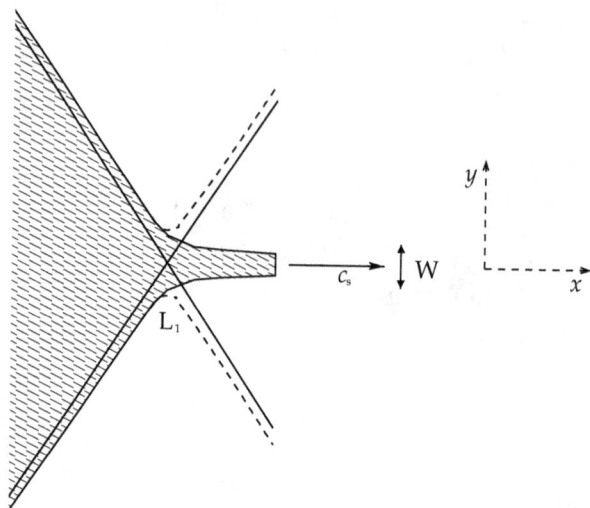

Fig. 9.10. Star 1 to the left overfills its Roche lobe. Away from the inner Lagrangian point L_1 material at its surface is stationary and lies on the equipotential surface shown as a dashed line. To the right of the L_1 point the material expands freely into the vacuum within the Roche lobe of star 2. Close to L_1 the stream passes through a sonic surface where it has a width W.

and the L_1 point is at $(x_1, 0, 0)$. Now at L_1

$$\frac{\partial^2 \Phi}{\partial y^2} = \frac{GM_1}{x_1^3} + \frac{GM_2}{(a - x_1^3)^3} - \Omega^2 > 0 \qquad (9.95)$$

$$\approx \Omega^2 \qquad (9.96)$$

because $\Omega^2 = GM/a^3$, by Kepler's third law, and $x_1 \approx a$. So

$$W \approx \frac{c_s}{\Omega}. \qquad (9.97)$$

The density at the sonic surface depends on how much the star is overfilling its Roche lobe,

$$\Delta R = R_1 - R_L. \qquad (9.98)$$

Consider a simple hydrostatic isothermal atmosphere, with $c_s^2 = P/\rho$, for which

$$\frac{1}{\rho}\frac{dP}{dr} = -\frac{GM}{R_1^2} = \frac{c_s^2}{H}, \qquad (9.99)$$

where the pressure scale-height

$$H = P\left(\frac{\mathrm{d}P}{\mathrm{d}r}\right)^{-1} \ll R \approx a. \tag{9.100}$$

The density at L_1 is related to the density ρ_* at the photosphere, where $r = R_1$, by $\rho_{L_1} = \rho_* \exp(\Delta R/H)$ and we may also make the approximation $\Omega^2 \approx c_s^2/HR_1$ and we note that $W \approx \sqrt{HR_1}$. Combining these equations we deduce that

$$\dot{M}_1 \approx -\rho_* c_s H R_1 e^{\Delta R/H}. \tag{9.101}$$

At the photosphere of the Sun $\rho_* \approx 2 \times 10^7 \,\mathrm{g\,cm^{-3}}$, $c_s \approx 10^5 \,\mathrm{cm\,s^{-1}}$ and $H \approx 270\,\mathrm{km} \approx 0.04\,R_\odot$ so that

$$\dot{M}_1 \approx 6 \times 10^{-10} \,M_\odot \,\mathrm{yr^{-1}} e^{(25\Delta R/R)} \tag{9.102}$$

for a star similar to Sun, but with an isothermal atmosphere, overfilling its Roche lobe by ΔR. This gives a time-scale for mass transfer

$$\tau_{\dot{M}} \approx 1.7 \times 10^9 \,\mathrm{yr} \ll \tau_{\mathrm{nuc}} \tag{9.103}$$

even when $\Delta R = 0$ and which grows exponentially with ΔR. With a somewhat more involved calculation for adiabatic flow from a polytrope of $n = 3/2$, a good model for Roche-lobe overflow from a red giant, we find that

$$\dot{M}_1 \approx -10 \frac{M_1}{M_\odot} \left(\frac{\mathrm{yr}}{P}\right) \left(\frac{\Delta R}{R_1}\right)^3 M_\odot \,\mathrm{yr^{-1}}, \tag{9.104}$$

for an orbital period P. From this we can deduce that a typical $1\,M_\odot$ giant filling its Roche lobe in a system with orbital period of 1 yr that must transfer mass on a typical evolution time-scale of 10^8 yr need only overfill its Roche lobe by

$$\Delta R \approx 0.001\,R_1. \tag{9.105}$$

Indeed as long as the time-scale on which a star expands $\tau_{\mathrm{expand}} \gg \tau_{\mathrm{dyn}}$, its dynamical time-scale, we can expect

$$R \approx R_{\mathrm{L}} \quad \mathrm{and} \quad \dot{R} \approx \dot{R}_{\mathrm{L}} \tag{9.106}$$

as the mass transfer proceeds.

9.7.2. *The stream*

Once inside the Roche lobe of the companion star the stream is rapidly accelerated by the combined stellar and centrifugal potential Φ together with the Coriolis force. The speed of the overflowing material quickly exceeds its sound speed and so the flow is supersonic and ballistic to a very good approximation. It swings past the companion at a closest approach d and then heads back towards the other star though it cannot escape its new Roche lobe again.[3] If the radius of the companion $R_2 > d$ the stream directly impacts its surface and is brought to rest rapidly in a shock creating a hot spot. This is the case for the archetypal system Algol (Sec. 9.9.1). If $R_2 < d$ the stream misses the companion and collides with itself. A ring of material forms in a Keplerian orbit about the companion. The ring expands viscously to form an accretion disc in which angular momentum is transported outwards and mass falls inwards on to the companion. The disc is truncated at the out edge where its angular momentum is returned to the binary orbit. This is the typical configuration of cataclysmic variables (Sec. 9.9.2).

9.7.3. *Stability of mass transfer*

To examine the stability of mass transfer define three derivatives of radii with respect to the mass of the lobe-filling star. The first is the rate of change of the Roche-lobe radius R_L for conservative mass transfer in which the angular momentum of the system J and the total mass M are conserved. Any material lost by star 1 is accreted by star 2.

$$\zeta_L = \left(\frac{\partial \log R_{L_1}}{\partial \log M_1} \right)_{M,J}. \qquad (9.107)$$

This can be approximated by $\zeta_L = 2.13q - 1.67$ and we see that it is positive for $M_1 > 0.78 M_2$ so that the Roche lobe shrinks in response to mass transfer from star 1 to star 2 if $q > 0.78$ and otherwise it

---

[3]For a more detailed discussion the reader is directed to the classic paper by Lubow and Shu (1975).

expands. The initial response of the star to mass loss is adiabatic as it regains hydrostatic equilibrium and loses thermal equilibrium in the process. So we define a second derivative at constant entropy s and composition of each isotope X_i throughout the star

$$\zeta_{ad} = \left(\frac{\partial \log R_1}{\partial \log M_1}\right)_{s,\{X_i\}}. \qquad (9.108)$$

For stars with radiative envelopes $\zeta_{ad} > 0$ so they shrink on mass loss while for stars with convective envelopes $\zeta_{ad} < 0$ and they expand on mass loss. On a thermal time-scale the star regains full equilibrium at its new mass but still with constant composition. A third derivative

$$\zeta_{eq} = \left(\frac{\partial \log R_1}{\partial \log M_1}\right)_{\{X_i\}}. \qquad (9.109)$$

describes the rate of change of radius with mass in equilibrium. For main-sequence stars $\zeta_{eq} > 0$, while for red giants and stars crossing the Hertzsprung gap $\zeta_{eq} < 0$.

The rate at which mass transfer proceeds depends on which of these derivatives are larger. If $\zeta_L > \zeta_{ad}$ then the Roche lobe shrinks faster than the radius of the star in direct response to mass transfer. So ΔR increases and consequently $\dot{M}$ increases rapidly. There is positive feedback and the mass transfer is unstable:

$$\left|\frac{M_1}{\dot{M}_1}\right| \to \tau_{dyn} \approx 10 \text{ to } 100 \text{ yr}, \qquad (9.110)$$

and mass transfer proceeds on a dynamical time-scale. Star 2 often cannot accrete the material at such a high rate. Instead it expands itself and the transferred material ends up in a common envelope around the two stars (Sec. 9.9.3). This is typically the outcome when a giant fills its Roche lobe when in orbit with a less massive companion because the giant expands while its Roche lobe is shrinking. Positive feedback drives the mass transfer up to the dynamical rate.

If $\zeta_L < \zeta_{ad}$ but $\zeta_L > \zeta_{eq}$ then the star shrinks in its immediate response to mass transfer but then expands on its thermal time-scale

τ_{th} so that

$$\left| \frac{M_1}{\dot{M_1}} \right| \rightarrow \tau_{\text{th}} \approx 10^5 \text{ to } 10^6 \text{ yr} \qquad (9.111)$$

and mass itself transfer proceeds on a thermal time-scale. This is the case when a subgiant in the Hertzsprung gap with a radiative or thin convective envelope fills its Roche lobe.

If both $\zeta_{\text{ad}} > \zeta_{\text{L}}$ and $\zeta_{\text{eq}} > \zeta_{\text{L}}$ the star shrinks in response to mass transfer and does not expand again to fill its Roche lobe until driven to either by its own nuclear evolution or until some angular momentum loss mechanism causes the orbit to shrink sufficiently. Either

$$\left| \frac{M_1}{\dot{M_1}} \right| \rightarrow \tau_{\text{nuc}} \approx 10^7 \text{ to } 10^9 \text{ yr}, \qquad (9.112)$$

the case for main-sequence stars or red giants in present-day Algols (see Sec. 9.9.1), or

$$\left| \frac{M_1}{\dot{M_1}} \right| \rightarrow \tau_{\text{J}}, \qquad (9.113)$$

the time-scale on which angular momentum is lost from the system as for cataclysmic variables (Sec. 9.9.2).

9.8. Period Evolution of Binary Stars

When a component of a binary system loses mass or there is mass transferred between the components, the orbit changes. Here we consider the circular case. In fact for simple isotropic mass loss at a constant rate from one star on a time-scale much longer than the orbital period, it can be shown that eccentricity is an adiabatic invariant. That is $\dot{e} \approx 0$. We have already argued that Roche-lobe overflow only begins after the binary star has circularised but there are possibilities for mass to be captured from the wind of a companion and the rate at which this occurs varies over an eccentric orbit at a rate that depends on the structure of the wind. Exactly how the eccentricity varies in such a case has not yet been fully worked out. Even for a circular binary star the eccentricity of the orbit changes if mass loss takes place on a time-scale short compared with the orbital period. For instance if half or more of the mass of a circular system is instantaneously lost, perhaps in a supernova explosion, the system becomes unbound ($e \geq 1$, $E \geq 0$). For our circular binary star we allow for mass loss from star 1 at a rate $-\dot{M}_1$. Some of this is accreted by star 2 at a rate $\dot{M}_2$ and the remainder is lost from the system at a rate $-\dot{M}$. So

$$-\dot{M}_1 = -\dot{M} + \dot{M}_2, \tag{9.114}$$

with $\dot{M}_1 \leq 0$, $\dot{M} \leq 0$ and $\dot{M}_2 \geq 0$. We assume that the wind leaves the system isotropically and carries off only the intrinsic angular momentum of star 1 in its orbit. This approximation breaks down if the spin angular momentum of star 1 is significant but otherwise we have

$$\dot{j} = \dot{M}a_1^2\Omega < 0. \tag{9.115}$$

Usually the spin angular momentum of a star reaches only about a tenth of the orbital angular momentum when $R \approx R_L$ if $q = 1$ and can be ignored. Recall that the orbital angular momentum

$$J = \frac{M_1 M_2}{M}a^2\Omega. \tag{9.116}$$

Taking the logarithm and differentiating with respect to time we find

$$\frac{\dot{j}}{J} = \frac{\dot{M}_1}{M_1} + \frac{\dot{M}_2}{M_2} - \frac{\dot{M}}{M} + \frac{2\dot{a}}{a} + \frac{\dot{\Omega}}{\Omega}. \tag{9.117}$$

Similarly Kepler's third law gives

$$\frac{2\dot{P}}{P} = -\frac{2\dot{\Omega}}{\Omega} = \frac{3\dot{a}}{a} - \frac{\dot{M}}{M}. \tag{9.118}$$

Combining these last five equations we find

$$\frac{\dot{P}}{P} = -\frac{2\dot{M}}{M} + \frac{3(M_2 - M_1)}{M_1 M_2}\dot{M}_2. \tag{9.119}$$

When there is only mass loss in the wind from star 1 we have $\dot{M}_2 = 0$ so

$$\frac{\dot{P}}{P} = -\frac{2\dot{M}}{M} \tag{9.120}$$

and integrating we find

$$PM^2 = \text{const} \quad \text{and} \quad aM = \text{const.} \tag{9.121}$$

Both the period and the separation increase with mass loss. Indeed the planets in the solar system will move further from the Sun as it evolves and loses mass. Mercury will still be engulfed when the Sun becomes a red giant. Venus will raise tides on the Sun sufficient for its orbit to decay. The Earth however may just escape if tidal dissipation is on the weak side of expectations.

When mass transfer is fully conservative so that $\dot{M} = 0$ and $\dot{J} = 0$ we find instead that

$$\frac{\dot{P}}{P} = -\frac{3\dot{M}_1}{M_1} - \frac{3\dot{M}_2}{M_2} \tag{9.122}$$

and thence

$$P(M_1 M_2)^3 = \text{const} \quad \text{and} \quad a(M_1 M_2)^2 = \text{const.} \tag{9.123}$$

The product $M_1 M_2 = (M - M_2)M_2$ has a minimum when $M_1 = M_2$ so both P and a reach minima during mass transfer when the masses are equal. The orbit shrinks while the mass losing star is more massive and expands once it is less massive than its companion.

9.9. The Zoo of Binary Stars

We have described the basic physics of binary stars and their interactions. Coupling this with stellar evolution leads to a veritable zoo of different types of binary star. Observations do overlap with what we expect but often require the introduction of new physical processes such as common envelope evolution (Sec. 9.9.3) that are not fully understood. We shall illustrate this with three examples, the Algols as the prototypes, the cataclysmic variables as those studied in most detail and the type Ia supernovae that are used as standard candles to measure the structure and evolution of the Universe, before briefly describing effects on more massive stars.

9.9.1. *Algols*

As one of the brightest stars in the northern hemisphere Algol, or β Persei, has been known for a long time. It is an eclipsing SB2 and so yields a great deal of information about its current state. Its variability was first definitely recorded by Montanari in Bologna but the name Algol suggests that it may have been recognised much earlier. Algol is derived from the Arabian *Al Ghūl* which has been variously translated as demon or changing spirit. However Allen (1899) felt it is more likely that the name is derived from Ptolemy who referred to it as the brightest star in the Gorgon's head, a constellation recognised by the Greeks at the time and indeed generally until quite recently (Goodricke refers to it in 1783). The Hebrews called it *Rōsh-ha-Satan* or Satan's head and the Chinese *Tseih She* or the piled up corpses. Whether these names reflect the variability or not must be left to our imaginations because no actual record has been found.

Its eclipsing was not noted for over a century until John Goodricke (1783) sent a short letter to the Royal Society describing how he had spotted a periodicity in the light variations of Algol. He and his cousin, Edward Pigott, had by then already obtained a fairly accurate estimate of the period of 2 days and 21 hours. Goodricke in a short paragraph at the end of his letter went on to suggest that the cause of the variation might be either a dark object orbiting and eclipsing the star or a dark spot on its surface. Confirmation of his first hypothesis did not come for yet another century when Vogel (1890) observed radial velocity shifts in the spectrum of Algol and found the positions of minimum light to correspond to the conjunctions of the eclipse model.

Observations improved with time giving better photometric and spectroscopic measurements of Algol and a number of similar systems. It seems that it had been apparent that something was not quite right with Algol for some time before Hoyle (1955) recorded what he described as the Algol paradox. From the shapes of the eclipses it was apparent that the fainter star was larger in radius. Such a situation was known not to be possible according to the theory of stellar evolution. If both stars were on the main sequence then the brighter would be larger. In fact the fainter could only be larger if it had evolved off the main sequence and indeed Parenago (1950) had already claimed that the fainter components of Algols were in many cases sub-giants. Hoyle argued that, although it would be possible to pick the two stars from the H–R diagram, one on the main sequence and the other a much older sub-giant, all reasonable theories of the formation of binary stars suggested that the two components would have formed at the same time and would be of the same age now. Thus he had identified the paradox without the need to introduce the masses of the stars directly and went on to successfully explain it in terms of the initially brighter star evolving to such a size that its fainter companion gobbled up matter from its surface. This companion could then move up the main sequence and become the brighter of the two (Fig. 9.11). The Algol system demonstrates how the evolution of both components of a binary system can be substantially altered from what they would have experienced as single stars.

At the same time Crawford (1955) was also solving the same paradox though more specifically in terms of the limitations placed on the mass ratios by the spectroscopically determined mass functions and the assumption that the brighter component does in fact lie on the main sequence. Struve (1948) had already pointed out that these mass functions are low. Crawford also introduced the concept of the giant filling its Roche lobe. In fact Walter (1931) had already noted that the cool stars in Algols are close to the limit of dynamical stability but this had gone largely unnoticed. At that time giants were still thought to be contracting protostars burning the light reactive elements in the ppII- and ppIII-chains rather than a later stage of evolution so it was harder to conceive of the idea of a star growing to fill its Roche lobe.

This semi-detached nature of Algols provided mutual support for the hypothesis formulated by Struve (1949) that the existence of gaseous streams between the two stars in Algols could account for

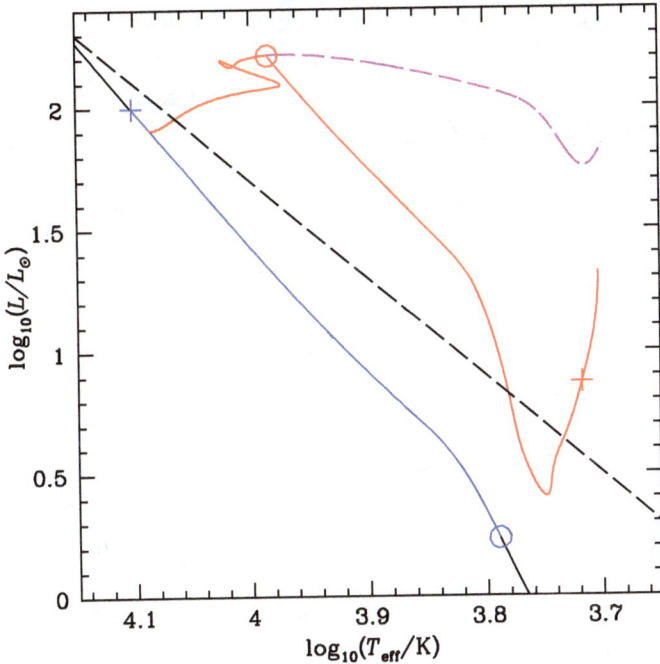

Fig. 9.11. The evolution of an Algol system in the H–R diagram. Initially the binary components are of 3 and $1\,M_\odot$ and the orbital period is 1.7 d. The more massive component (red track) evolves first and begins mass transfer shortly after the end of the main sequence at the point marked by the red circle. The dashed magenta track shows how its evolution would continue were it not to fill its Roche lobe. At that point its companion (blue track) has not evolved from its position on the zero-age main sequence marked by the blue circle. Initially the mass transfer is relatively rapid on a thermal time-scale and the donor loses about $2.2\,M_\odot$ in only a Myr. After that mass transfer slows to a nuclear time-scale and it loses another $0.1\,M_\odot$ in the following Myr. The donor is then at the point marked with the red cross and has a helium core of mass $M_c = 0.37\,M_\odot$. The companion has hardly evolved in this time but, by slightly non-conservative mass transfer, has moved up the zero-age main sequence to the point marked by the blue cross with a mass of $3.2\,M_\odot$. The black, dashed straight line is at constant radius and demonstrates that the more evolved star is now redder and larger but less luminous than its companion, typical of an Algol system. In this model mass transfer continues for another 8 Myr until the donor has lost its hydrogen envelope and cools as a helium white dwarf when its companion, still on the main sequence would appear as a massive blue straggler in a cluster.

an asymmetry in the radial velocity curve. Although the photometric light curve of U Cephei showed symmetric eclipses the radial velocity curve is asymmetric. Struve explained this in terms of the spectrum of a gaseous stream moving faster than the two stars superimposed on the symmetric curve of the star. Evidence had also been provided by Wood (1950) who had found that binaries with period fluctuations almost always have one star filling its Roche lobe.

With a fairly definite theory and the dawn of numerical stellar evolution the stage was set for the construction of theoretical models of these semi-detached systems. The first step was taken by Morton (1960). Concentrating on the initially more massive star, he examined the process of mass transfer. He pointed out that, because all observed Algols have the sub-giant component already less massive than its hot companion, the initial rate of mass transfer must have been much faster than that taking place now. It must have been sufficiently fast to make it unusual to observe a system in a state where the primary is still the more massive. Over the decade following Morton's work detailed models were made independently by others. Kippenhahn and Wiegert (1967) introduced the nomenclature of case A to indicate mass transfer before the exhaustion of central hydrogen burning and case B for mass transfer afterwards, when the star has evolved off the main sequence. In all of these models conservative mass transfer (all the matter lost by the primary being accreted by the secondary) was assumed but Paczyński and Zil'ołkowski (1967) showed that the resulting Algol systems are more realistic if half the mass lost by the primary is actually lost from the system.

To understand Algol evolution we start from the premise that the more massive star evolves to a giant first and so is the first to fill its Roche lobe. Let this more massive star be star 1 and the mass ratio $q = M_1/M_2$ so that at the onset of mass transfer $q > 1$. However in Sec. 9.7.3 we argued that if a giant fills its Roche lobe while $q > 1$ then mass transfer proceeds on a dynamical time-scale. It is only once the mass ratio has fallen below some critical q_{crit} that its Roche lobe can grow faster than a giant expands as it loses mass and mass transfer can slow to a nuclear time-scale. We can estimate q_{crit} by noting that the radius of the giant can be approximated by

$$R_1 = f(L)M_1^{-0.27}, \tag{9.124}$$

where $f(L)$ is an increasing function of luminosity, that depends only on conditions at the core, and the exponent of mass M_1 reflects the behaviour of the convective envelope which is similar to a polytrope of $n = 3/2$ for which $R \propto M^{-1/3}$. On time-scales much shorter than the nuclear time-scale $f(L) \approx$ const so for dynamical mass transfer we may write

$$\frac{\dot{R}_1}{R_1} = -0.27 \frac{\dot{M}_1}{M_1}. \tag{9.125}$$

Dynamical mass transfer ensues because of the positive feedback on the rate when a star grows directly as a result of mass transfer. This is the case when $\dot{R}_1 > \dot{R}_L$ when $R_1 = R_L$. Differentiating formula (9.91) we find

$$\frac{\dot{R}_L}{R_L} = \frac{\dot{M}_1}{3M_1} - \frac{\dot{M}}{M} + \frac{\dot{a}}{a} \tag{9.126}$$

so for conservative mass transfer ($\dot{M} = 0$ and $\dot{J} = 0$) and using Eqs. (9.117) and (9.118) we find such positive feedback when

$$-0.27 \frac{\dot{M}_1}{M_1} > \frac{\dot{M}_1}{M_1} \left\{ \frac{6M_1 - 5M_2}{3M_2} \right\}. \tag{9.127}$$

But $\dot{M}_1 < 0$ so mass transfer is dynamically unstable if

$$q > q_{\text{crit}} \approx 0.7. \tag{9.128}$$

It does indeed turn out that all observed Algol systems have $q < q_{\text{crit}}$. U Cep actually has a mass ratio only just less than q_{crit} and as a result has a large rate of change of orbital period, in fact the only period change that can be unequivocally attributed to mass transfer.

In fact mass transfer on a dynamical time-scale is much too fast for the companion to accrete the transferred mass which rather builds up rapidly around it to form a common envelope. In Sec. 9.9.2 we shall argue that this is the route to cataclysmic variables rather than classical Algols. The simplest way around this is that all Algols must have begun mass transfer before the most massive star has evolved on to the giant branch unless it has suffered sufficient mass loss that $q < q_{crit} \approx 0.7$ and the Roche lobe expands faster than the star. The classical solution to this is that all present-day Algols actually began their mass transfer while the more massive star was evolving across the Hertzsprung gap and feedback maintained mass transfer on a thermal time-scale. This is still some thousand times faster than the nuclear rate so we still expect to see almost all systems with $q < q_{crit}$ particularly given that we know of only about 400. A more radical, but sometimes necessary, possibility is that the presence of a companion can enhance mass loss from a giant and thereby reduce the mass ratio below q_{crit} before the onset of mass transfer. Such enhanced mass loss may well be driven by the fact that the giant, tidally locked to its orbit, is forced to spin much faster than if it were isolated. It is also possible for material lost from a giant in a relatively slow moving wind to be accreted by its companion. Evidence for this is found in barium stars, all of which are stars on the first ascent of the giant branch in, often eccentric, binary systems, but which show s-process elements, that are not formed before the asymptotic giant branch, in their spectra. All their companions are likely to be white dwarf remnants of former AGB stars that transferred mass by accretion from a slow wind rather than Roche-lobe overflow.

9.9.2.   *Cataclysmic variables*

Cataclysmic variables are very close binary stars in which the primary component is a white dwarf which is accreting material transferred from its Roche-lobe filling companion. Figure 9.12 illustrates the basic components. The companion to the white dwarf is always less massive, often substantially, and is typically a low-mass main-sequence star for which the Roche-lobe filling state dictates an orbital period of a few hours and a separation of about a solar radius. The nuclear evolution time-scale of such stars is generally much greater than the age of the Galaxy so they remain internally inert so the evolution of the system is driven by angular momentum loss in a magnetically locked wind or by gravitational radiation. In a very

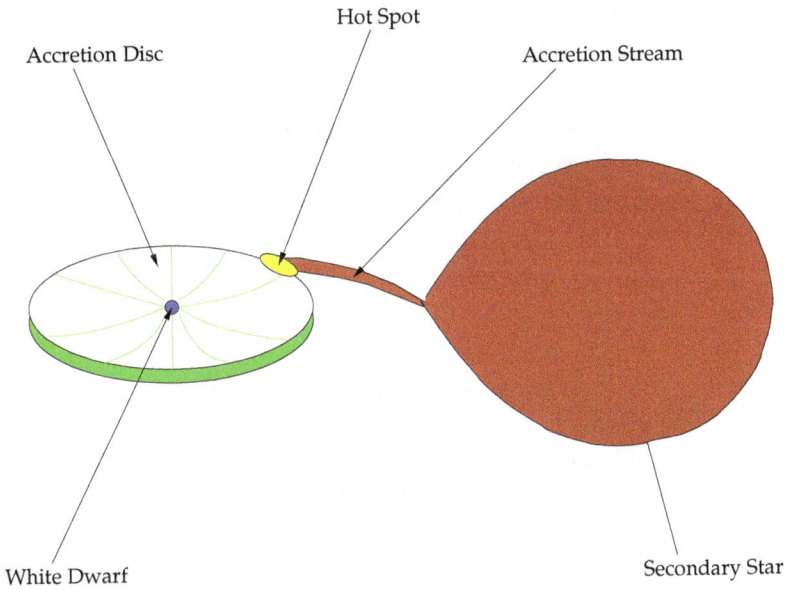

Fig. 9.12. A schematic diagram of a cataclysmic variable with the major observable components marked.

few systems the secondary star can be slightly evolved. For example GK Per, the widest system classified as a cataclysmic variable, has an orbital period of 47 h and its white dwarf has a subgiant companion. The nuclear, or in some cases mass-loss, time-scales of evolved companions can be relatively short and their nature is therefore fundamentally different from those systems with unevolved low-mass secondaries. Most importantly the mass transfer rates are higher. These systems, particularly those with very large red giant or supergiant secondaries, are classified as symbiotic stars. At the other extreme the companion can be totally inert yet more evolved. It can be another white dwarf of lower mass than the primary. AM CVn, with a period of 89 min, is the prototype of this class of cataclysmic variables.

Observationally, in addition to the two stars, a third component, an accretion disc, is important and often dominates the light from the cataclysmic variable. It is formed because material overflowing from the companion at the inner Lagrangian point L_1 has too much angular momentum to fall directly on to the white dwarf. Viscous dissipation allows the majority of the matter to accrete slowly through the disc on to the white dwarf while angular momentum is carried outwards until it can be tidally returned to the orbit. Many cataclysmic variables are observationally very clean systems in which the light variations and spectra of each of the three main components can be separated out. Often the signature of the high velocity accretion stream and the hot spot where it impacts the edge of the disc can also be identified.[4]

Two instabilities gave cataclysmic variables their name and were responsible for their early observation. The first gives rise to the classical novae which had again been known since ancient times. The first formal record was made by Don Anthelme in 1670 who observed the new star Nova Vel at 2nd magnitude only for it to fade again within months. The eruption arises because hydrogen rich material transferred to the white dwarf from its companion builds up in a degenerate layer on the surface. When the base of this layer becomes dense enough the hydrogen ignites in a thermal nuclear runaway that leads to a large increase in brightness and probably the ejection of most of the accreted material. The second observable cataclysmic behaviour is due to an instability in the accretion disc. Under some conditions material can accumulate and fall through a disc in bursts. The quasi-periodic increase in brightness of the disc makes them visible as dwarf novae that were first noticed when the comet hunter John Russell Hind (1856) reported that U Gem had appeared at ninth magnitude on 15th December 1855 but had gone by 31st December that year. It was seen to brighten again by Norman Pogson in March of 1856. There are yet other systems which have never displayed either of these phenomena and others that are dominated by magnetic fields so that the accretion disc is completely disrupted.

---

[4]An excellent, detailed and very readable review of the observations from early times to the present day forms a substantial part of the book *Cataclysmic Variables* by Warner (1995) to which the interested reader is encouraged to turn.

9.9.3. *Common envelope evolution*

The white dwarfs in cataclysmic variables must have originally
formed as the cores of giants which must have had room to grow
to 100 or even $1000\,R_\odot$ before interaction. However their orbital
separation is now only a few solar radii. The generally accepted
route by which a binary reduces its period is common-envelope
evolution proposed by Paczyński (1976). Following dynamical mass
transfer from the giant, the pair becomes a common-envelope system
(Fig. 9.13) in which the degenerate core of the original giant and
the relatively dense red dwarf are orbiting within the low-density
envelope of the giant that now engulfs both stars. From here on what
happens is as much plausible conjecture as fact. By some frictional
process the two cores are supposed to spiral together towards the
centre of the envelope. During this process the orbital energy released
is transferred to the envelope which it drives away in a strong wind.
Because the orbital energy of the cores and the binding energy of
the envelope are of the same order it can be envisaged that in some
cases the balance is just such that the entire envelope is blown away
when the cores reach a separation of a few solar radii. If more energy
is transferred the envelope is lost while the orbit is still quite wide.
If less energy is transferred the cores coalesce before the envelope is
lost. In practice coalescence most likely occurs when the red dwarf
reaches a depth in the envelope where it has comparable density with
the envelope or when it is tidally disrupted by the white dwarf.

In the simplest models we define a parameter α_{CE} to be the
fraction of the orbital energy released, during the spiralling-in, that
goes into driving away the envelope. Knowing α_{CE} and the binding
energy of the envelope, we can calculate the final orbital separation
from the initial. Note that the binding energy of the envelope is
calculated differently by different modellers. The most significant
discrepancy is whether we use the binding energy of the single-star
giant envelope before the common envelope forms or that of the
common envelope itself on the assumption that it has swollen up to
the size of the orbit. The magnitude of α_{CE} is expected to be less than
one because at least part of the released energy should be radiated
away. However population synthesis models that recreate sufficient

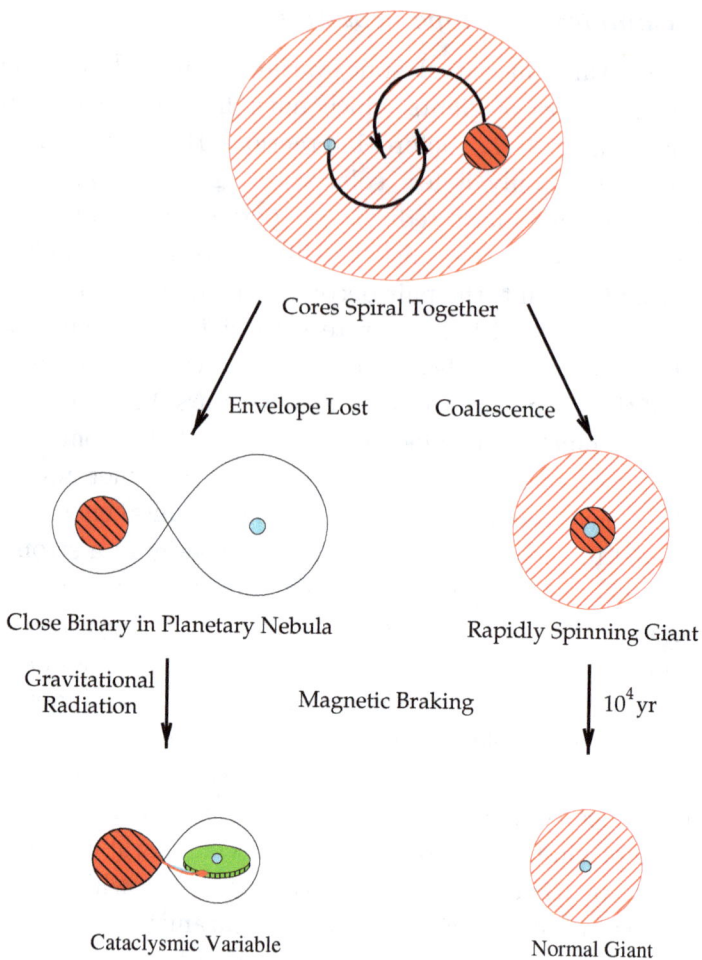

Fig. 9.13. Example common-envelope evolution. After dynamical mass transfer from a giant, a common envelope enshrouds the relatively dense companion and the core of the original giant. These two spiral together as their orbital energy is transferred to the envelope until either the entire envelope is lost or they coalesce. In the former case a close white-dwarf and main-sequence binary is left, initially as the core of a planetary nebula. Magnetic braking or gravitational radiation may shrink the orbit and create a cataclysmic variable. Coalescence results in a rapidly rotating giant which will very quickly spin down by magnetic braking.

numbers of cataclysmic variables and other close systems, such as X-ray binaries and the progenitors of type Ia supernovae, indicate that large α_{CE} are required. Typically about three times the energy released seems to be required.

Sources other than the orbital energy are available but it is not yet established exactly how they might be tapped. There is always ongoing nuclear burning around the giant's core and indeed this energy is important if it is assumed that the common envelope expands to fill the orbit as it forms and so is included surreptitiously when the radius of the common envelope is taken to be the orbital separation rather than the radius of the Roche-lobe filling giant. In general this requires that the time-scale for common-envelope evolution be comparable with or longer than the thermal time-scale of the envelope so that the nuclearly generated energy is comparable with the envelope binding energy. It also requires an efficient means of converting this nuclear luminosity to the kinetic energy of mass loss to avoid radiation. For instance Han, Podsiadlowski and Eggleton (1994) proposed that we ought to include the ionisation energy in the binding energy of the envelope. This greatly reduces what is required but to such an extent that the envelopes of many normal AGB star models are unbound. It is also difficult to see how this energy can be tapped in an envelope that is hot enough to remain fully ionised. Because it is an inherently three dimensional hydrodynamical process with a very wide range of fluid densities full theoretical models of common envelope evolution are still in their infancy. In the current state of the art (Ivanova *et al.*, 2013) it seems that the cores very quickly spiral in, over a few orbits, without ejecting much of the envelope which may well take considerably longer to leave. We can hope for significant progress over the coming decades.

9.9.4. *Type Ia supernovae*

Luminous type Ia supernovae are amongst the brightest objects in the Universe and their use as standard candles by cosmologists has elevated the need to understand their progenitors. The major energy source of type Ia supernovae is the decay of ^{56}Ni to ^{56}Fe and the total energy released in a type Ia supernova is consistent with the decay of approximately a solar mass of ^{56}Ni. These facts strongly implicate the thermonuclear explosion of a white dwarf though the actual explosion mechanism is not fully understood. White dwarfs may be divided into three major types: (i) helium white dwarfs, composed almost entirely of helium, form as the degenerate cores of low-mass red giants which lose their hydrogen envelope before helium can ignite; (ii) carbon/oxygen white dwarfs, composed of about 20% carbon and 80% oxygen, form as the cores of asymptotic giant branch stars or naked helium-burning stars that lose their envelopes before carbon ignition; and (iii) oxygen/neon white dwarfs, composed of heavier combinations of elements, form from giants that ignite carbon in their cores but still lose their envelopes before the degenerate centre collapses to a neutron star.

In binary systems, mass transfer can increase the mass of a white dwarf. Close to the Chandrasekhar mass ($M_{\text{Ch}} \approx 1.44\,M_\odot$) degeneracy pressure can no longer support the star which collapses releasing its gravitational energy. Oxygen–neon white dwarfs lose enough energy in neutrinos and collapse sufficiently, before oxygen ignites, to avoid explosion (accretion induced collapse). Carbon–oxygen white dwarfs, on the other hand, reach densities early enough, at about $1.38\,M_\odot$, during collapse for carbon fusion to set off a thermonuclear runaway under degenerate conditions and release enough energy to create a type Ia supernova. Accreting helium white dwarfs reach sufficiently high densities to ignite helium well below M_{Ch} ($M \approx 0.7\,M_\odot$) but an explosion under these conditions is expected to be considerably fainter than a type Ia supernova because only about $0.1\,M_\odot$ of material is converted to ^{56}Ni. They are now thought to have been observed as faint transient phenomena.

The process is further complicated by the nature of the accreting material. If it is hydrogen-rich, accumulation of a layer of only $10^{-4}\,M_\odot$ or so leads to ignition of hydrogen burning sufficiently violent to eject most, if not all of or more than, the accreted layer in the novae outbursts of cataclysmic variables. So the white dwarf mass does not significantly increase and ignition of its interior is avoided. However if the accretion rate is high $\dot{M} > 10^{-7}\,M_\odot\,\text{yr}^{-1}$ hydrogen can burn as it is accreted, bypassing nova explosions, and allowing the white dwarf mass to grow. Though, if it is not much larger than this, $\dot{M} > 3 \times 10^{-7} M_\odot \, \mathrm{yr}^{-1}$, hydrogen cannot burn fast enough and accreted material builds up a giant-like envelope around the core and burning shell which can rapidly lead to more drastic interaction with the companion and the end of the mass transfer episode. Rates in the narrow range for steady burning are found only when the companion is in the short-lived phase of thermal time-scale expansion as it evolves from the end of the main sequence to the base of the giant branch. Super-soft X-ray sources are probably in such a state but, without invoking some special feedback mechanism, such as disc winds, cannot be expected to remain in it for very long and white dwarf masses very rarely increase sufficiently to explode as type Ia supernovae.

At first sight, a more promising scenario might be mass transfer from one white dwarf to another. In a very close binary orbit gravitational radiation can drive two white dwarfs together until the less massive fills its Roche lobe. If both white dwarfs are CO and their combined mass exceeds M_{Ch} enough mass could be transferred to set off a type Ia supernova. However if the mass ratio $M_{\mathrm{donor}}/M_{\mathrm{accretor}}$ exceeds 0.628 mass transfer is dynamically unstable because a white dwarf expands as it loses mass. Based on calculations at somewhat lower, steady accretion rates, Nomoto and Iben (1985) claimed that the ensuing rapid accretion of material allows carbon to burn in mild shell flashes, converts the white dwarf to oxygen and neon and ultimately leads to accretion induced collapse and not a type Ia supernova. They found a limit of one-fifth of the Eddington accretion rate was necessary to avoid igniting carbon non-degenerately. Even for stable mass transfer driven by gravitational radiation this is exceeded. We are still searching for the progenitors of type Ia supernovae from among the diverse binary systems in the stellar zoo.

9.9.5. *Massive stars, neutron stars and black holes*

The effect on the more massive component, the first to evolve, of a close binary companion is generally to accelerate its mass loss. For low- and intermediate-mass stars the result is a less massive white dwarf than expected from isolated evolution. Indeed mass transfer beginning before helium ignition is responsible for most observed helium white dwarfs in our neighbourhood, because few single stars of low enough mass were born long enough ago to have evolved in the lifetime of our Galaxy. Similarly increased mass loss from massive stars changes the nature of the remnants they leave behind but also the nature of any associated supernovae. In particular the first instance mass transfer only strips most of the hydrogen envelope to leave a naked helium star with a thin hydrogen-rich envelope which is easily expelled or burnt. Such a star can evolve to explode as a type Ib/c supernova rather than the type II experienced by a similar initial mass single star. Indeed evidence in the form of both the relative rates of different types of supernovae and images of supernova progenitor star fields indicate that most hydrogen-free supernovae are the result of binary interaction rather than the end of massive single stars.

The effects on the initially lower-mass companion are more convoluted because it can often accrete substantial mass, particularly when its thermal response time-scale is shorter than the mass-transfer time-scale. Otherwise the transfer is likely to be non-conservative and may lead to common envelope evolution. For stars with convective cores the increased total mass is reflected by growth of the convective core which mixes in fresh hydrogen, effectively rejuvenating the star. The transferred material carries with it angular momentum acquired from the orbit and this can often spin up the accretor to near break-up spin. The spin in turn drives strong mixing that further rejuvenates the star. Even accounting for rejuvenation, sufficient mass transfer can drive the accretor to evolve and explode as a supernova before its originally more massive companion has a chance to evolve much further itself. Indeed the accretor in a binary system in which neither star was originally massive enough to reach a supernova can often acquire sufficient mass to explode.

The sudden loss of much of the entire mass of a binary system in a supernova almost always leads to its disruption, particularly when the supernova is asymmetric and kicks the remnant to speeds of as much as $500 \, \mathrm{km \, s^{-1}}$. However X-ray binary stars with a neutron star or black hole component are known and so, not uncommonly, both the mass expelled and the kick must be small enough to leave a bound system. Evolution, often including a common envelope phase, can lead to mass transfer on to the compact object. Because of its small radius accretion is always through a disc, the inner parts of which are heated to temperatures at which they radiate X-rays. If the accreting compact object merges with its companion even more exotic stars, such as Thorne–Żytkow objects and quasistars, giants with neutron degenerate and black hole cores might form but as yet none have been unambiguously recognised. When the second component of a binary system itself ends up massive enough it too can evolve through a supernova to leave a second neutron star or black hole and if close enough these can merge by emission of gravitational radiation in the lifetime of the Universe. The gravitational wave signatures of merging neutron stars and black holes in distant parts of the Universe have now been detected and represent an excellent test of the application of general relativity over large length-scales.

In such a brief chapter there are many exciting aspects of binary star evolution on which we have not touched. We hope nevertheless that we have presented enough of a basic understanding to inspire the curiosity of our readers and so we provide a number of suggestions for further exploration in the bibliography. Our understanding of the evolution of binary stars is still rapidly changing but, nevertheless, it is already possible to put forward plausible evolutionary histories for any system of two stars at any separation. Algorithms have been developed to evolve large populations of millions of binary systems with suitable choices for initial mass and period distributions. Whenever we model the effects of stellar evolution in clusters of stars or galaxies to look at, for example, their overall spectra, nucleosynthesis or supernovae rates, it is essential to consider the effects of binary stars carefully. They can rarely be ignored completely.

9.10.   Questions

1. A binary star system with a period of 4 yr lies at 200 pc from the Sun. The projected orbit of each component holding the other fixed is an ellipse. For each star the semi-major axis of the ellipse passes

through the companion and the apastron separation is four times the periastron separation. The semi-major axis subtends 20 mas and the semi-minor axis 8 mas. Find the eccentricity, inclination, semi-major axis and total mass of the system.

2. Two stars of equal surface temperature form an eclipsing binary system. The orbit is circular. If the stars are spherical and limb-darkening can be ignored, show that both eclipses are the same depth in magnitudes and that this depth cannot exceed a certain amount. Find this amount.

3. A total of N stars are placed at random on the celestial sphere. Show that the probability distribution for the angular separation of nearest neighbours is

$$P(\theta)\, d\theta = \frac{N-1}{2^{N-1}} (\sin \theta)(1 + \cos \theta)^{N-2}\, d\theta.$$

The restricted bright star catalogue contains 4908 objects brighter than $V = 6$. Of these 114 are doubly bright visual systems with separations less than $3.5\,\mu$rad. What would be the most likely angular separation of any star's nearest neighbour if all the stars were single and placed at random? What is the probability that all 114 double stars are simply random superpositions?

4. A red giant of mass M_1 is in a binary system with a main-sequence star of mass M_2. The red giant is losing mass in a stellar wind at a rate $\dot{M} < 0$ so that, if the intrinsic angular momentum of the stars is neglected, this wind carries off orbital angular momentum at a rate

$$\frac{\dot{J}_{\text{orb}}}{J_{\text{orb}}} = \frac{M_2 \dot{M}}{M_1 M},$$

where $M = M_1 + M_2$.

   On a short time-scale the radius of the giant responds according to

$$R \propto M_1^{-n} \quad 0 < n < 1$$

and the radius of its Roche lobe is approximated by

$$\frac{R_L}{a} = 0.426 \left(\frac{M_1}{M}\right)^{\frac{1}{3}}.$$

Now suppose that the giant is filling its Roche lobe ($R = R_L$) and that wind mass loss is taking place on a time-scale much shorter than the nuclear time-scale. Show, by differentiating $\log R/R_L$ or otherwise, that mass transfer is driven by the wind if

$$q = \frac{M_1}{M_2} < \frac{1 + 3n}{3(1 - n)}. \tag{†}$$

What happens otherwise?

Show further that, when (†) is satisfied, the rate of mass transfer to the main-sequence star

$$\dot{M}_2 = -\frac{1 + 3n - 3(1 - n)q}{(1 + q)(5 - 3n - 6q)}\dot{M}.$$

What is the physical consequence if $6q > 5 - 3n$?

5. A supergiant of C/O core mass M_c and envelope mass M_{env}, of which the binding energy may be expressed as

$$E_{bind} \approx -\frac{2GM_{env}^2}{R_G},$$

where R_G is a fiducial radius defined by

$$\frac{R_G}{R_\odot} \approx 1000 \left(\frac{M_c}{M_\odot}\right)^2 \left(\frac{M_{env}}{M_\odot}\right)^{-\frac{1}{3}},$$

is in a binary with a C/O white dwarf of mass M_{wd}.

The giant fills its Roche lobe and dynamical mass transfer leading to common-envelope evolution ensues. Show that, if the common envelope efficiency is α_{ce}, the final separation of the cores when the envelope has been lost is a_f where

$$\frac{a_f}{R_G} \approx \frac{\alpha_{ce}}{2} \frac{M_c M_{wd}}{M_{env}^2},$$

in the limit $a_f \ll a_i$, where a_i is the initial separation.

The radius of a white dwarf of mass $M_{\rm wd}$ can be approximated by

$$\frac{R_{\rm wd}}{R_\odot} \approx 0.01 \left(\frac{M_\odot}{M_{\rm wd}}\right)^{\frac{1}{3}}$$

and the radius of the hot giant core by

$$R_{\rm c}\,(M_{\rm c}) \approx 5 R_{\rm wd}(M_{\rm c}).$$

The spiralling cores coalesce if $a_{\rm f} \leq 3\max\,(R_{\rm wd}, R_{\rm c})$. Estimate the minimum envelope mass $M_{\rm crit}$ required for the cores to coalesce if $M_{\rm c} = 0.6\,M_\odot$, $M_{\rm wd} = 0.9\,M_\odot$ and $\alpha_{\rm ce} = 1.0$.

Suppose the process of coalescence heats the degenerate white dwarf and the supergiant core to a temperature at which carbon burning $^{12}{\rm C} + {}^{12}{\rm C} \rightarrow {}^{24}{\rm Mg}$ (13.93 MeV per reaction) can ignite. Estimate the total nuclear energy that can be released and compare it with the binding energy of the white dwarf which may be modelled as an $n = 3/2$ polytrope.

Comment on the result for $M_{\rm env}$ in the range from well below to well above $M_{\rm crit}$.

Bibliography

General

A number of texts on the structure and evolution of stars have appeared and we list here those that we have used most extensively when preparing the lectures on which this text is based. Schwarzschild's book is a good straightfoward introduction, Clayton's is excellent for nuclear physics, Kippenhahn and Wiegert's provides a modern overview and Cox and Guili's two volumes a comprehensive overview of the physics.

Clayton, D. D. (1968). *Principles of Stellar Evolution and Nucleosynthesis* (McGraw-Hill, New York).

Cox, J. P. and Guili, R. T. (1968). *Principles of Stellar Structure, Vol. 1: Physical Principles, Vol. 2: Applications to Stars* (Gordon and Breach, New York).

Kippenhahn, R. and Weigert, A. (1996). *Stellar Structure and Evolution* (Springer, Berlin).

Schwarzschild, M. (1958). *Structure and Evolution of the Stars* (Princeton University Press, Princeton).

For those interested in the historical development of the subject we direct you to two fascinating texts written by those actually involved as the theory developed.

Eddington, A. S. (1926). *The Internal Constitution of the Stars* (Cambridge University Press, Cambridge).

Gamow, G. (1964). *A Star Called the Sun* (Viking Press, New York).

Chapter 1

For an excellent overview of the general history of Astronomy see

Ferris, T. (1988). *Coming of Age in the Milky Way* (William Morrow & Co., New York).

Other works referred to in Chapter 1 are

Bellini, A. (2009). Ground-based CCD astrometry with wide field imagers. III. WFI@2.2m proper-motion catalog of the globular cluster ω Centauri, *Astron. Astrophys.* **493**, pp. 959–978.

Bessel, F. W. (1838). On the parallax of 61 Cygni, *Mon. Not. R. Astron. Soc.* **4**, pp. 152–161.

Cannon, A. J. and Pickering, E. C. (1901). Spectra of bright southern stars photographed with the 13-inch Boyden telescope as part of the Henry Draper memorial, *Ann. Harvard Coll. Obs.* **28**, pp. 129–263.

Cannon, A. J. and Pickering, E. C. (1912). Classification of 1,477 stars by means of their photographic spectra, *Ann. Harvard Coll. Obs.* **56**, pp. 65–114.

Cousins, A. W. J. (1975). VRI photometry at the S.A.A.O, *Mon. Not. R. Astron. Soc. Southern Africa* **34**, pp. 68–71.

Eldridge, J. J. and Relaño, M. (2011). The red supergiants and Wolf–Rayet stars of NGC 604, *Mon. Not. R. Astron. Soc.* **411**, pp. 235–246.

Fellgett, P. (1995). Simple stars, *The Observatory* **115**, p. 93.

Morgan, W. W. and Keenan, P. C. (1973). Spectral classification, *Ann. Rev. Astron. Astrophys.* **11**, pp. 29–50.

Parsons, S. G. *et al.* (2017). Testing the white dwarf mass–radius relationship with eclipsing binaries, *Mon. Not. R. Astron. Soc.* **470**, pp. 4473–4492.

Payne, C. (1925). Astrophysical data bearing on the relative abundance of the elements, *Pro. Natl. Acad. Sci. USA* **11**, pp. 192–198.

Pickering, E. C. (1890). The Draper catalogue of stellar spectra photographed with the 8-inch Bache telescope as a part of the Henry Draper memorial, *Ann. Harvard Coll. Obs.* **27**, pp. 1–388.

Pogson, N. M. (1856a). Magnitudes of thirty-six of the minor planets for the first day of each month of the year 1857, *Mon. Not. R. Astron. Soc.* **17**, pp. 12–15.

Prša, A. *et al.* (2016). Nominal values for selected solar and planetary quantities: IAU 2015 resolution B3, *Astron. J.* **152**, pp. 41–47.

Russell, H. R. (1929). On the composition of the Sun's atmosphere, *Astrophys. J.* **70**, pp. 11–82.

Rutherford, E. (1929). Origin of actinium and age of the earth, *Nature* **123**, pp. 313–314.

Saha, M. N. (1921). On a physical theory of stellar spectra, *Proc. R. Soc. London* **99**, pp. 135–153.

Southworth, J. (2015). DEBCat: A catalog of detached eclipsing binary stars, in *Living Together: Planets, Host Stars, and Binaries*, ASP Conference Series, Vol. 496 (Astronomical Society of the Pacific), pp. 164–165.

Westera, P., Lejeune, T., Buser, R., Cuisinier, F. and Bruzual, G. (2002). A standard stellar library for evolutionary synthesis. III. Metallicity calibration. *Astron. Astrophys.* **381**, pp. 524–538.

Chapter 3

Data for the equation of state Fig. 3.8 is calculated according to

Pols, O. R., Tout, C. A., Eggleton, P. P. and Han, Z. (1996). Approximate input physics for stellar modelling, *Mon. Not. R. Astron. Soc.* **274**, pp. 964–974.

Chapter 4

Opacity data for Fig. 4.8 is taken from

Alexander, D. R. and Ferguson, J. W. (1994). Low-temperature Rosseland opacities, *Astrophys. J.* **437**, pp. 879–891.

Iglesias, C. A., Rogers, F. J. and Wilson, B. G. (1992). Spin–orbit interaction effects on the Rosseland mean opacity, *Astrophys. J.* **397**, pp. 717–728.

Chapter 5

For a historical view see

Wooley, R. v. d. R. and Stibbs, D. W. N. (1953). *The Outer Layers of a Star* (Oxford University Press, Oxford).

For an observational view see

Gray, F. D. (1992). *The Observation and Analysis of Stellar Photospheres* (Cambridge University Press, Cambridge).

Chapter 6

An interesting account of the attempts to explain the energy source of the Sun in the nineteenth century, including the first conception of the death of the Sun, are described in lucid detail by Helge Kragh.

Kragh, H. (2016). The source of solar energy, ca. 1840–1910: From meteoric hypothesis to radioactive speculations, *European Phys. J. H* **41**, pp. 365–394.

For an account of the solar neutrino problem and its solution see

Pallavicini, M. (2015). Solar neutrinos: Experimental review and prospectives, *J. Phys.: Conf. Ser.* **598**, 012007.

Gamow's seminal paper on quantum mechanical tunnelling in α-decay can be found at

Gamow, G. (1928). Zur Quantentheorie des Atomkernes, *Z. Phys.* **51**, pp. 204–212.

For discussion, including the historical development, of the triple-α reaction see

Tout, C. A. (2006). The triple-α process and the origin of the elements, *Contemporary Phys.* **47**, pp. 145–155.

Chapter 7

Henyey, L. G., Wilets, L., Böhm, K. H., Lelevier, R. and Levee, R. D. (1959). A method for automatic computation of stellar evolution, *Astrophys. J.* **129**, pp. 628–636.

For a discussion of the Vogt–Russell theorem see

Kähler, H. (1978). The Vogt–Russell theorem, and new results on an old problem, in *The HR diagram — The 100th Anniversary of Henry Norris Russell*, IAU Symposia, Vol. 80 (Springer), pp. 303–311.

Chapter 8

The following references, in chronological order, describe the origin and the development of the Eggleton code to the Cambridge STARS code that we use to construct our stellar models.

Eggleton, P. P. (1971). The evolution of low mass stars, *Mon. Not. R. Astron. Soc.* **151**, pp. 351–364.

Eggleton, P. P. (1972). Composition changes during stellar evolution, *Mon. Not. R. Astron. Soc.* **156**, pp. 361–376.

Eggleton, P. P. (1973). A numerical treatment of double shell source stars, *Mon. Not. R. Astron. Soc.* **163**, pp. 279–284.

Eggleton, P. P., Faulkner, J. and Flannery, B. P. (1971). An approximate equation of state for stellar material, *Astron. Astrophys.* **23**, pp. 325–330.

Eldridge, J. J. and Tout, C. A. (2004). A more detailed look at the opacities for enriched carbon and oxygen mixtures, *Mon. Not. R. Astron. Soc.* **348**, pp. 201–206.

Pols, O. R., Tout, C. A., Eggleton, P. P. and Han, Z. (1995). Approximate input physics for stellar modelling, *Mon. Not. R. Astron. Soc.* **274**, pp. 964–974.

Schröder, K.-P., Pols, O. R. and Eggleton, P. P. (1997). A critical test of stellar evolution and convective core 'overshooting' by means of zeta Aurigae systems, *Mon. Not. R. Astron. Soc.*, **285**, pp. 696–710.

Stancliffe, R. J., Tout, C. A. and Pols, O. R. (2004). Deep dredge-up in intermediate-mass thermally pulsing asymptotic giant branch stars, *Mon. Not. R. Astron. Soc.* **352**, pp. 984–992.

Other works referred to in Chapter 8 are

Böhm-Vitense, E. (1958). Über die Wasserstoffkonvektionszone in Sternen verschiedener Effekivtemperaturen und Leuchtkräfte, *Z. Astroph.* **46**, pp. 108–143.

de Jager, C., Nieuwenhuijzen, H. and van der Hucht, K. A. (1988). Mass-loss Rates in the Hertzsprung–Russell diagram, *Astron. Astrophys. Supp. Ser.* **72**, pp. 259–289.

Nugis, T. and Lamers, H. J. G. L. M. (2000). Mass-loss rates of Wolf-Rayet stars as a function of stellar parameters, *Astron. Astrophys.* **360**, pp. 227–244.

Reimers, D. (1975). Circumstellar absorption lines and mass loss from red giants, *Mem. Soc. R. Sci. Liège* **8**, pp. 369–382.

Schönberg, M. and Chandrasekhar S. (1942). On the evolution of the main-sequence stars, *Astrophys. J.* **96**, pp. 161–172.

Schwarzschild, M. and Härm, R. (1965). Thermal instability in non-degenerate stars, *Astrophys. J.* **142**, pp. 855–867.

Smartt, S. J. (2015). Observational constraints on the progenitors of core-collapse supernovae: The case for missing high-mass stars, *Pub. Astron. Soc. Australia* **32** id. e016.

Vassiliadis, E. and Wood, P. R. (1993). Evolution of low- and intermediate-mass stars to the end of the asymptotic giant branch with mass loss, *Astrophys. J.* **413**, pp. 641–657.

Vink, J. S., de Koter, A. and Lamers, H. J. G. L. M. (2001). Mass-loss predictions for O and B stars as a function of metallicity, *Astron. Astrophys.* **369**, pp. 574–588.

Woosley, S. E., Heger, A. and Weaver, T. A. (2002). The evolution and explosion of massive stars, *Rev. Mod. Phys.* **74**, pp. 1015–1071.

Chapter 9

For a good introduction to the subject consult the monograph

Pringle, J. E. and Wade, R. A. (1985). *Interacting Binary Stars* (Cambridge University Press, Cambridge).

And for a more complete treatise with historical discussion see

Kopal, Z. (1959). *Close Binary Systems* (Chapman & Hall, London).

Other works referred to in Chapter 9 are

Allen, R. H. (1899), *Star Names and Their Meanings* (Stechert, New York).

Crawford, J. A. (1955). On the subgiant components of eclipsing binary systems, *Astrophys. J.* **121**, pp. 71–76.

Darwin, G. H. (1879). VIII. The determination of the secular effects of tidal friction by a graphical method, *Proc. R. Soc. London* **29**, pp. 168–181.

Eggleton, P. P. (1983). Approximations to the radii of Roche lobes, *Astrophys. J.* **268**, pp. 368–369.

Eggleton, P. P. (2001). *Evolutionary Processes in Binary and Multiple Stars* (Cambridge University Press, Cambridge).

Goodricke Jr., J. (1783). A series of observations on, and a discovery of, the period of the variation of the light of the bright star in the head of Medusa, called Algol, *Phil. Trans. R. Soc. London* **73**, pp. 474–482.

Han, Z., Podsiadlowski, P. and Eggleton, P. P. (1994). A possible criterion for envelope ejection in asymptotic giant branch or first giant branch stars, *Mon. Not. R. Astron. Soc.* **270**, pp. 121–130.

Herschel, W. (1803). Account of the changes that have happened, during the last twenty-five years, in the relative situation of double-stars; with an investigation of the cause to which they are owing, *Phil. Trans. R. Soc. London* **93**, pp. 339–382.

Hind, J. R. (1856). On a new variable star, *Mon. Not. R. Astron. Soc.* **16**, p. 56.

Hoyle, F. (1955). *Frontiers of Astronomy* (Heinemann, London).

Ivanova, N. *et al.* (2013). Common envelope evolution: Where we stand and how we can move forward, *Ann. Rev. Astron. Astrophys.* **21**, id. 59.

Kippenhahn, R. and Wiegert, A. (1967). Entwicklung in engen Doppelstern-systemen I. Massenaustausch vor und nach Beendigung des zentralen Wasserstoff–Brennens, *Z. Astrophys.* **65**, pp. 251–273.

Lubow, S. H. and Shu, F. H. (1975). Gas dynamics of semidetached binaries, *Astrophys. J.* **198**, pp. 383–405.

Michell, J. (1767). An inquiry into the probable parallax, and magnitude of the fixed stars, from the quantity of light which they afford us, and the particular circumstances of their situation, *Phil. Trans. R. Soc. London* **57**, pp. 234–264.

Morton, D. C. (1960). Evolutionary mass exchange in close binary systems, *Astrophys. J.* **132**, pp. 146–161.

Nomoto, K. and Iben Jr., I. (1985). Carbon ignition in a rapidly accreting degenerate dwarf — A clue to the nature of the merging process in close binaries, *Astrophys. J.* **297**, pp. 531–537.

Paczyński, B. (1971). Evolutionary processes in close binary systems, *Ann. Rev. Astron. Astrophys.* **9**, pp. 183–208.

Paczyński, B. (1976). Common envelope binaries, in *Structure and Evolution of Close Binary Systems, Proc. IAU Symp.* **73** (Reidel, Dordrecht) pp. 75–80.

Paczyński, B. and Ziółkowski, J. (1967). Evolution of close binaries. III, *Acta Astron.* **17**, pp. 7–14.

Parenago, P. P. (1950). Uber die Massen von Bedeckungsveranderlichen mit bekannter Radialgeschwindigkeit nur des Hauptsterns, *Astron. Zhur.* **27**, pp. 41–47.

Pogson, N. (1856b). On variable stars, *Mon. Not. R. Astron. Soc.* **17**, pp. 23–26.

Struve, O. (1948). The masses and mass-ratios of close binary systems, *Ann. Astrophys.* **11**, pp. 117–123.

Struve, O. (1949). Spectroscopic binaries (George Darwin Lecture), *Mon. Not. R. Astron. Soc.* **109**, pp. 487–506.

Vogel, N. C. (1890). Spectrographische Beobachtungen an Algol, *Astron. Nachr.* **123**, p. 289.

Warner, B. (1995). *Cataclysmic Variable Stars* (Cambridge University Press, Cambridge).

Wood, F. B. (1950). On the change of period of eclipsing variables stars, *Astrophys. J.* **112**, pp. 196–206.

Index

A

α Centauri, 16
α-isotopes, 156
absolute magnitude, 13
absorption, 91
    atmosphere, 122
    free-free, 94
absorption line, 19
accretion induced collapse, 318
adiabat
    Sun, 108
Alfvén radius, 8
Algol, 307
Algol paradox, 307
AM Canum Venaticorum, 313
angular momentum
    binary star, 278
apparent magnitude, 12
apsidal constant, 290
apsidal motion, 290

B

barium stars, 312
barrier energy, 162
Bayer catalogue, 11
Betelgeuse, 11
big bang nucleosynthesis, 153

binary star, 271
    number, 272
    observation, 274
binding energy per nucleon, 148
black body, 50
blue straggler, 24
    formation, 309
bolometric magnitude, 14
Boltzmann distribution, 52
boson, 51
boundary conditions, 193
Bremsstrahlung, 94

C

carbon burning, 157
case A mass transfer, 310
case B mass transfer, 310
cataclysmic variable, 312
centrifugal force, 295
centrifugal potential, 161
Cepheid instability, 228
Chandrasekhar mass, 69
chromosphere
    Sun, 22
circularisation, 286
CNO cycle, 153
colour, 17
colour–magnitude diagram, 22

common envelope, 315
conditioning, 200
conduction, 98
    importance, 82
conservative mass transfer, 302
contact binary, 296
convection, 98
    diffusion, 195
convective overshooting, 111
    STARS code, 221
cooling track, 258
core breathing pulses, 229
Coriolis force, 295
corona, 8
Coulomb barrier, 160
critical mass ratio, 310
cross-section factor, 165

D

Darwin instability, 288
degeneracy, 64
density of states
    electron, 64
    photons, 51
    warm degenerate matter, 72
detailed balance, 125
diffusion, 194
    coefficient, 194
    ions, 84
Doppler broadening, 21
Doppler shift, 21
double diffusion, 116
dredge up
    first, 227
    second, 230
    third, 236
dynamical mass transfer, 303
dynamical time-scale
    Sun, 9

E

eccentricity, 275
eclipsing binary star, 283
Eddington closure, 131
Eddington luminosity, 133

Eddington's first approximation, 130
Eddington's quartic, 59
Einstein coefficients, 123
electron scattering, 90
electron screening, 174
ellipse, 275
emission, 123
emission line, 21
energy
    binary star, 279
    generation, 142
    net of a star, 43
energy generation
    time dependent, 144
energy generation rate, 142
    CNO cycle, 172
    pp-chain, 171
    triple-α reaction, 173
enthalpy, 54
entropy
    constant, 56
equation of state, 47
    composite, 74
equation of transfer, 125
equations of stellar evolution, 192, 194
Euler momentum equation, 32
evolution
    $1\,M_\odot$ star, 239
    $5\,M_\odot$ star, 222
    $7\,M_\odot$ star, 244
    $15\,M_\odot$ star, 247
    $40\,M_\odot$, 249
    massive stars, 246

F

fate of the Earth, 243
Fermi energy, 65
Fermi momentum, 65
filter, 17
Flamsteed catalogue, 12

G

Gamow window, 169
Gamow energy, 169

gas constant, 48
globular cluster, 24
gravitational contraction, 144
    ideal gas, 146
gravitational settling, 84
gravitational wave, 321
grey atmosphere, 127

H

Härm–Schwarzschild instability,
    231
habitable zone, 240
half-life, 265
Hayashi track, 209
heat capacity, 54
helioseismology, 6
helium
    Sun, 5
helium burning, 155
helium core flash, 241
Henry Draper catalogue, 12
Henyey method, 199
Hertzsprung–Russell diagram, 22
homology, 201
    convective stars, 207
    evolution, 211
    main sequence, 203
    variables, 201
horizontal branch, 24, 261
hot bottom burning, 237
hydrogen
    Sun, 5
hydrogen burning, 149
hydrostatic equilibrium, 34

I

identical particles, 159
implicit integration, 200
interpulse, 235
ionisation, 60
    helium, 62
    hydrogen, 62
irradiance
    Sun, 3

K

Kelvin–Helmholtz time-scale
    Sun, 9
Kepler's laws, 275
    first, 280
    second, 278
    third, 281
Kirchoff's law, 124

L

Lagrangian derivatives, 191
Lagrangian point, 296
Lane–Emden equation, 35
Laplace–Runge–Lenz vector, 279
Ledoux criterion, 113
light year, 15
limb darkening, 137
lithium production, 152
local thermodynamic equilibrium, 79
luminosity class, 21
luminosity gradient, 143
luminous blue variable, 255

M

magic number, 180
magnitude, 12
main sequence, 22
main-sequence
    hook, 225
    lifetime, 207
mass conservation, 31
mass function, 282
mass loss, 252
mass transfer, 294
    rate, 299
    stability, 302
Maxwell stress, 33
Maxwell–Boltzmann distribution, 158
mean free path
    electron, 81
    neutrinos, 262
    photon, 80, 88
mean molecular weight, 49
    electron, 65
meridional circulation, 256

metal, 5, 49
metallicity, 50
  stars, 251
  Sun, 5
Mira variable, 253
mixing, 194
mixing length, 106
mixing length theory, 104
molecular band, 21
molecules, 73
Morgan–Keenan system, 19

N

Navier–Stokes equation, 32
neon burning, 157
neon-22, 156
neutrino losses, 184
  nuclear reactions, 150
neutrino oscillations, 152
neutron capture, 179
neutron drip line, 181
neutron source, 181
  AGB stars, 238
neutron star, 70
Newton's laws, 277
Newton–Raphson iteration, 199
nova
  classical, 314
  dwarf, 314
nuclear binding energy, 147
nuclear radius, 160
nuclear reaction rates, 158
nuclear spin, 160
nuclear statistical equilibrium
  (*see also* NSE), 176

O

on phase, 236
opacity, 89
  electron scattering, 90
  H$^-$, 95
  Kramers, 94
  molecules, 95
  photodisintegration, 158
  power laws, 97

optical depth, 123
orbit, 275
orbital elements, 281
orbital period evolution, 304
oxygen burning, 158

P

p-process, 184
P Cygni profile, 137
pair production, 184
parallax, 15
parsec, 15
partition function
  excited states, 93
  ions, 61
penetration factor, 161
pep reaction, 153
photodisintegration, 157
photosphere, 132
physics of stars, 1
plane-parallel atmosphere, 128
polytrope, 34
  convective, 108
  Eddington, 59
  white dwarf
    non-relativistic, 67
    relativistic, 69
polytropic index, 35
population III
  evolution, 251
populations I, II and III, 6
power down, 236
pressure, 32, 43, 47
  electron, 66
  gas, 48
  ion, 73
  radiation, 50
    importance, 57
pressure broadening, 21
pressure ionisation, 63
primary carbon, 237
primary nitrogen, 237
prominence, 8
proper motion, 15
proton capture, 154

proton to neutron ratio, 178
proton-proton chains, 150
pulsation, 34, 44

Q

quantum tunnelling, 161

R

r-process, 181
radiation constant, 52
radiative transfer, 84
Rayleigh–Jeans tail, 18
reaction equilibria, 175
reaction rate
    non-resonant, 169
    power-law, 172
    resonant, 172
red clump, 24
red giant, 210
reddening, 18
rejuvenation, 320
relaxation, 198
resonance, 167
Reynolds number, 32
    convection, 110
Reynolds stress, 32
Roche lobe, 297
Roche-lobe overflow, 297
Roche-lobe radius, 298
Roche potential, 295
Rosseland mean opacity, 89
rotating coordinates, 295
rotation, 34
RR Lyrae variables, 229

S

δ Scuti stars, 229
s-process, 180
Saha equation
    ionisation, 60
    molecules, 73
    nuclear statistical equilibrium,
        176
scattering, 126
Schwarzschild criterion, 102

Schwarzschild radius, 71
semi-detached binary, 297
semi-latus rectum, 275
semi-major axis, 275
semiconvection, 111
shooting, 197
silicon burning, 158
solar calibration, 239
    radius, 108
solar eclipse, 7
solar flare, 8
solar neutrino problem, 152
solar system expansion, 306
solar wind, 252
specific intensity, 121
spectral lines, 19
    formation, 135
spectral type, 20
spectroscopic binary star,
    282
star formation, 44
star names, 11
STARS code, 219
Stefan–Boltzmann constant, 52
stellar masses, 27
stellar wind, 252
stimulated emission, 123
stream, 302
Sun, 2
    age, 4
    centre, 8
    composition, 4
    luminosity, 3
    mass, 3
    mean density, 4
    radius, 3
sunspot, 7
super wind, 253
super-AGB star, 245
super-soft X-ray source, 319
superadiabatic gradient, 105
supernova
    core-collapse, 264
    kick, 320
    type Ia, 317

type Ib/c, 320
type II, 264
supernova classification, 261
symbiotic star, 313
synchronisation, 286

T

tachocline, 7
technetium, 180
temperature, 43
temperature gradient
    convective, 107
    radiative, 89
thermal mass transfer, 303
thermal time-scale
    Sun, 9
thermohaline mixing, 116
thermostatic control, 174
Thorne–Żytkow object, 321
tidal potential, 288
tides, 285
    equilibrium, 286
triple-α reaction, 155
turbulence, 32

U

U Cephei, 311
Urca process, 185

V

virial of Clausius, 41
virial theorem, 39
viscosity, 32
visual binary star, 281

W

white dwarf, 67
    formation, 258
WKB approximation, 163
Wolf–Rayet wind, 255

X

X-ray binary, 320

Z

Zeeman splitting, 21
zero-age main sequence, 220